Assessment Resources

HOLT, RINEHART AND WINSTON

A Harcourt Classroom Education Company

Austin • New York • Orlando • Atlanta • San Francisco • Boston • Dallas • Toronto • London

To the Teacher

Geometry Assessment Resources contains blackline masters that provide the teacher with both traditional and alternative assessments, giving teachers a variety of choices to best suit their needs.

- **Quick Warm-Up: Assessing Prior Knowledge** and **Lesson Quiz** (one page per lesson) contains a short set of prerequisite exercises for each lesson followed by a quiz consisting of free-response questions.
- **Mid-Chapter Assessment** (one page per chapter) contains multiple-choice questions and free-response questions that assess the instruction in the first half of each chapter.
- **Chapter Assessment** (two forms per chapter)

  Form A is a two-page multiple-choice test that includes assessment questions for every lesson in the chapter.

  Form B is a two-page free-response test that includes questions for every lesson in the chapter.
- **Alternative Assessment** (two forms per chapter)

  Form A is a one-page authentic-assessment activity or performance-assessment activity that involves concepts from the first half of the chapter.

  Form B is a one-page authentic-assessment activity or performance-assessment activity that involves concepts from the second half of the chapter.

Printed in the United States of America

ISBN 0-03-054313-4

1 2 3 4 5 6 7 066 03 02 01 00

Table of Contents

Chapter 1	Quick Warm-Ups and Lesson Quizzes	**1**
	Mid-Chapter Assessment	**5**
	Chapter Assessments	**9**
	Alternative Assessments	**13**
Chapter 2	Quick Warm-Ups and Lesson Quizzes	**15**
	Mid-Chapter Assessment	**18**
	Chapter Assessments	**21**
	Alternative Assessments	**25**
Chapter 3	Quick Warm-Ups and Lesson Quizzes	**27**
	Mid-Chapter Assessment	**31**
	Chapter Assessments	**36**
	Alternative Assessments	**40**
Chapter 4	Quick Warm-Ups and Lesson Quizzes	**42**
	Mid-Chapter Assessment	**46**
	Chapter Assessments	**51**
	Alternative Assessments	**55**
Chapter 5	Quick Warm-Ups and Lesson Quizzes	**57**
	Mid-Chapter Assessment	**61**
	Chapter Assessments	**66**
	Alternative Assessments	**70**
Chapter 6	Quick Warm-Ups and Lesson Quizzes	**72**
	Mid-Chapter Assessment	**75**
	Chapter Assessments	**79**
	Alternative Assessments	**83**
Chapter 7	Quick Warm-Ups and Lesson Quizzes	**85**
	Mid-Chapter Assessment	**89**
	Chapter Assessments	**93**
	Alternative Assessments	**97**
Chapter 8	Quick Warm-Ups and Lesson Quizzes	**99**
	Mid-Chapter Assessment	**103**
	Chapter Assessments	**106**
	Alternative Assessments	**110**
Chapter 9	Quick Warm-Ups and Lesson Quizzes	**112**
	Mid-Chapter Assessment	**115**
	Chapter Assessments	**119**
	Alternative Assessments	**123**

Chapter 10 Quick Warm-Ups and Lesson Quizzes **125**
Mid-Chapter Assessment **128**
Chapter Assessments **133**
Alternative Assessments **137**

Chapter 11 Quick Warm-Ups and Lesson Quizzes **139**
Mid-Chapter Assessment **143**
Chapter Assessments **147**
Alternative Assessments **151**

Chapter 12 Quick Warm-Ups and Lesson Quizzes **153**
Mid-Chapter Assessment **156**
Chapter Assessments **159**
Alternative Assessments **163**

Answers **165**

NAME ________________ CLASS ________ DATE ________

Quick Warm-Up: Assessing Prior Knowledge

1.1 The Building Blocks of Geometry

Find three objects in your classroom that can be represented by each geometric figure.

1. point ________________
2. line segment ________________
3. angle ________________
4. line ________________
5. plane ________________

Lesson Quiz

1.1 The Building Blocks of Geometry

Use symbolic notation to name the following:

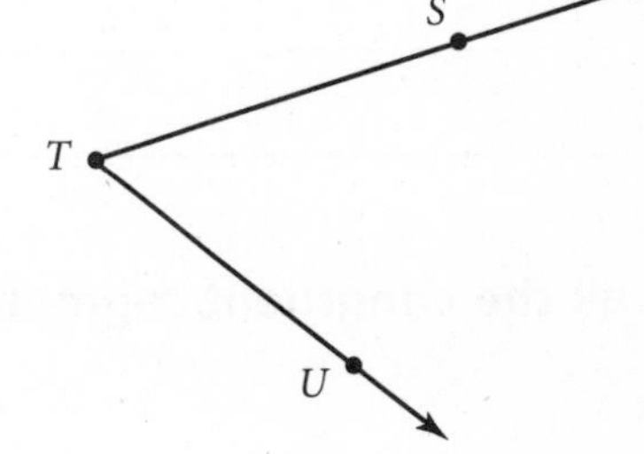

1. Name two segments.

____

2. Name two rays.

____

3. Name the angle in three different ways.

____

4. Draw a point in the interior of ∠STU.

Identify whether each object is best modeled by a point, a line, or a plane.

5. a pen ________________
6. a piece of paper ________________
7. the side of a house ________________
8. a distant star ________________

9. Where a segment begins and ends are its ________________.
10. An angle is formed by two rays with a common endpoint called a ________________.
11. The intersection of two ________________ is a line.

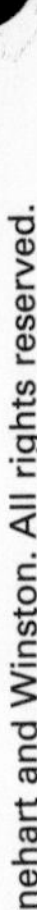

NAME ______________________ CLASS ____________ DATE ____________

Quick Warm-Up: Assessing Prior Knowledge

1.2 Measuring Length

Identify an angle, a segment, a ray, a line, and a point in the figure at right.

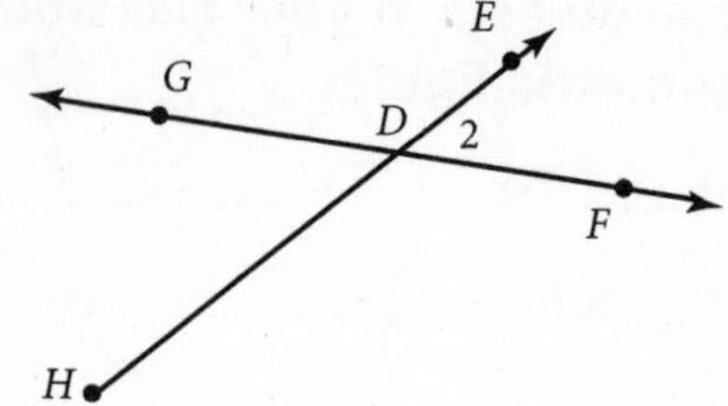

____

____

Lesson Quiz

1.2 Measuring Length

Find the lengths of the segments determined by the points on the number line below.

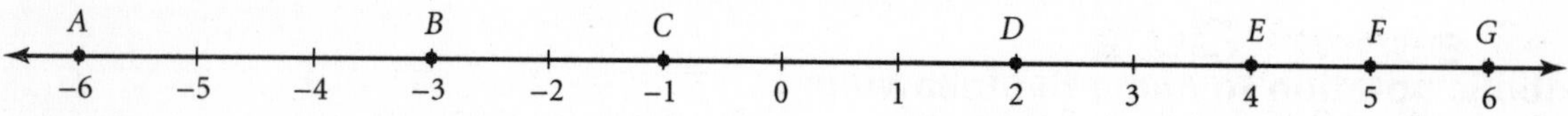

1. AE ______________ 2. CF ______________

3. BF ______________ 4. EF ______________

Name all the congruent segments in each figure.

5.

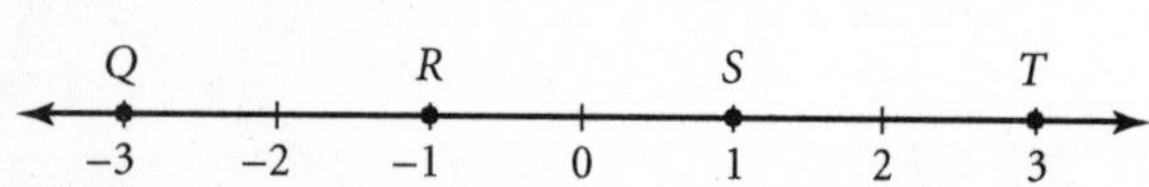

6.

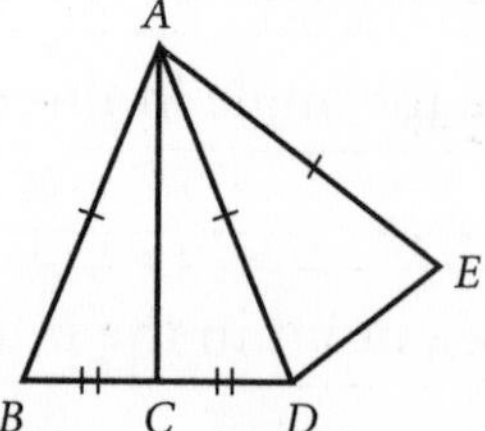

______________________________ ______________________________

7. Write congruence and equivalence statements for the segments in Exercise 5.

8. Write congruence and equivalence statements for the segments in Exercise 6.

NAME ______________________ CLASS ____________ DATE ____________

Quick Warm-Up: Assessing Prior Knowledge

1.3 Measuring Angles

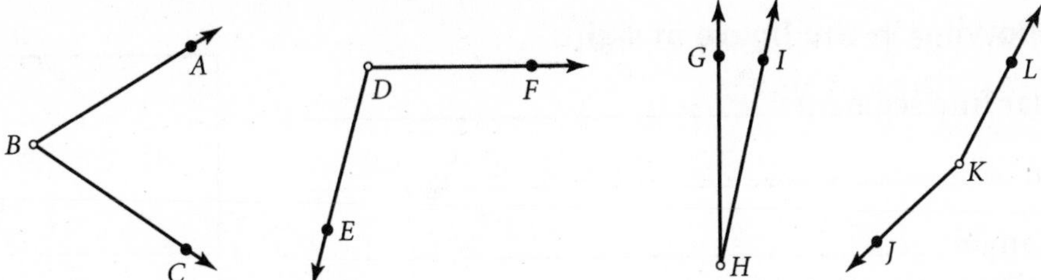

1. Order the angles from largest to smallest. ______________________

2. Which angles are larger than a right angle? Which are smaller?

____

3. Does changing the length of the rays that make up an angle change the measure of the angle? ____________

Lesson Quiz

1.3 Measuring Angles

Find the measure of each angle in the diagram at right.

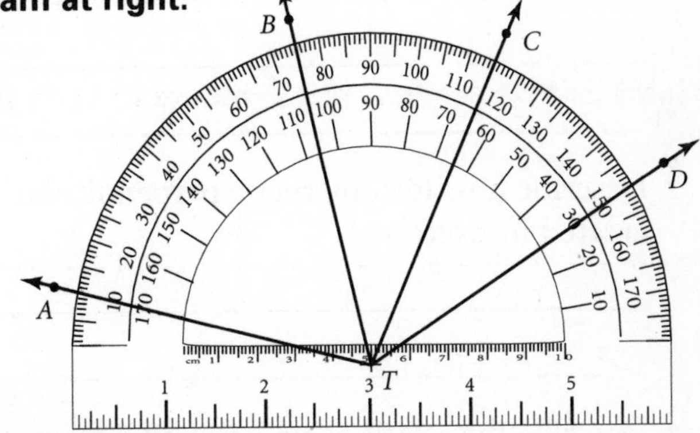

1. m∠*DTC* ____________
2. m∠*DTB* ____________
3. m∠*DTA* ____________
4. m∠*CTB* ____________
5. m∠*ATC* ____________

Find the indicated measures in the figure at right.

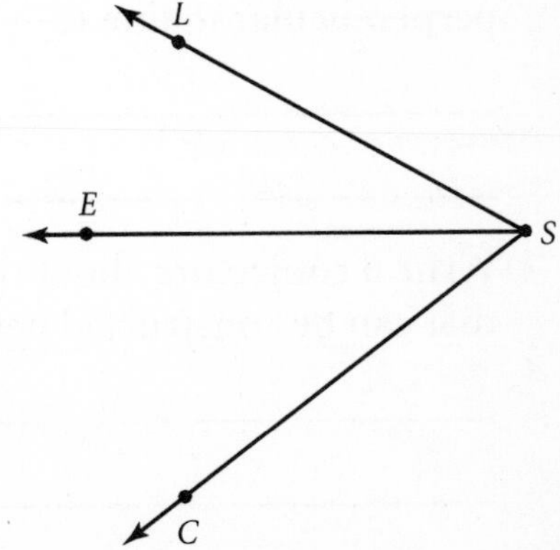

6. m∠*CSE* = 30°, m∠*ESL* = 50°, m∠*CSL* = ________
7. m∠*CSE* = 110°, m∠*CSL* = 155°, m∠*ESL* = ________
8. m∠*CSE* = ________, m∠*ESL* = 36°, m∠*CSL* = 90°
9. m∠*CSE* = ________, m∠*ESL* = 59°, m∠*CSL* = 123°
10. m∠*CSE* = $\frac{3}{5}x$, m∠*ESL* = $\frac{2}{5}x$, , m∠*CSL* = ________
11. m∠*CSE* = ________, m∠*ESL* = 37°, m∠*CSL* = x

NAME ______________________ CLASS ____________ DATE ____________

Quick Warm-Up: Assessing Prior Knowledge

1.4 Exploring Geometry by Using Paper Folding

Identify the following in the figure at right:

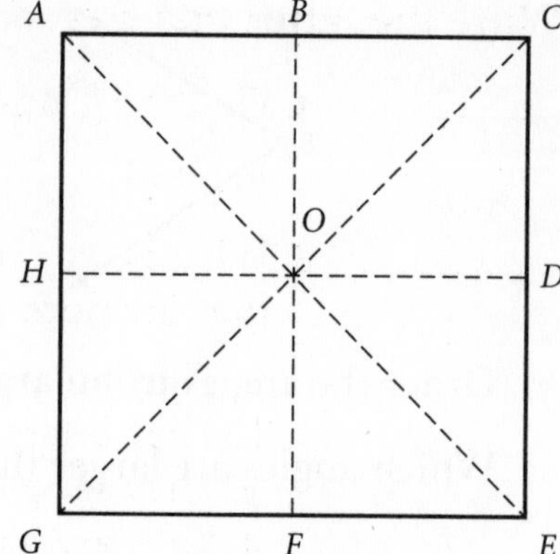

1. perpendicular line segments ______________________
2. a right angle ______________________
3. a 45 degree angle ______________________
4. a line segment that divides a right angle in half ____________

Lesson Quiz

1.4 Exploring Geometry by Using Paper Folding

Construct the geometric figures below by folding a sheet of paper.

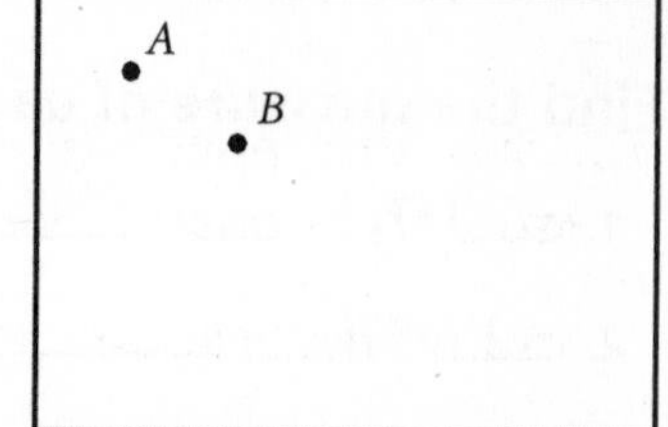

1. Describe how to construct line ℓ through points A and B.

____

____

2. Describe how to construct a perpendicular line to line ℓ created in Exercise 1.

____

____

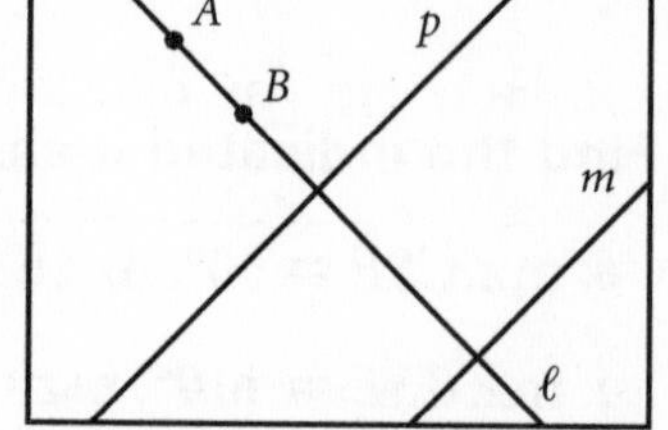

3. Describe the relationship between lines m and p that are both perpendicular to line ℓ.

____

____

4. Write a conjecture about the number of lines that can be constructed perpendicular to line ℓ.

____

____

NAME ________________ CLASS ________ DATE ________

Mid-Chapter Assessment

Chapter 1 (Lessons 1.1–1.4)

Write the letter that best answers the question or completes the statement.

______ 1. Find the geometric figure that best models a paintbrush.

a. point b. line c. plane d. ray

______ 2. What is the relationship between lines m and p if they are both perpendicular to ℓ?

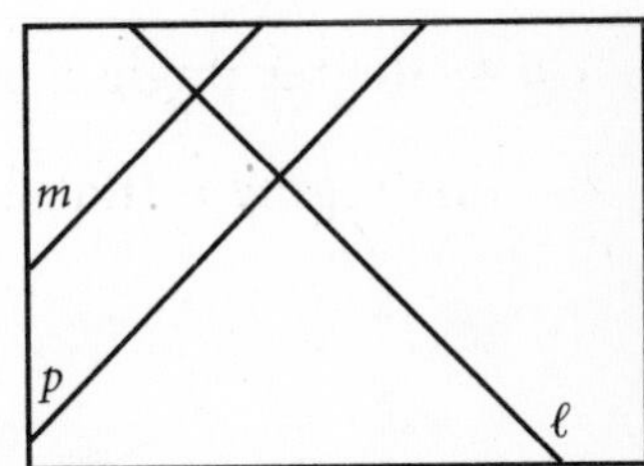

a. They are perpendicular.
b. They intersect.
c. They are parallel.
d. They are planes.

______ 3. If a point lies on the perpendicular bisector of a segment, then the point is ______ equidistant from the endpoints of the segment.

a. always b. sometimes
c. never d. cannot be determined

______ 4. What is the value of x if $QS = 60$?

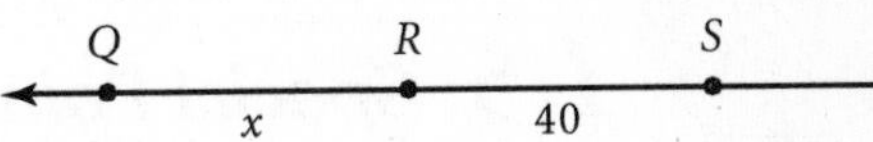

a. 5 b. 10
c. 15 d. 20

Tell whether each statement is true or false, and explain your reasoning in each false case.

5. A line starts at one point and goes on forever. ______

6. Two rays form an angle. ______

7. How can you determine whether a given line is the angle bisector of an angle?

______

In the diagram at right, $m\angle PSR = 60°$, $m\angle PSQ = (x + 5)°$, and $m\angle QSR = (x + 15)°$.

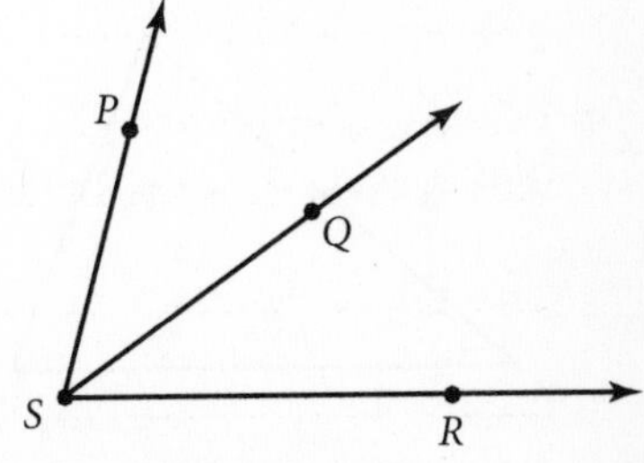

8. Find $m\angle PSQ$. ______

9. Find $m\angle QSR$. ______

NAME ______________________ CLASS ______________ DATE ______________

Quick Warm-Up: Assessing Prior Knowledge
1.5 Special Points in Triangles

Draw and label an example of each of the following kinds of triangles: right triangle, equilateral triangle, isosceles triangle, acute triangle, obtuse triangle, and scalene triangle.

Lesson Quiz
1.5 Special Points in Triangles

Describe how to draw each of the following:

1. the inscribed circle of a triangle ______________________________

____

____

2. the circumscribed circle of a triangle ______________________________

____

____

Draw each of the following:

3. the inscribed circle of $\triangle ABC$

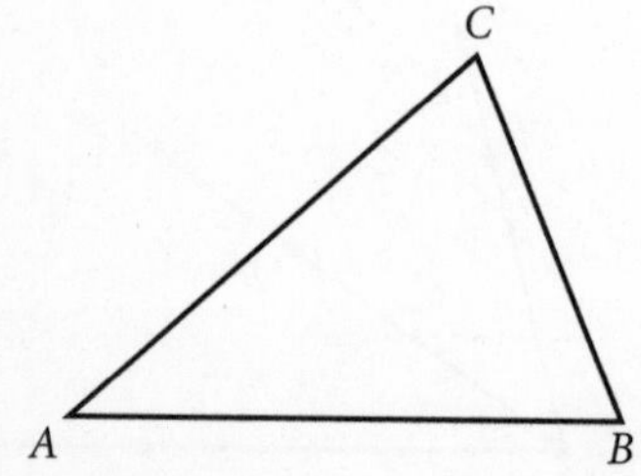

4. the circumscribed circle of $\triangle BOP$

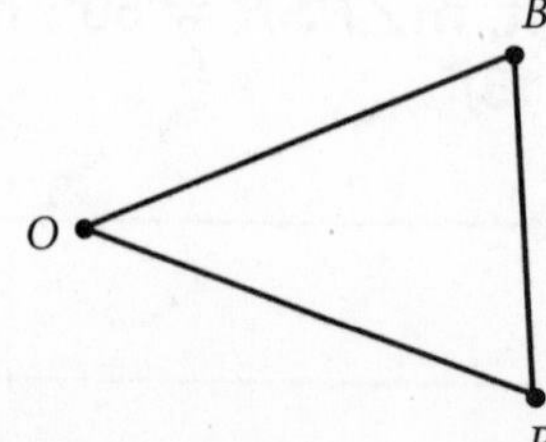

NAME ______________________ CLASS ____________ DATE ____________

Quick Warm-Up: Assessing Prior Knowledge

1.6 Motion in Geometry

Create a pair of drawings to show each of these motions.

1. a translation or slide

2. a reflection or flip

3. a rotation or turn

4. a line segment that divides a right angle in half

Lesson Quiz

1.6 Motion in Geometry

Identify each rigid motion as a reflection, translation, or neither.

1.

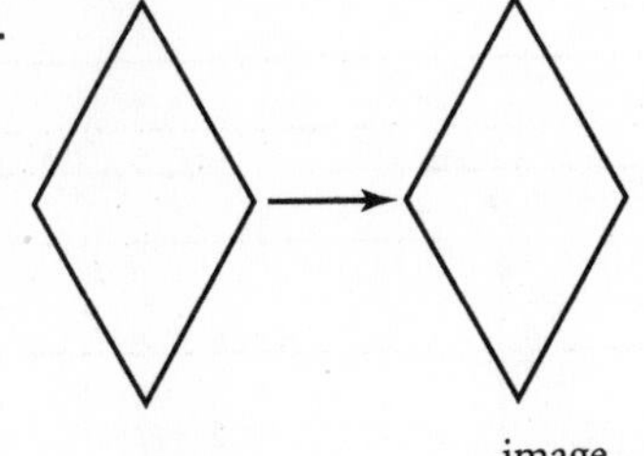

____

2.

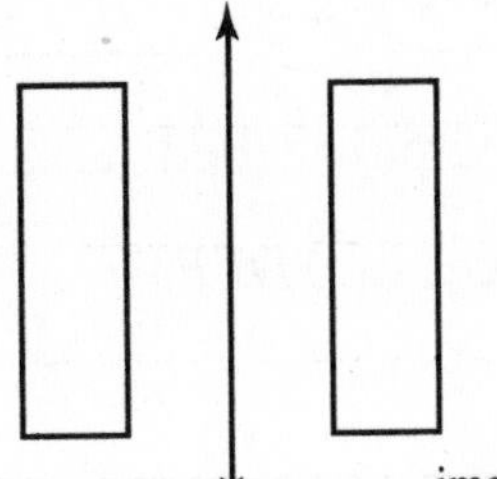

____

3.

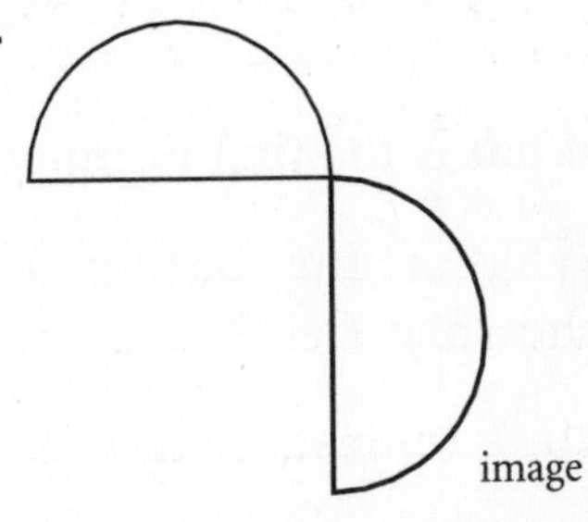

____

Reflect each figure across the given line.

4.

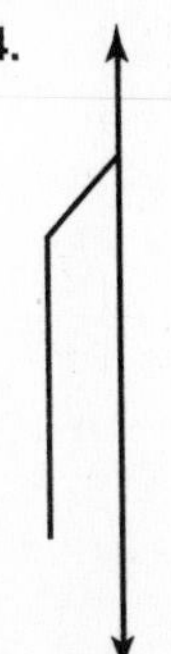

5.

6.

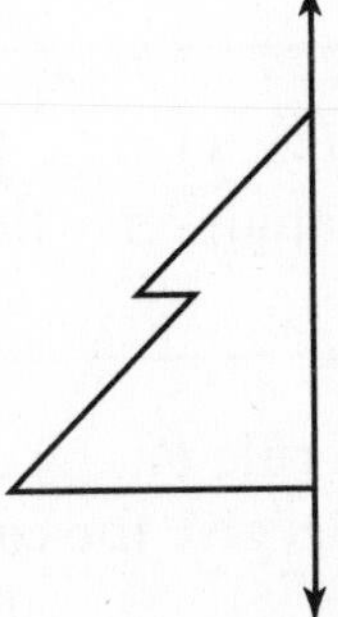

Quick Warm-Up: Assessing Prior Knowledge
1.7 Motion in the Coordinate Plane

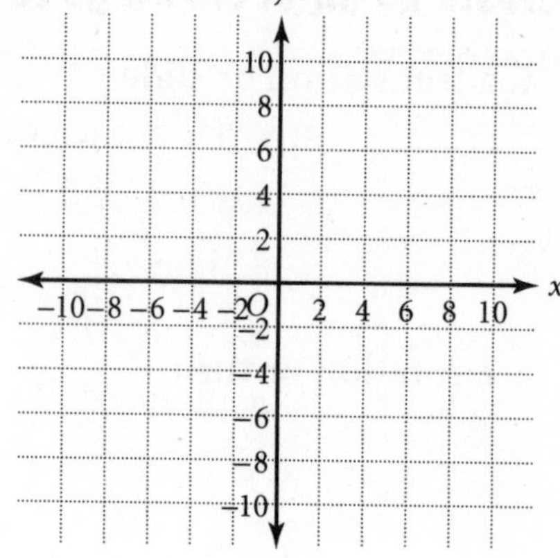

1. Find and draw the set of all ordered pairs in which the first number is −3.

2. Graph a square that has (2, 3) as one of its vertices. List and compare the *x*- and *y*-coordinates of the vertices.

____

____

Lesson Quiz
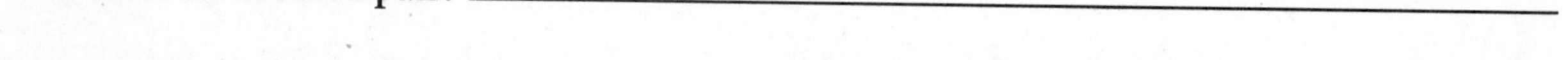
1.7 Motion in the Coordinate Plane

1. What is an ordered pair? ______________________________

____

2. What is the first number in an ordered pair called? ______________

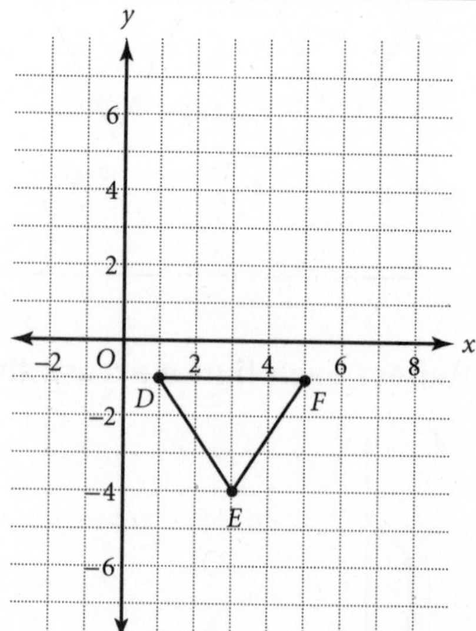

3. What are the coordinates of the vertices of ΔDEF, shown at right?

____

4. Draw the reflection of ΔDEF across the *x*-axis on the grid provided.

5. What are the coordinates of the vertices after the reflection across the *x*-axis?

____

6. If the rule $F(x, y) = (x + 3, y - 1)$ is applied to the original triangle, give the coordinates of its image.

____

7. If the same rule $F(x, y)$ above is applied again to the new triangle, give the coordinates of its image.

____

NAME ________________ CLASS ________ DATE ________

Chapter Assessment

Chapter 1, Form A, page 1

Write the letter that best answers the question or completes the statement.

______ 1. What plane contains points *M*, *N*, and *O*?

a. plane *B*
b. plane *X*
c. plane *MN*
d. plane *MNO*

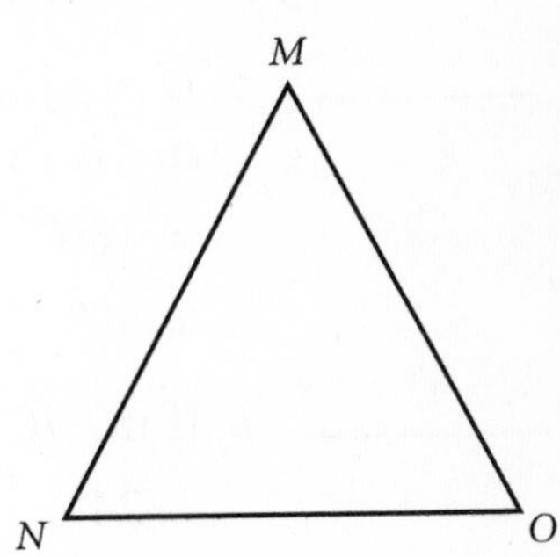

______ 2. How many rays in the figure have *A* as an endpoint?

a. 5 b. 10
c. 3 d. 15

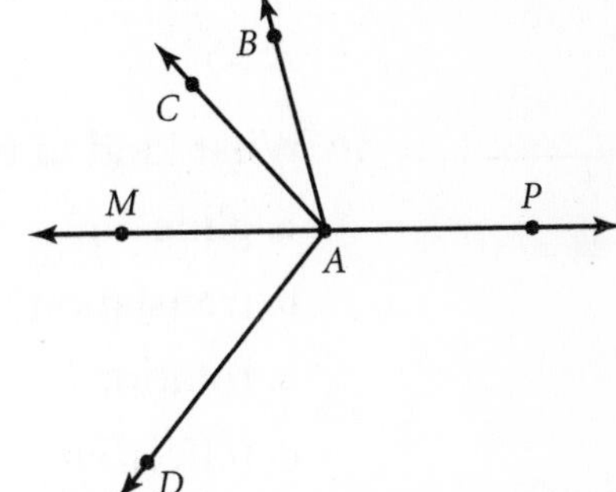

______ 3. Every point on the ______ of a segment is equidistant from the endpoints of the segment.

a. angle bisector b. median
c. perpendicular bisector d. plane

______ 4. Which segments are congruent?

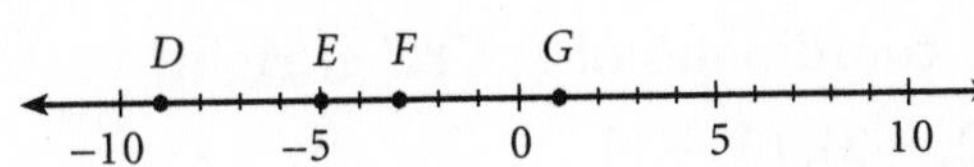

a. $\overline{DE}$ and $\overline{FG}$ b. $\overline{DE}$ and $\overline{EF}$
c. $\overline{EF}$ and $\overline{FG}$ d. $\overline{DG}$ and $\overline{EG}$

______ 5. *H* is between points *Q* and *R*. $QH = 23$ and $HR = 12$. What is the length of *QR*?

a. 12 b. 11 c. 35 d. 48

______ 6. At how many points do the perpendicular bisectors of the sides of a triangle intersect?

a. 0 b. 1 c. 2 d. 3

Chapter Assessment

Chapter 1, Form A, page 2

Use the figure to answer Exercises 7–9.

______ 7. If m∠JCK = 58° and m∠KCP = 35°, what is m∠JCP?

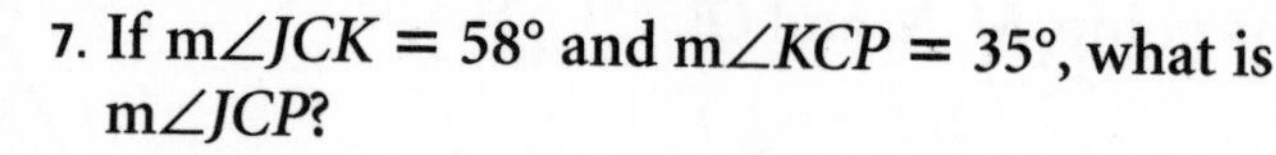

a. 180° b. 93°

c. 90° d. 83°

______ 8. If m∠JCK = 92° and m∠JCP = 116°, what is m∠KCP?

a. 24° b. 208° c. 90° d. 92°

______ 9. If m∠KCP = 35° and m∠JCP = 145°, what is m∠JCK?

a. 35° b. 45° c. 190° d. 110°

______ 10. What kind of transformation changes $\overline{ED}$ to $\overline{E'D'}$?

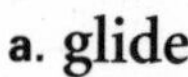

a. glide

b. translation

c. rotation

d. reflection

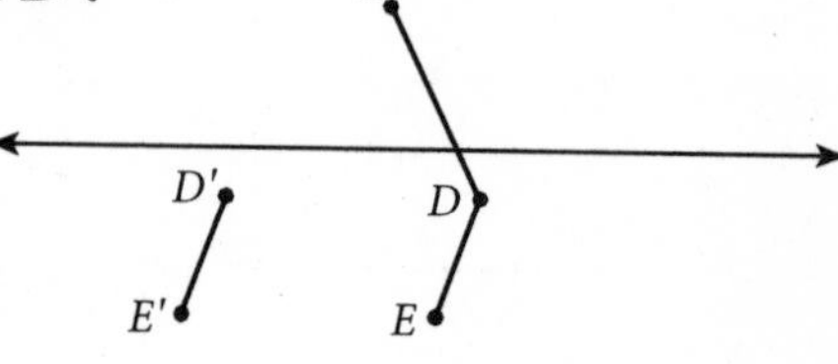

______ 11. What kind of transformation is shown in the figure at right?

a. slide

b. translation

c. rotation

d. reflection

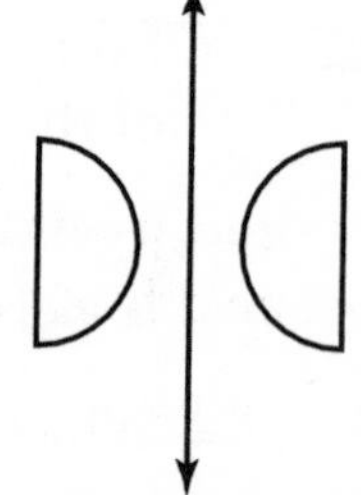

______ 12. What are the coordinates of △XYZ at right?

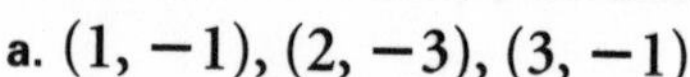

a. (1, −1), (2, −3), (3, −1)

b. (−1, 1), (−3, 2), (−1, 3)

c. (1, 1), (2, 2), (3, 3)

d. (−1, 1), (−2, −2), (−3, −3)

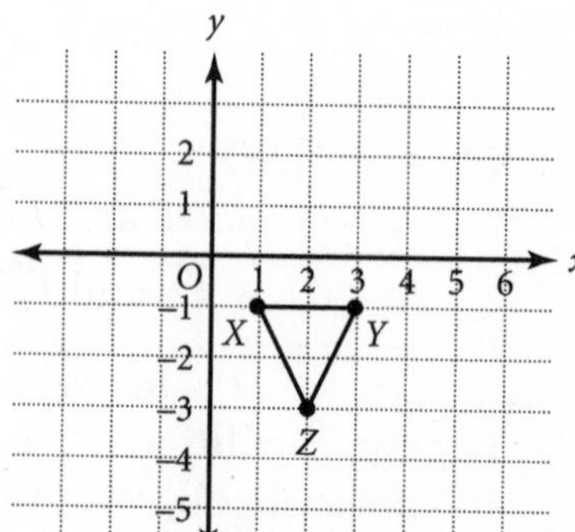

______ 13. What would be the coordinates of △$X'Y'Z'$, the image of △XYZ reflected across the y-axis?

a. (1, 1), (2, 3), (3, 1)

b. (1, −1), (2, 1), (3, −1)

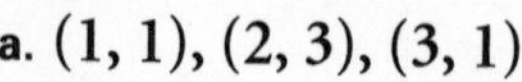

c. (1, −1), (−1, −1), (0, −3)

d. (−1, −1), (−3, −1), (−2, −3)

NAME ____________________ CLASS __________ DATE __________

Chapter Assessment

Chapter 1, Form B, page 1

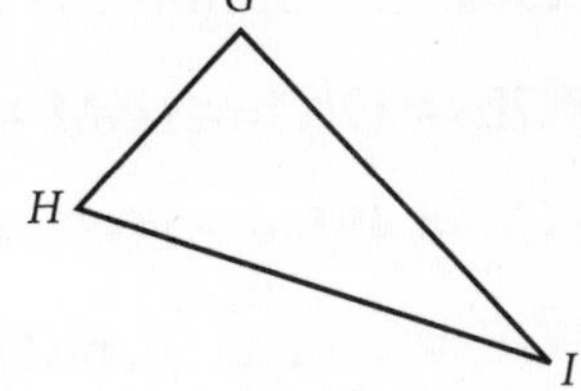

1. Name the segments in the triangle at right.

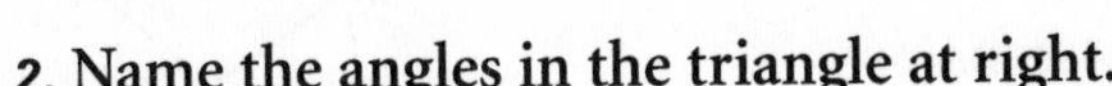

2. Name the angles in the triangle at right.

____

3. Name the plane that contains the points *G*, *H*, and *I*.

____

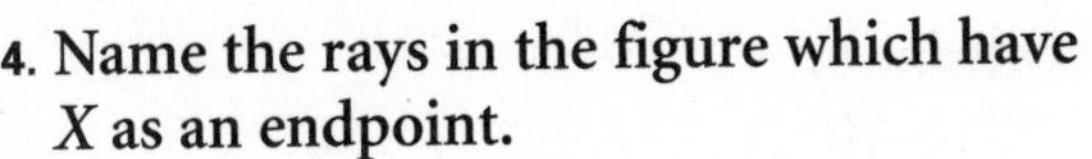

4. Name the rays in the figure which have *X* as an endpoint.

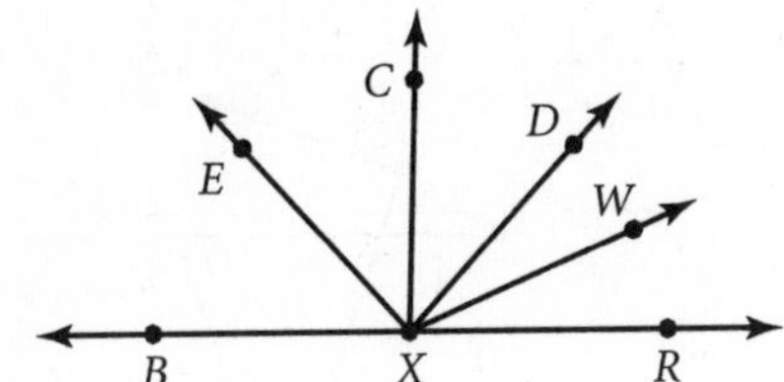

____

5. Describe the relationship between lines ℓ and *m*.

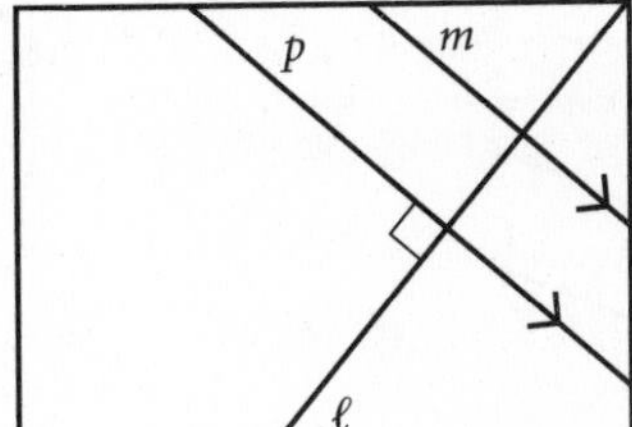

____

S is between points B and D. Sketch each figure in the space provided and find the missing measure.

6. $BS = 49$, $SD = 5$, $BD =$ ________

7. $BS =$ ________, $SD = 13.9$, $BD = 30$

8. $BS = 115$, $SD =$ ________, $BD = 180$

9. For which of the above exercises, if any, are $\overline{BS}$ and $\overline{SD}$ congruent? ________

10. Explain the difference between the inscribed circle and circumscribed circle of a triangle.

____

____

____

NAME ______________________ CLASS ____________ DATE ____________

Chapter Assessment

Chapter 1, Form B, page 2

In Exercises 11–13, find the missing measures.

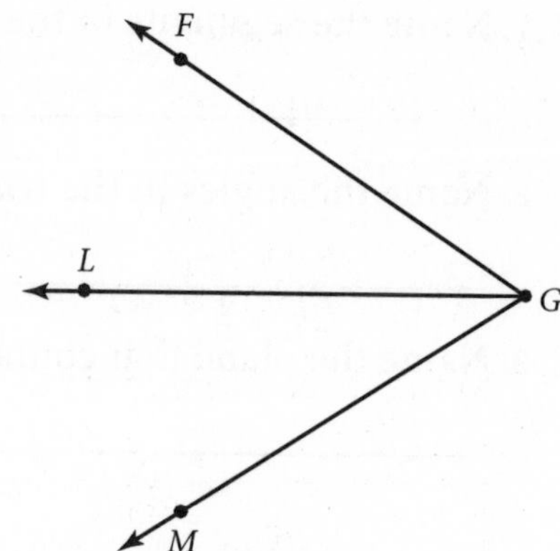

11. $m\angle FGL = 17°$, $m\angle LGM = 32°$, $m\angle FGM =$ ________

12. $m\angle FGL = 48°$, $m\angle LGM =$ ________, $m\angle FGM = 96°$

13. $m\angle FGL =$ ________, $m\angle LGM = 65°$, $m\angle FGM = 95°$

14. For which of the above exercises, if any, are $\angle FGL$ and $\angle LGM$ congruent?

____

Identify each transformation as a translation, rotation, or reflection.

15.

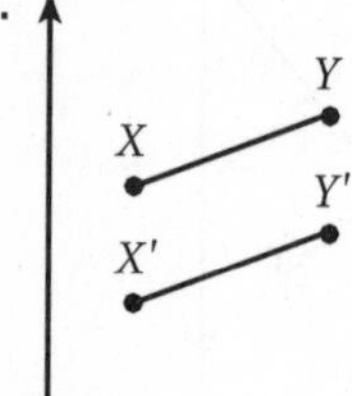

16.

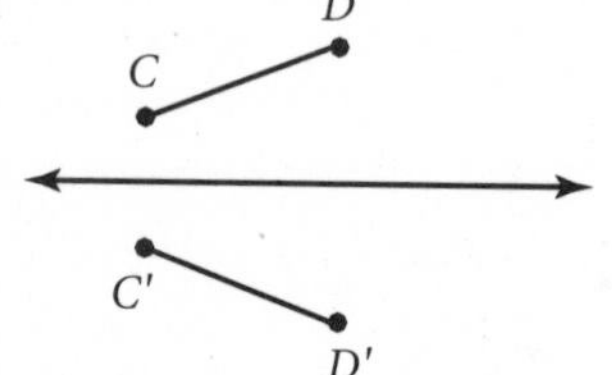

17.

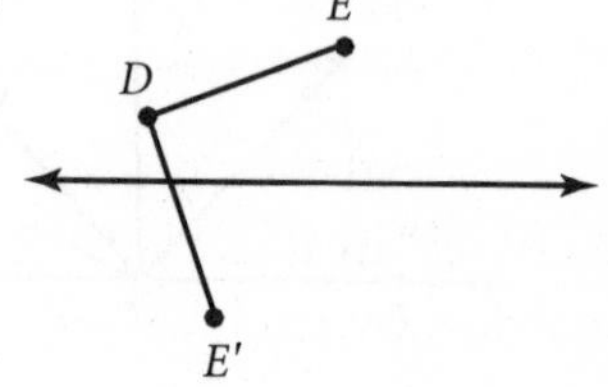

______________ ______________ ______________

18. Using the rule $T(x, y) = (x + 3, y + 3)$, transform $\triangle ABC$. Draw $\triangle A'B'C'$ on the grid provided, and give the coordinates of its vertices.

____

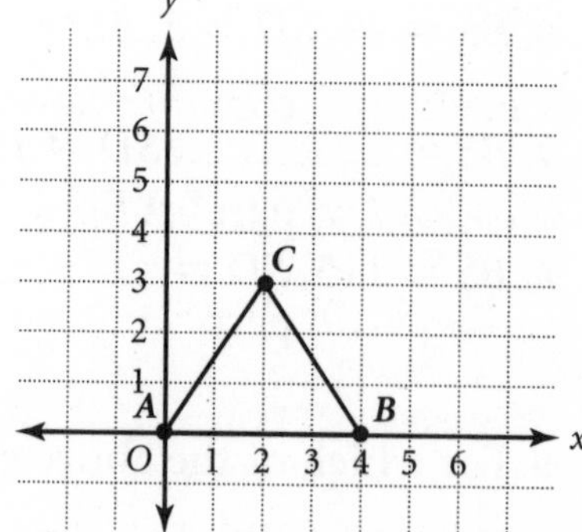

19. What type of transformation is shown in Exercise 18?

____

20. Write the rule to transform the new triangle back to $\triangle ABC$'s original coordinates.

____

21. Use the rule you wrote in Exercise 20 to transform $\triangle ABC$. Give the coordinates of the new vertices.

____

NAME ______________________ CLASS ______________ DATE ______________

Alternative Assessment

Modeling Geometric Figures, Chapter 1, Form A

TASK: Identify real-world models for geometric figures.

HOW YOU WILL BE SCORED: As you work through the task, your teacher will be looking for the following:

- whether you can find real-world models for geometric figures
- how well you describe the models mathematically

1. Name three different real-world objects that can be modeled by a point.

____

2. Name three different real-world objects that can be modeled by a line.

____

3. Name three different real-world objects that can be modeled by a plane.

____

4. Name a real-world object that contains all three geometric figures discussed in Exercises 1–3. List the parts of the object that can be modeled by each of the following figures:

Point: ______________ Line: ______________ Plane: ______________

______________ ______________ ______________

For Exercises 5–9, use the object you named in Exercise 4.

5. Name three collinear points. ______________________

6. Name two coplanar lines. ______________________

7. Name two parallel lines. ______________________

8. Name two perpendicular lines. ______________________

9. Describe a real-world object that contains a model of parallel planes. Use your object from Exercise 4, or another object, to illustrate your answer.

____

SELF-ASSESSMENT: What other objects or situations could answer Exercises 4–9?

NAME ______________________ CLASS ______________ DATE ______________

Alternative Assessment

Constructions, Chapter 1, Form B

TASK: Describe different constructions in a triangle and different situations in which those constructions would be useful.

HOW YOU WILL BE SCORED: As you work through the task, your teacher will be looking for the following:

- whether all constructions are explained clearly and mathematically
- how well you describe the situations in which the constructions would be useful

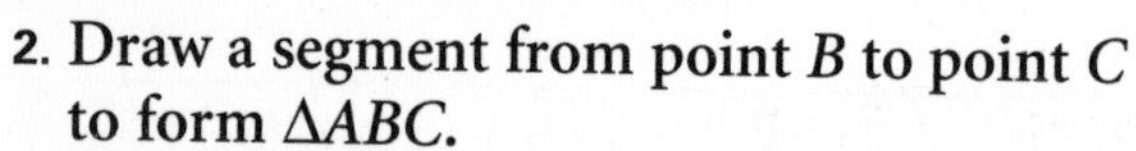

1. Using the coordinate grid provided, reflect $\overline{AB}$ across the y-axis. Label the reflection of point B as point C.

2. Draw a segment from point B to point C to form ΔABC.

3. Describe the process for constructing the medians of ΔABC.

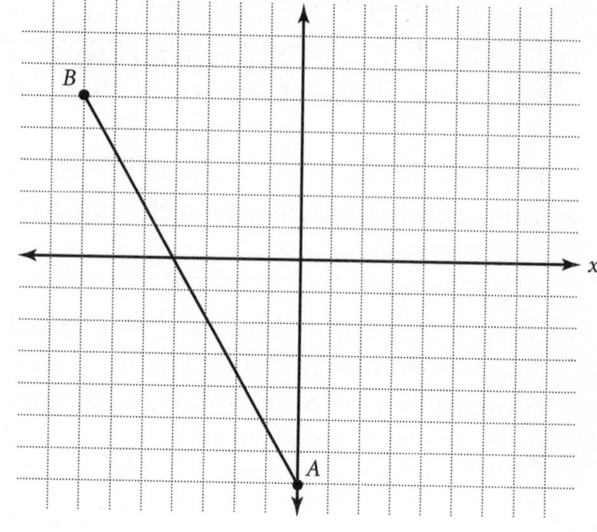

____

____

____

____

4. What point is created by the intersection of the three medians? ______________

5. The intersection point from Exercise 4 is the center of gravity for the triangle. Describe a situation in which it would be useful to find this point.

____

6. Describe the process for determining the center of the inscribed circle, the incenter of ΔABC.

____

7. Describe a situation in which it would be useful to find the incenter.

____

8. Are there any congruent angles created in the process of determining the incenter of ΔABC? Explain.

____

SELF-ASSESSMENT: Describe a situation in which all four of the special points of a triangle would be useful.

NAME ______________________ CLASS ______________ DATE ______________

Quick Warm-Up: Assessing Prior Knowledge

2.1 An Introduction to Proofs

1. The length of a square is 5 cm. What is its width? ______________

2. What is the sum of the first 5 counting numbers? ______________

3. Solve the equation $2x + 1 = 5$. ______________

Lesson Quiz

2.1 An Introduction to Proofs

1. The line and table below show that one line segment, $\overline{DE}$, can be formed by two points on a line. Add points to each line until you have the total number of points, n, listed in the table. Record in the table the number of segments created and their names.

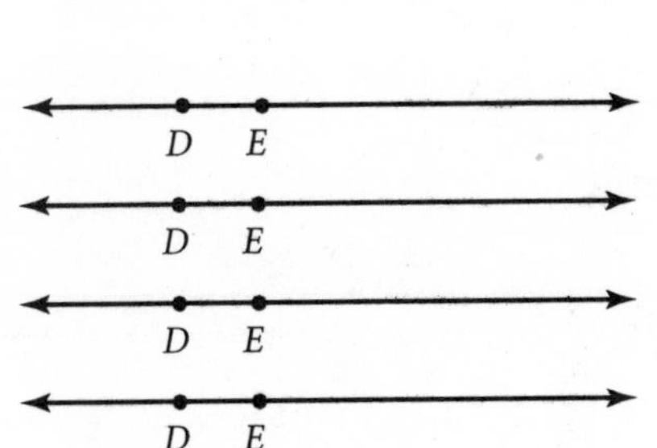

n	Segment names	Number of segments
2	$\overline{DE}$	1
3		
4		
5		

2. Write an expression in terms of n for the number of segments formed by n points on the line. ______________

3. Devise a method to find the sum of the first n odd numbers. Then use your method to write an expression in terms of n for the sum of the first n odd numbers.

____

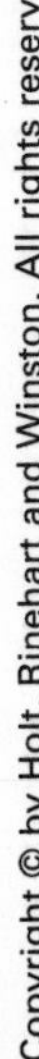

NAME ______________________ CLASS ____________ DATE ____________

Quick Warm-Up: Assessing Prior Knowledge
2.2 An Introduction to Logic

For Exercises 1–3, classify each statement as true or false.

1. A statement is a sentence that is either true or false. ____________
2. If Tom lives in Boston, then he lives in Massachusetts. ____________
3. If Tom lives in Massachusetts, then he lives in Boston. ____________
4. What is an Euler diagram? ____________

Lesson Quiz
2.2 An Introduction to Logic

Refer to the following statement to answer Exercises 1–4:

All TV comedy programs are fictional.

1. Rewrite the statement as a conditional.

____

2. Identify the hypothesis and the conclusion of the statement.

____

3. Draw an Euler diagram that illustrates the statement.

4. Write the converse of the conditional statement. Is the converse true or false? Explain.

____

____

NAME ______________________ CLASS __________ DATE __________

Quick Warm-Up: Assessing Prior Knowledge
2.3 Definitions

1. Write the definition of a triangle as a conditional. ______________________

2. What is the converse of "If *A*, then *B*"? ______________________

3. If a conditional is a true statement, will its converse be true? ______________________

Lesson Quiz
2.3 Definitions

Write the given statement in the specified forms. For the conclusion, state whether the resulting biconditional is a definition.

1. A rose is a red flower.

   Conditional: ______________________

   Converse: ______________________

   Biconditional: ______________________

   Conclusion: ______________________

2. Phoenix is a city in Arizona.

   Conditional: ______________________

   Converse: ______________________

   Biconditional: ______________________

   Conclusion: ______________________

3. A high school senior is in the twelfth grade.

   Conditional: ______________________

   Converse: ______________________

   Biconditional: ______________________

   Conclusion: ______________________

4. Is the following statement a definition? Explain. "A 150° angle is an obtuse angle." ______________________

____

NAME ______________________ CLASS ______________ DATE ______________

Mid-Chapter Assessment

Chapter 2 (Lessons 2.1–2.3)

Write the letter that best answers the question or completes the statement.

______ 1. What are the next three terms in this sequence?

$$\frac{1}{5}, \frac{1}{10}, \frac{1}{20}, \frac{1}{40}, \ldots$$

a. $\frac{1}{50}, \frac{1}{60}, \frac{1}{70}$ b. $\frac{1}{60}, \frac{1}{20}, \frac{1}{240}$ c. $\frac{1}{80}, \frac{1}{160}, \frac{1}{320}$ d. $\frac{1}{60}, \frac{1}{80}, \frac{1}{100}$

______ 2. Which conditional statement represents the claim "All birds can fly"?

a. If it flies, then it is a bird. b. If it is a bird, then it can fly.

c. Anything that flies is a bird. d. If it has feathers, then it is a bird.

Tell whether each statement is true or false.

3. In the statement "If it is cloudy, then it will rain" the hypothesis is "it will rain". ______

4. Adjacent angles are angles in a plane that have a common vertex and a common side but no common interior points. ______

Refer to the following statement to answer Exercises 5–7:

All rabbits are white.

5. Rewrite the statement as a conditional.

____

6. Identify the hypothesis and the conclusion of the statement.

____

7. Write the converse of the conditional statement. Is the converse true or false? Explain.

____

____

8. Write the statement below as a biconditional. Is the statement true or false? Explain.

All baseball players are athletes.

____

____

NAME ______________________ CLASS ____________ DATE ____________

Chapter Assessment

Chapter 2, Form A, page 1

Write the letter that best answers the question or completes the statement.

______ 1. What conjecture can you make about the sequence's relation to zero?

$$\frac{1}{3}, \frac{1}{6}, \frac{1}{12}, \frac{1}{24}, \ldots$$

a. The denominator is doubled, so the terms get further from zero.
b. The denominator is increasing, so the terms get closer to zero.
c. The denominators are multiples of 3, so the terms get closer to zero.
d. The numerator is always 1, so the terms do not change their distance from zero.

______ 2. What is the converse of the statement below?

If a triangle is equilateral, then it is isosceles.

a. A triangle is equilateral if and only if it is isosceles.
b. All equilateral triangles are isosceles.
c. A triangle is isosceles if and only if it is equilateral.
d. If a triangle is isosceles, then it is equilateral.

______ 3. What conclusion can you draw from the two statements below?

If an animal is a dog, it has four legs.
Murphy is a dog.

a. Murphy is an animal.
b. Murphy is not an animal.
c. Murphy has four legs.
d. Murphy does not have four legs.

______ 4. Choose which of the following is a counterexample for the *converse* of the statement "If an animal is a dog, then it has four legs."

a. Brandy the cat has four legs.
b. Spot the dog has four legs.
c. Murphy the dog is an animal.
d. Polly the parrot has two legs.

______ 5. Which proposed definition is true?

a. Two lines are parallel if and only if they intersect to form a right angle.
b. Two lines are perpendicular if and only if they are coplanar and do not intersect.
c. Two lines are parallel if and only if they are coplanar and do not intersect.
d. Two lines are parallel if and only if they intersect and do not form right angles.

______ 6. Which statement is the biconditional of "All rings are round"?

a. If it is a ring, then it is round.
b. If it is a ring, then it is not round.
c. If it is round, then it is a ring.
d. It is a ring if and only if it is round.

NAME ______________________ CLASS ______________ DATE ______________

Chapter Assessment

Chapter 2, Form A, page 2

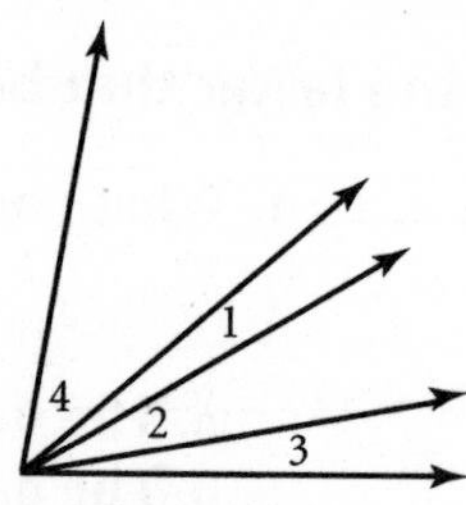

______ 7. Which angles form an adjacent pair?

a. $\angle 1$ and $\angle 3$
b. $\angle 3$ and $\angle 4$
c. $\angle 2$ and $\angle 3$
d. $\angle 2$ and $\angle 4$

______ 8. What property justifies the conclusion of the statement below?

If $\overline{BC} \cong \overline{DE}$, then $\overline{DE} \cong \overline{BC}$.

a. Symmetric Property of Congruence
b. Addition Property of Equality
c. Reflexive Property of Congruence
d. Transitive Property of Congruence

______ 9. What property justifies the conclusion?

$x + 7 = 21$ ← Given
$x + 7 - 7 = 21 - 7$
$x = 14$ ← conclusion

a. Symmetric Property of Equality
b. Addition Property of Equality
c. Subtraction Property of Equality
d. Reflexive Property of Equality

______ 10. What is the measure of $\angle ABC$?

a. 105°
b. 75°
c. 15°
d. 90°

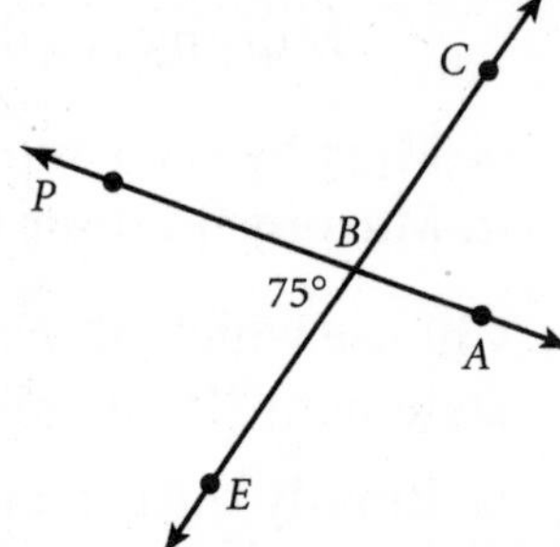

______ 11. Which of the following is the name given to a statement that has been proven by deductive reasoning?

a. conjecture
b. postulate
c. property
d. theorem

______ 12. What property justifies the conclusion of the statement below?

$m\angle ABC = m\angle CBA$

a. Symmetric Property of Equality
b. Addition Property of Equality
c. Reflexive Property of Equality
d. Transitive Property of Equality

NAME ______________________ CLASS ____________ DATE ____________

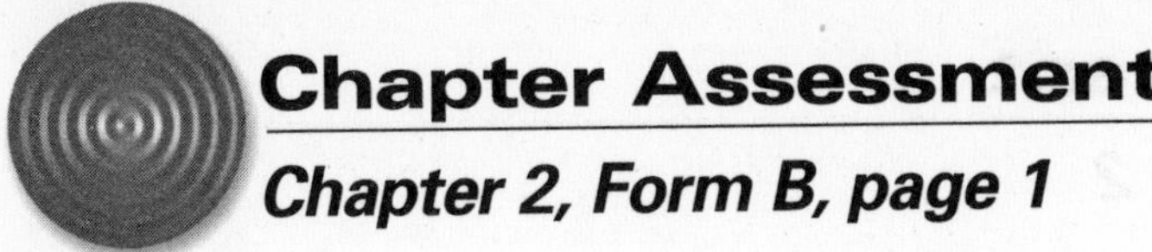

Chapter Assessment

Chapter 2, Form B, page 1

1. What are the next four terms of the sequence below?

$$\frac{1}{400}, \frac{1}{350}, \frac{1}{300}, \frac{1}{250}, \ldots$$

____

2. What conjecture can you make about the sequence's terms?

____

Refer to the following statement to answer Exercises 3 and 4:

All Knicks fans live in New York.

3. Rewrite the statement as a conditional.

____

4. Identify the hypothesis and the conclusion of the statement.

____

5. Arrange the three statements to form a logical chain. Then write the conditional statement proved by the logical chain.

If it is hot, then it is summer.
If people go swimming, then it is hot.
If the sun is shining, then people go swimming.

____

____

____

6. Write the given statement in the specified forms. Is the biconditional true or false? Explain.

A 45° angle is an obtuse angle.

Conditional: ______________________________

Converse: ______________________________

Biconditional: ______________________________

Conclusion: ______________________________

Chapter Assessment

Chapter 2, Form B, page 2

Tell whether each statement is true or false.

7. When you interchange the hypothesis and the conclusion of a statement, the new conditional is the converse of the original conditional. ________
8. A, B, and C are collinear points with B between A and C. Thus, $AB + BC = AC$. ________
9. In the statement "If it is cold, then it will snow" the conclusion is "it is cold." ________
10. For all real numbers x, y and z, if $x = y$ and $y = z$, then $x = z$. ________
11. If two lines intersect, then the adjacent angles formed are congruent. ________

Complete each proof.

Given: $\overrightarrow{NM} \perp \overrightarrow{NO}$
$m\angle MNO = m\angle RQP$

Prove: $\overrightarrow{QR} \perp \overrightarrow{QP}$

Statements	Reasons
$\overrightarrow{NM} \perp \overrightarrow{NO}$	Given
$m\angle MNO = 90°$	12. ________
$m\angle MNO = m\angle RQP$	13. ________
$m\angle RQP = 90°$	14. ________
$\overrightarrow{QR} \perp \overrightarrow{QP}$	15. ________

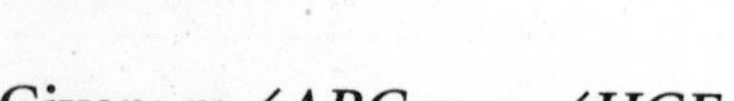

Given: $m\angle ABC = m\angle HGF$
$m\angle 1 = m\angle 3$
Prove: $m\angle 2 = m\angle 4$

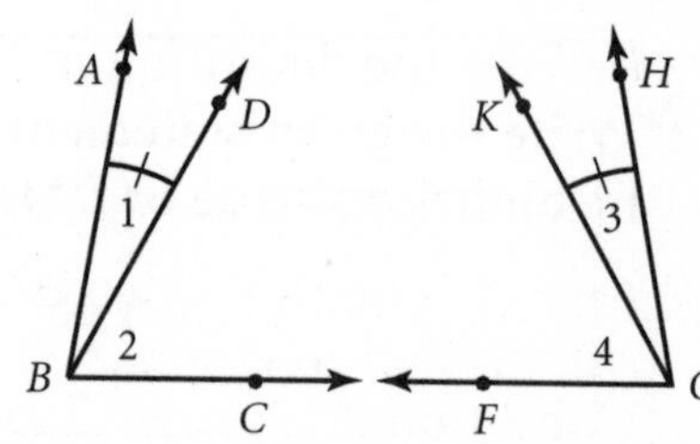

Statements	Reasons
$m\angle ABC = m\angle HGF$ $m\angle 1 = m\angle 3$	Given Given
$m\angle ABC - m\angle 1 = m\angle 2$ $m\angle HGF - m\angle 3 = m\angle 4$	16. ________
$m\angle ABC - m\angle 1 = m\angle HGF - m\angle 3$	17. ________
$m\angle 2 = m\angle 4$	18. ________

NAME ______________________ CLASS __________ DATE __________

Alternative Assessment

Conditionals and Definitions, Chapter 2, Form A

TASK: Write conditionals and definitions describing real-world situations.

HOW YOU WILL BE SCORED: As you work through the task, your teacher will be looking for the following:

- whether you are able to write conditional statements
- how well you write definitions and translate them into other kinds of statements

1. Describe the tardiness policy for your school or a particular class.

____

____

2. Using your description of the tardiness policy, write a conditional statement that describes what happens when a student is late to class.

____

3. Write the converse of the tardiness conditional statement.

____

4. Is the converse true? Why or why not?

____

5. If your original conditional statement from Exercise 1 were converted into a biconditional statement, would it still be true? Why or why not?

____

6. Define the term *congruent angles.* ______________________

7. Rewrite the definition of congruent angles by using a biconditional statement.

____

8. Choose a math term and write its definition in the form of a conditional statement.

____

9. Is the converse of your definition true? Why or why not?

____

SELF-ASSESSMENT: What conditional or biconditional statements do you use when speaking to others? For example, "I will ______ if ______."

NAME ______________________ CLASS ____________ DATE ____________

Alternative Assessment

Proofs, Chapter 2, Form B

TASK: Prove statements true or false.

HOW YOU WILL BE SCORED: As you work through the task, your teacher will be looking for the following:

- whether your proof is accurate
- how well you construct your argument

In each of the following situations, prove or disprove the statement by using the given information. You may use either a two-column or paragraph proof.

1. Statement: Anthony can still make a B in his class if he makes 86% or higher on the last test.

   Given information: Anthony is taking a beginning biology class at a local college. The professor weighs each of the five tests equally in the calculation of students' final grades. Anthony's first four test scores were 90%, 65%, 78%, and 81%. To make a B, his average test grade must be at least 80%.

   Proof:

2. Statement: Carol was 45 years old in 1995.

   Given information: In 1995, Carol was twice as old as her daughter Monique. In the year 2000, Monique was 28 years old.

   Proof:

SELF-ASSESSMENT: Using a real-world situation, write a statement and given information like those in Exercises 1 and 2. Then prove your statement.

NAME ________________ CLASS ________ DATE ________

Quick Warm-Up: Assessing Prior Knowledge

3.1 Symmetry in Polygons

Refer to the diagram to answer the questions.

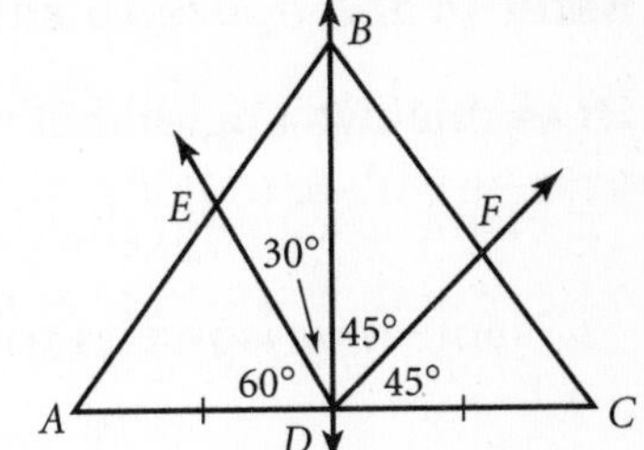

1. Name all of the rays that bisect $\overline{AC}$. ________
2. Name the perpendicular bisector of $\overline{AC}$. ________
3. Name the bisector of $\angle CDB$. ________

Lesson Quiz

3.1 Symmetry in Polygons

Determine whether each figure has reflectional symmetry, rotational symmetry, both, or neither.

1. 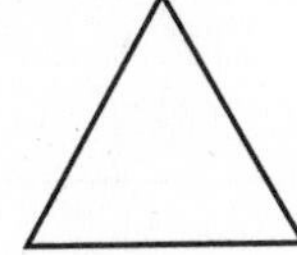

____

2. 

____

3. 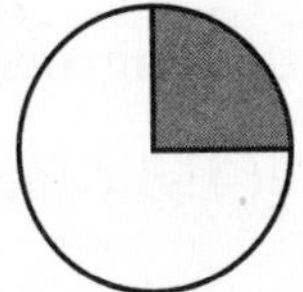

____

Each figure shows part of a shape with reflectional symmetry. Complete each figure.

4. 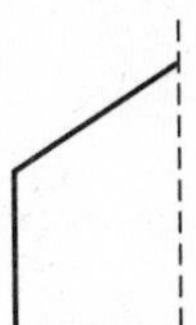

5. 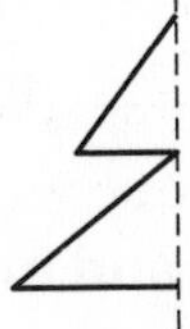

Find the measure of the central angle for each of the following.

6. square ________
7. regular hexagon ________
8. regular decagon ________

9. Draw and label 3 triangles: an equilateral, an isosceles, and a scalene.

NAME ______ CLASS ______ DATE ______

Quick Warm-Up: Assessing Prior Knowledge

3.2 Properties of Quadrilaterals

Refer to the figure to answer the questions.

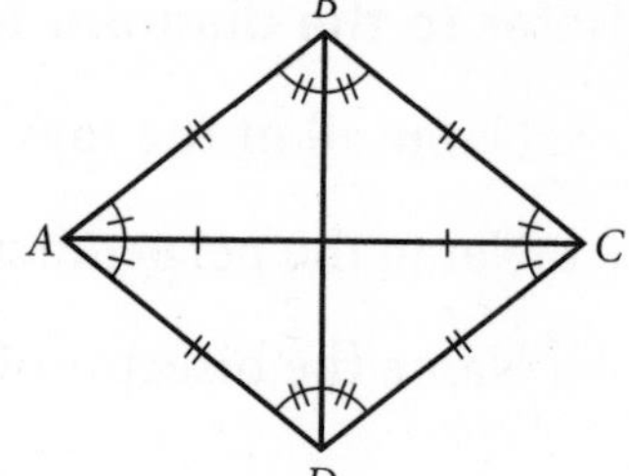

1. Name two angles that are bisected by $\overline{AC}$. ______
2. Name a segment that is bisected by $\overline{BD}$. ______
3. Name four isosceles triangles. ______

Lesson Quiz

3.2 Properties of Quadrilaterals

Identify the type of quadrilateral *ABCD* is.

1. *ABCD* has one and only one pair of parallel sides. ______
2. *ABCD* has four congruent sides and a 60° angle ______
3. *ABCD* has four right angles and four congruent sides. ______
4. *ABCD* has exactly two pairs of congruent sides and four right angles. ______

***EFGH* is a parallelogram with m∠*EFG* = 120°, m∠*EHF* = 58°, *EF* = 8, and *FO* = 5. Find each length or angle measure.**

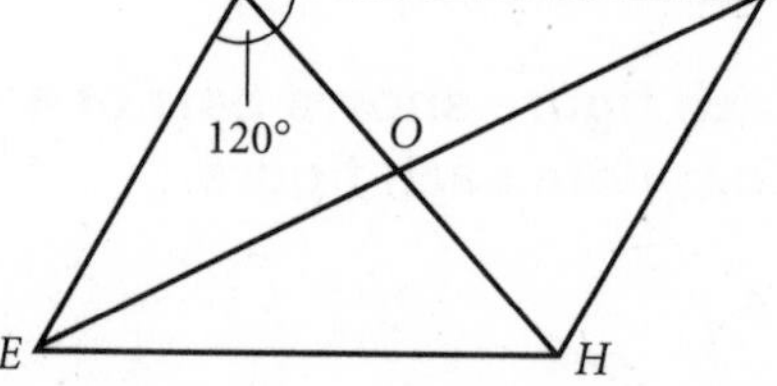

5. m∠*EHG* ______
6. *GH* ______
7. m∠*FHG* ______
8. *HO* ______
9. *FH* ______
10. m∠*GFH* ______

***JKLM* is a rectangle with *JL* = 12 and m∠*JKM* = 75°. Find each length or angle measure.**

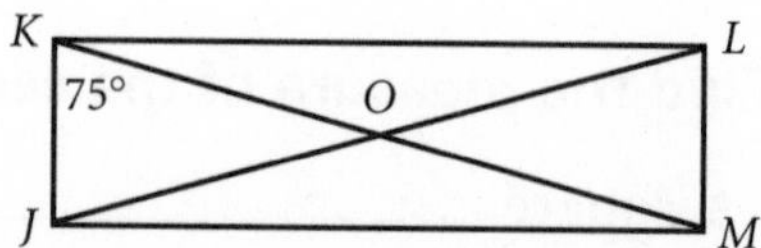

11. *OL* ______
12. m∠*MKL* ______
13. *KM* ______
14. m∠*KMJ* ______
15. m∠*JML* ______
16. *OM* ______

Find each length or angle measure if *JL* = 19 and m∠*JKM* = 63°.

17. *OL* ______
18. m∠*MKL* ______
19. *KM* ______
20. m∠*KMJ* ______
21. m∠*KML* ______
22. *OM* ______

NAME ______________________ CLASS ______________ DATE ______________

Quick Warm-Up: Assessing Prior Knowledge

3.3 Parallel Lines and Transversals

Refer to the diagram to answer the questions.

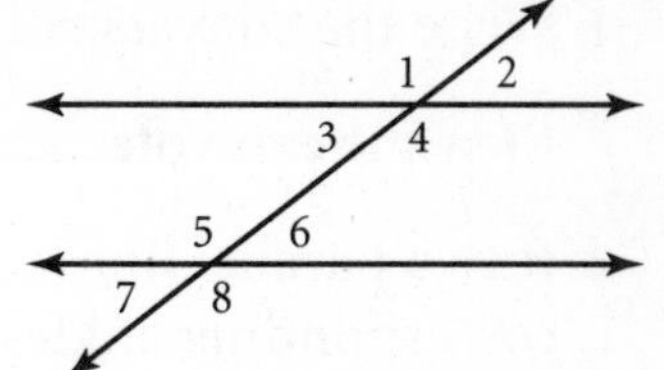

1. Is ∠4 congruent to ∠1? If so, what postulate or theorem explains why? ______________________

2. If ∠5 is congruent to ∠1, what can you conclude about ∠1 and ∠8? ______________________

3. What is the relationship between ∠3 and ∠4?

____

Lesson Quiz

3.3 Parallel Lines and Transversals

Based on the given figure, identify the type of special angle pair.

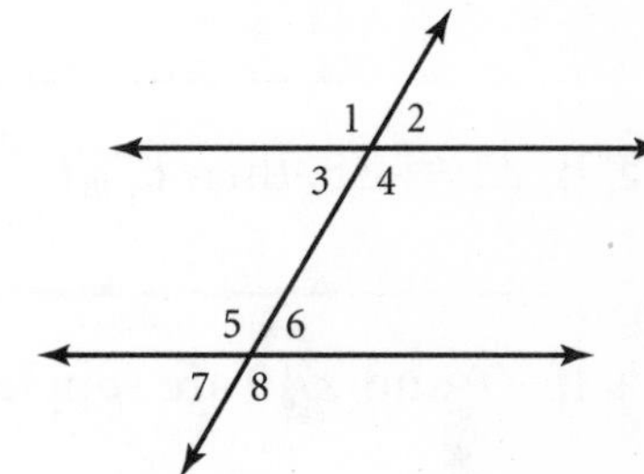

1. angles 4 and 5 ______________________

2. angles 2 and 6 ______________________

3. angles 3 and 5 ______________________
4. angles 1 and 8 ______________________

If $\ell_1 \parallel \ell_2$ and m∠5 = 65°, find each angle measure in the figure at right.

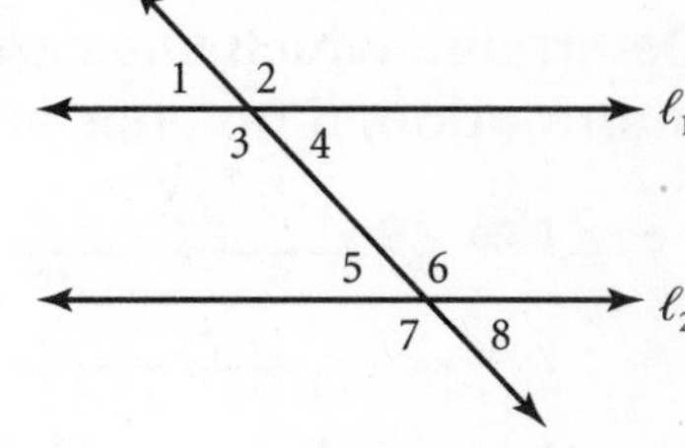

5. m∠1 ______ 6. m∠3 ______ 7. m∠7 ______

8. m∠4 ______ 9. m∠8 ______ 10. m∠2 ______

If $\ell_3 \parallel \ell_4$, m∠3 = $(2x - 10)°$, and m∠7 = $(x + 70)°$, find each angle measure in the figure at right.

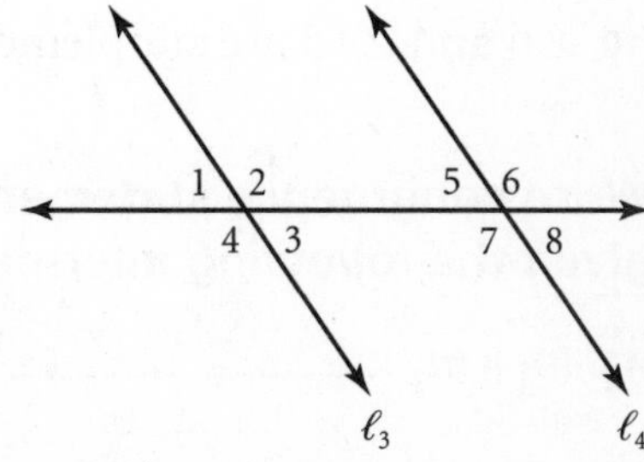

11. m∠4 ______ 12. m∠1 ______ 13. m∠8 ______

14. m∠2 ______ 15. m∠5 ______ 16. m∠6 ______

Quick Warm-Up: Assessing Prior Knowledge

3.4 Proving That Lines Are Parallel

1. Write the converse of this conditional: If Rosalia is 18 years old, then she can vote. ______________________

2. If two parallel lines are cut by a transversal, how many pairs of corresponding angles are formed? ____________

3. If two parallel lines are cut by a transversal, then what is the relationship between alternate interior angles? ____________

Lesson Quiz

3.4 Proving That Lines Are Parallel

Use the diagram to complete each statement.

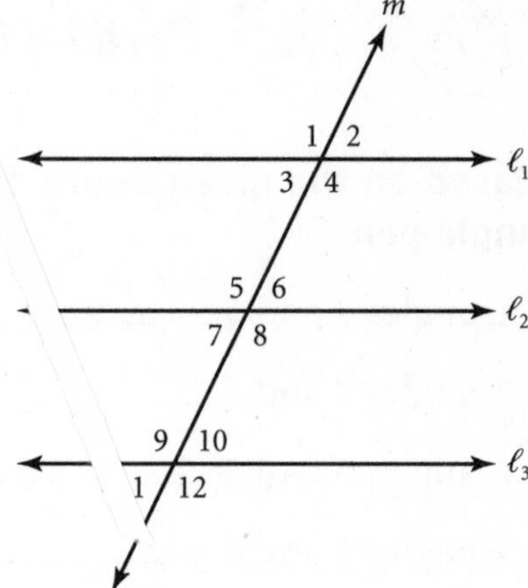

1. If $\angle 3 \cong \angle 10$, then $\ell_1 \parallel \ell_3$ because ______________________.

2. If $\angle 1 \cong \angle 5$, then $\ell_1 \parallel \ell_2$ because ______________________.

3. If $\angle 8$ and $\angle 10$ are supplementary, then ℓ_2, $\parallel \ell_3$ because ______________________.

Determine which lines are parallel given the following information. If no lines are parallel, write *none*.

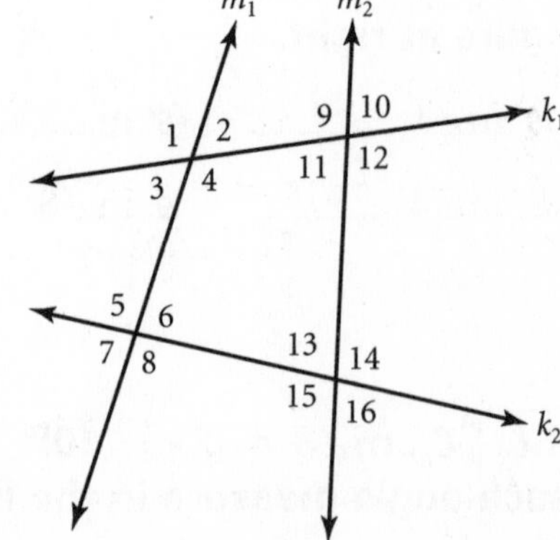

4. $\angle 1 \cong \angle 9$ ____________

5. $\angle 11 \cong \angle 14$ ____________

6. $\angle 1 \cong \angle 7$ ____________

7. $\angle 6 \cong \angle 15$ ____________

8. $\angle 12 \cong \angle 16$ ____________

9. $\angle 4 \cong \angle 13$ ____________

10. $\angle 6$ and $\angle 15$ are supplementary. ____________

Write congruence statements for alternate exterior angles, given the following information.

11. $m_1 \parallel m_2$ ______________________

12. $k_1 \parallel k_2$ ______________________

NAME ______________ CLASS ______________ DATE ______________

Mid-Chapter Assessment

Chapter 3 (Lessons 3.1–3.4)

Write the letter that best answers the question or completes the statement.

______ 1. Which of the following figures has only reflectional symmetry?

a. 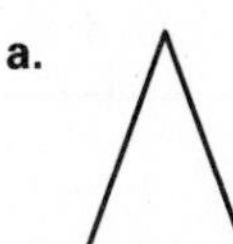 b. 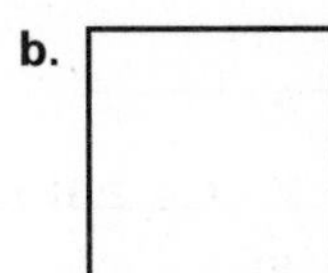 c. 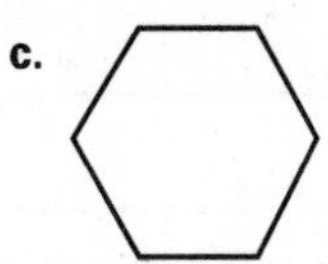 d. 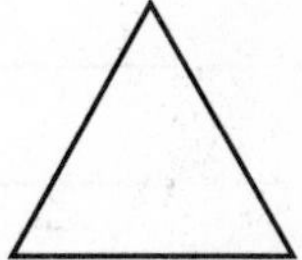

______ 2. Find the rotational symmetry of a regular hexagon.

a. 30°, 60°, 90°, 120°, 150°, 180° b. 45°, 90°, 135°, 180°, 225°, 270°
c. 60°, 120°, 180°, 240°, 300°, 360° d. 90°, 180°, 270°, 360°, 450°, 540°

______ 3. What is a parallelogram with four congruent sides called?

a. trapezoid b. rhombus c. square d. rectangle

______ 4. If $\ell_1 \parallel \ell_2$ and $m\angle 3 = 75°$, then $m\angle 5 =$ ______.

a. 75° b. 105°
c. 150° d. 180°

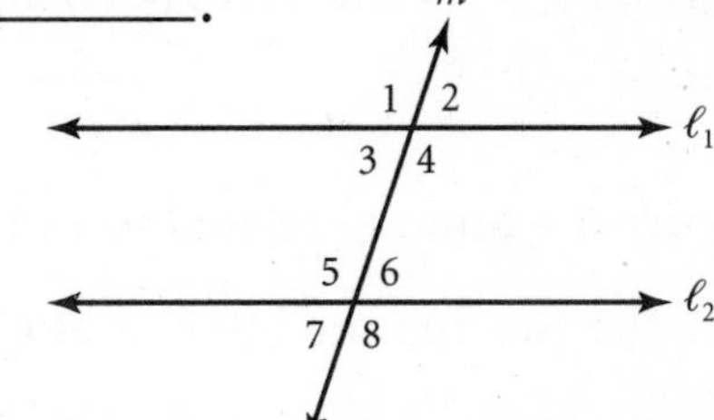

ABCD is a rhombus. Find each length or angle measure.

5. BC ______________ 6. $m\angle BOC$ ______________

7. $m\angle ADC$ ______________ 8. $m\angle ABO$ ______________

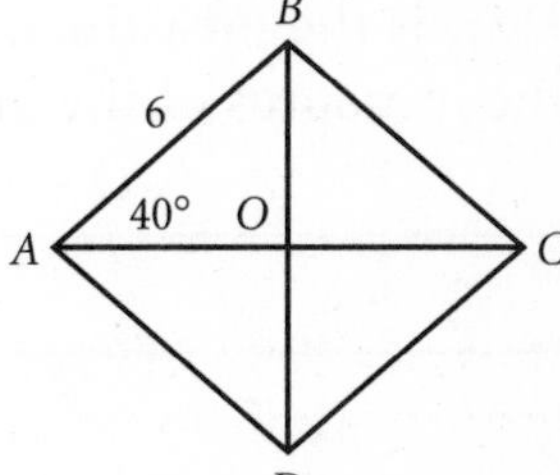

If $k_1 \parallel k_2$, $m\angle 4 = (3x - 10)°$, and $m\angle 5 = (x + 70)°$, find each measure.

9. $m\angle 8$ __________ 10. $m\angle 6$ __________

11. $m\angle 3$ __________ 12. $m\angle 7$ __________

13. How would you transform line n above to be a line of symmetry of k_1, and k_2?

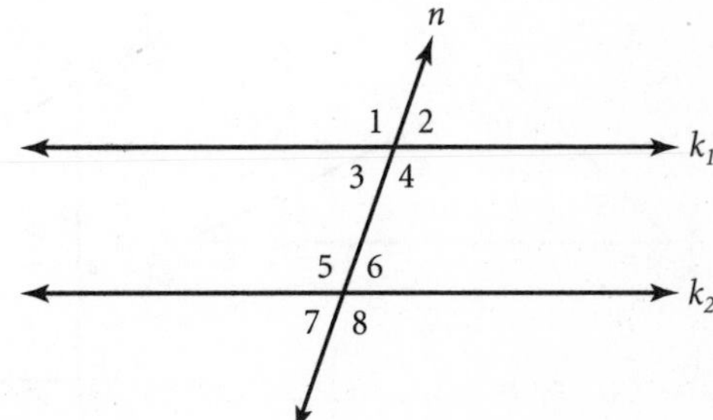

____

____

NAME ______________________ CLASS ____________ DATE ____________

Quick Warm-Up: Assessing Prior Knowledge

3.5 The Triangle Sum Theorem

Refer to the diagram to answer the questions.

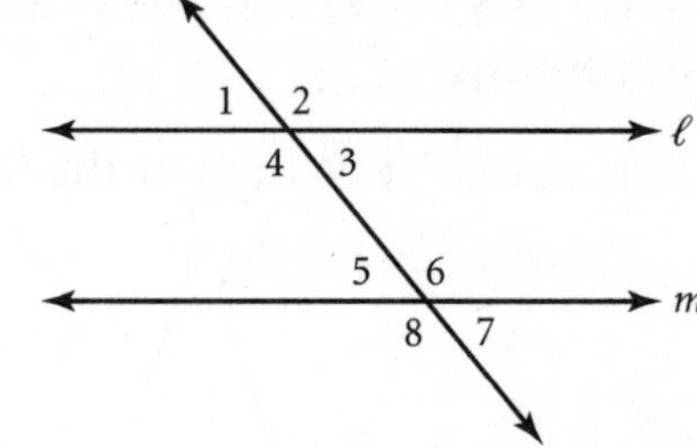

1. If ∠1 and ∠5 have different measures, are lines ℓ and *m* parallel? Why or why not?

   ____

   ____

2. If m∠3 + m∠6 = 180, what can you conclude about lines ℓ and *m*?

   ____

Lesson Quiz

3.5 The Triangle Sum Theorem

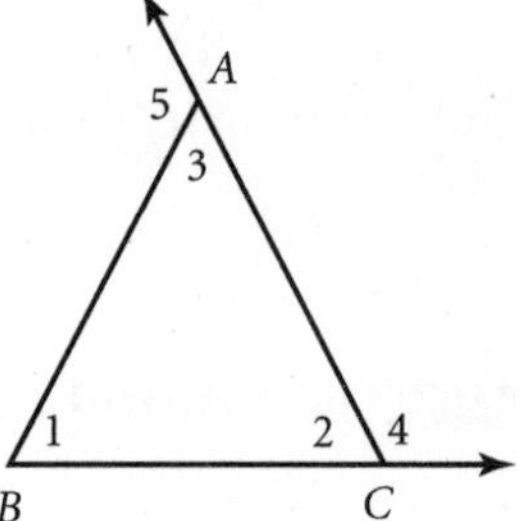

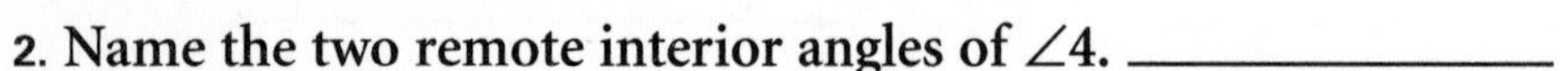

1. How are ∠4 and ∠5 related to △*ABC*? ______________

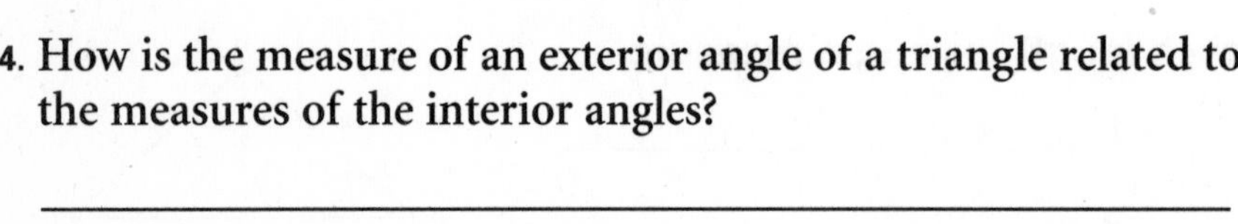

2. Name the two remote interior angles of ∠4. ______________
3. Name the two remote interior angles of ∠5. ______________
4. How is the measure of an exterior angle of a triangle related to the measures of the interior angles?

   ____

   ____

Find the indicated angle measure.

5.

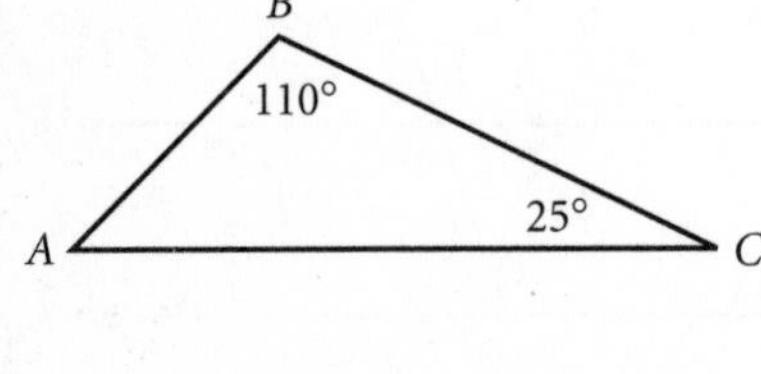

m∠*A* ______________

6.

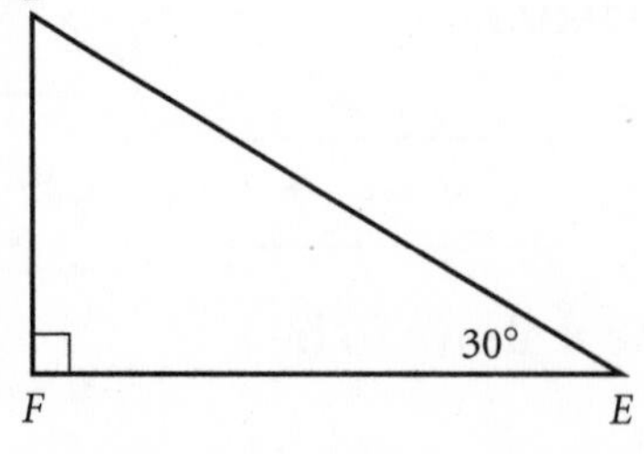

m∠*D* ______________

7.

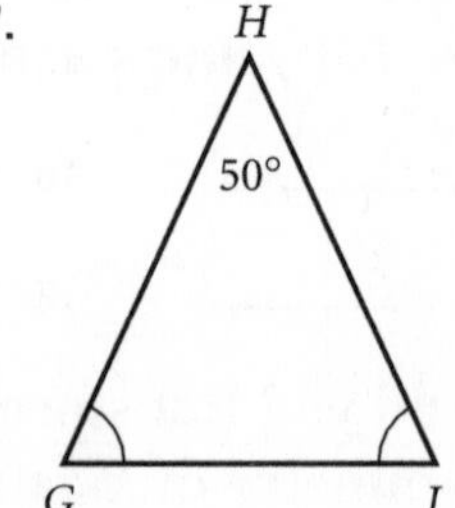

m∠*G* ______________

NAME ____________________ CLASS ____________ DATE ____________

Quick Warm-Up: Assessing Prior Knowledge

3.6 Angles in Polygons

1. If $\angle A$ in $\triangle ABC$ has a measure of 70°, then what is the measure of $m\angle B + m\angle C$? ____________

2. If $\triangle XYZ$ is a triangle all of whose angles are congruent, then what is the measure of $\angle X$?

Lesson Quiz

3.6 Angles in Polygons

1. What is the sum of the measures of the interior angles of a pentagon? ____________
2. What is the measure of one interior angle of a regular octagon? ____________
3. What is the sum of the measures of the exterior angles of a heptagon? ____________
4. What is the measure of one exterior angle of a regular hexagon? ____________

Find each angle measure.

5.

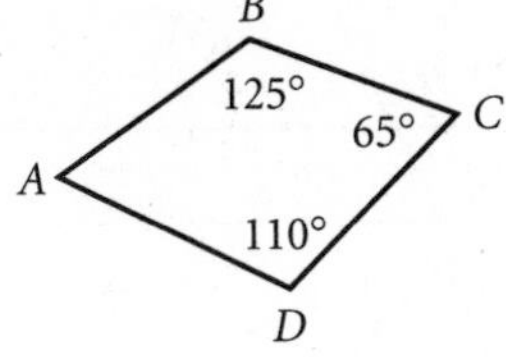

$m\angle A$ ____________

6.

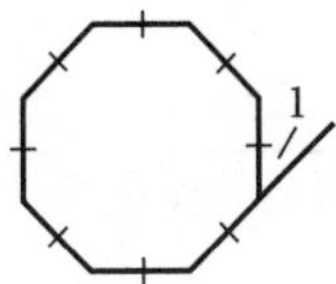

$m\angle 1$ ____________

7.

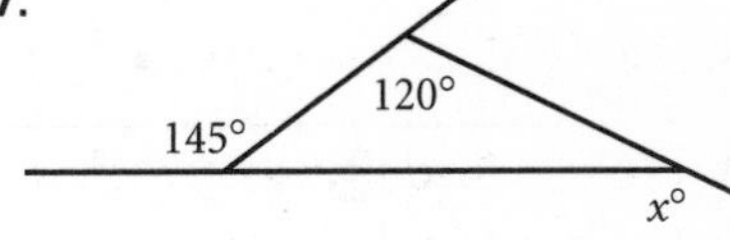

x ____________

8.

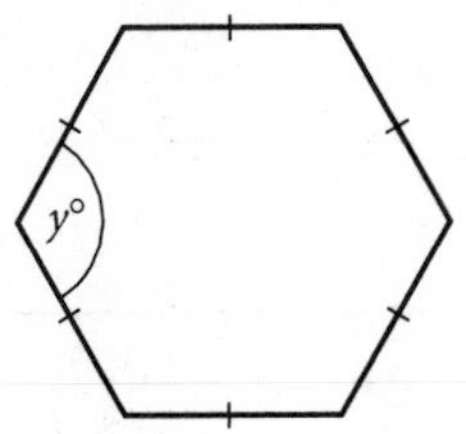

y ____________

9.

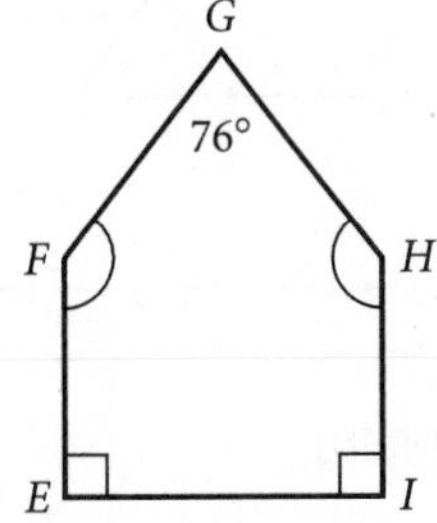

$m\angle F$ ____________

10.

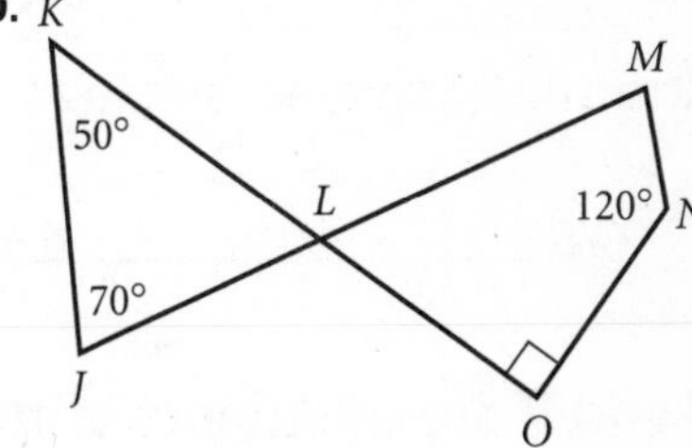

$m\angle M$ ____________

11. An exterior angle of a regular polygon is 24°. Find the number of sides in the polygon. ____________

NAME ______________________ CLASS ____________ DATE ____________

Quick Warm-Up: Assessing Prior Knowledge

3.7 Midsegments of Triangles and Trapezoids

1. If M is the midpoint of $\overline{AB}$ and $MA = 7$, then $AB = \underline{?}$ and $MB = \underline{?}$. ____________
2. What is the average of 15, 27, and 36? ____________
3. If the average of x and 20 is 16.25, what is x? ____________

Lesson Quiz

3.7 Midsegments of Triangles and Trapezoids

Find each measure, if it is possible. If not, write *cannot be determined.*

1. 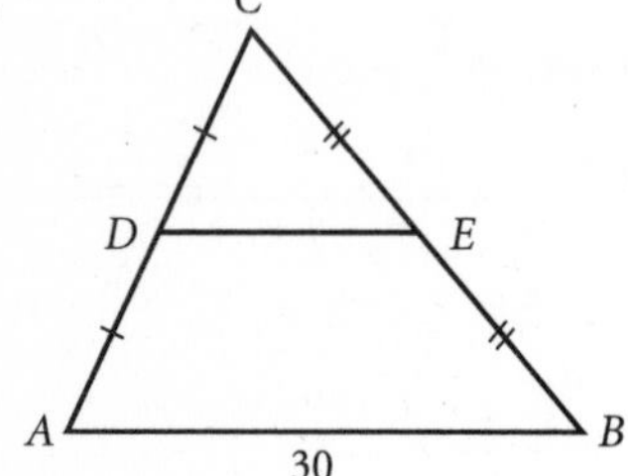

DE ____________

2. 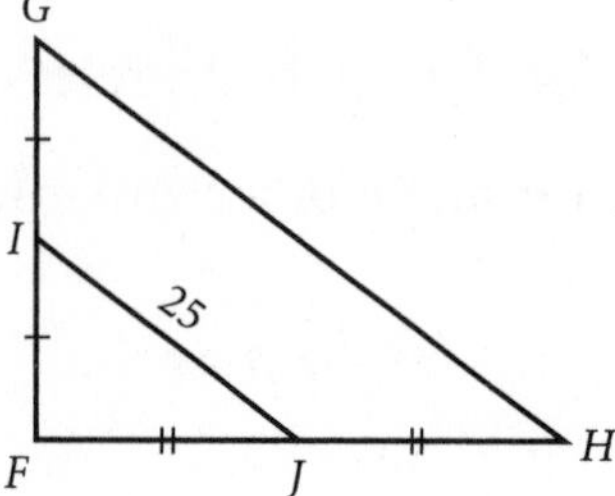

GH ____________

3. 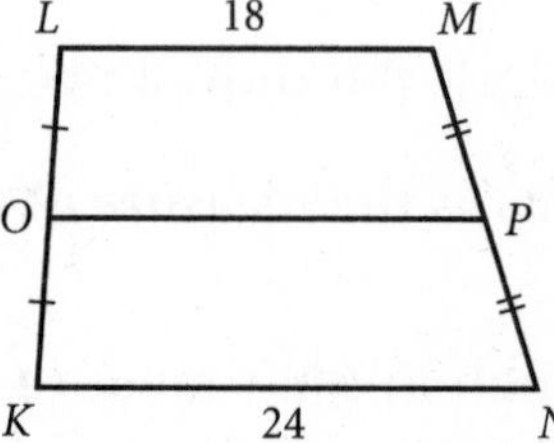

OP ____________

4. 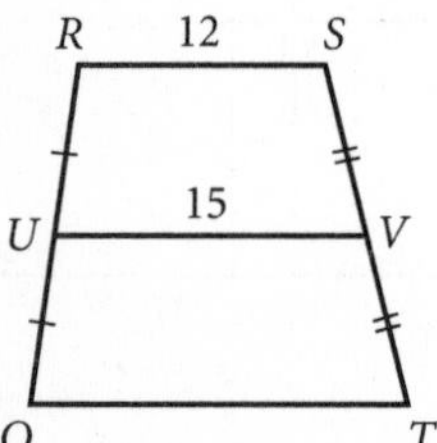

QT ____________

5. 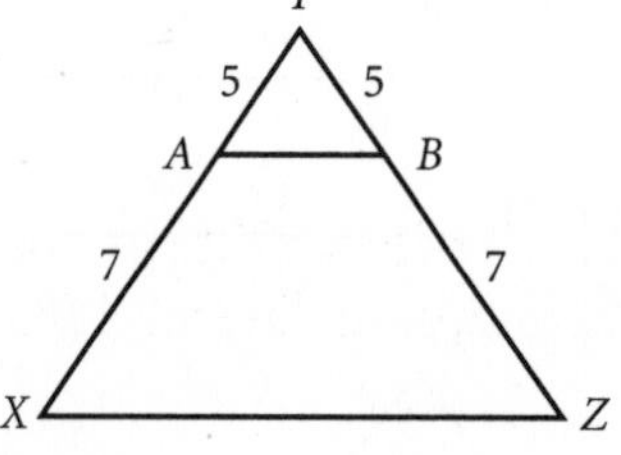

AB ____________

6. 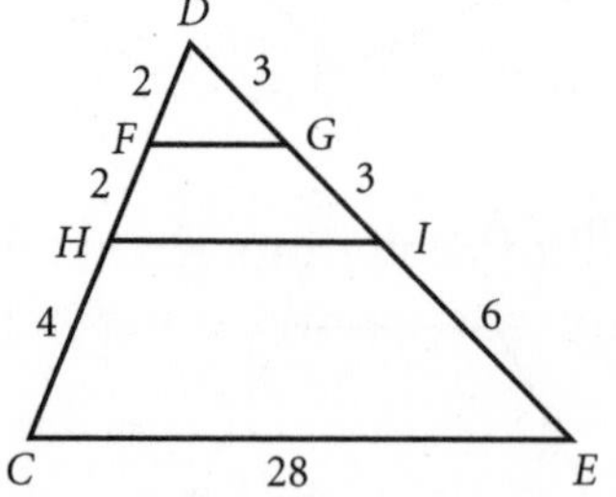

FG ____________

***J* and *K* are midpoints of the sides of $\triangle LMN$.**

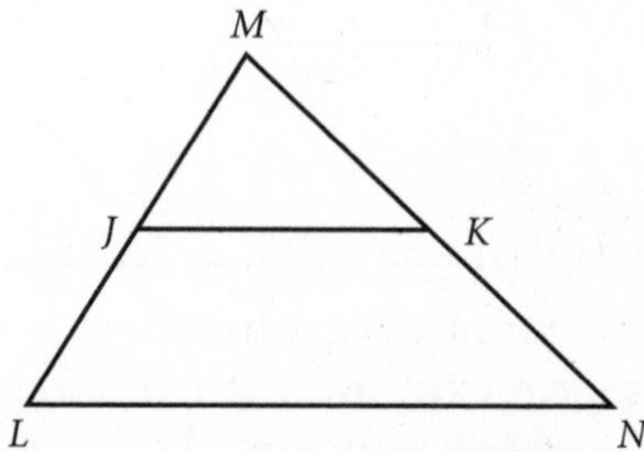

7. If $JK = 2x$ and $LN = x + 30$, find JK. ____________
8. If $JK = x - 5$ and $LN = x + 15$, find LN. ____________
9. If $LN = 15$ and $JK = 3x - 1.5$, find x. ____________
10. If $JK = MK = 7x$, and $MN = 70$, find x. ____________

Quick Warm-Up: Assessing Prior Knowledge

3.8 Analyzing Polygons with Coordinates

1. In a coordinate plane, to go from (2, 5) to (8, 16) you must move right $\underline{?}$ units and up $\underline{?}$ units. ______

2. Evaluate $\frac{d-c}{b-a}$ for $a = -3$, $b = 10$, $c = 0$, and $d = 39$. ______

3. If you multiply any nonzero real number by its reciprocal, the product is $\underline{?}$. ______

Lesson Quiz

3.8 Analyzing Polygons with Coordinates

1. Find the slope of the segment with endpoints at (1, 5) and (−3, 1). ______

2. If $\ell_1 \parallel \ell_2$ and the slope of ℓ_1 is −3, what is the slope of ℓ_2? ______

3. If $k_1 \perp k_2$ and the slope of k_1 is $\frac{1}{2}$, what is the slope of k_2? ______

$\overline{AB}$ and $\overline{CD}$ have the following endpoints. Determine if the segments are parallel, perpendicular, or neither.

4. $A(3, 0)$, $B(-2, 2)$, $C(-1, 4)$, $D(-3, -1)$ ______

5. $A(2, 3)$, $B(0, 4)$, $C(1, 2)$, $D(2, 3)$ ______

6. $A(4, -3)$, $B(1, 1)$, $C(2, 1)$, $D(5, -3)$ ______

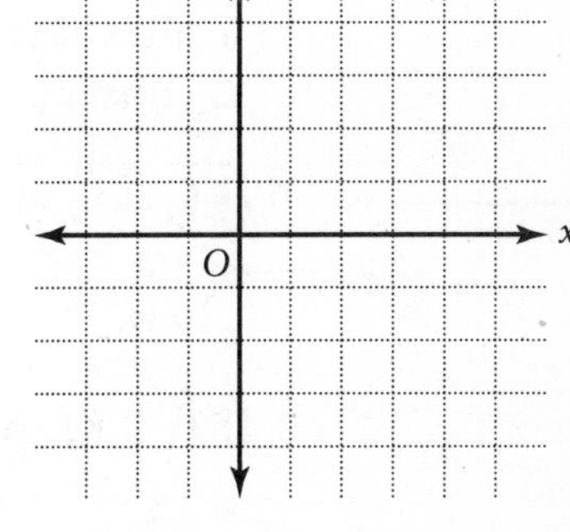

7. Plot the points $A(-1, -1)$, $B(1, 3)$, $C(4, 3)$, $D(2, -1)$ on the grid at right. Connect them to form quadrilateral $ABCD$.

Find the slope of each line segment in your graph.

8. $\overline{AB}$ ____ 9. $\overline{BC}$ ____ 10. $\overline{CD}$ ____ 11. $\overline{AD}$ ____

12. Identify quadrilateral $ABCD$ and justify your answer.

______

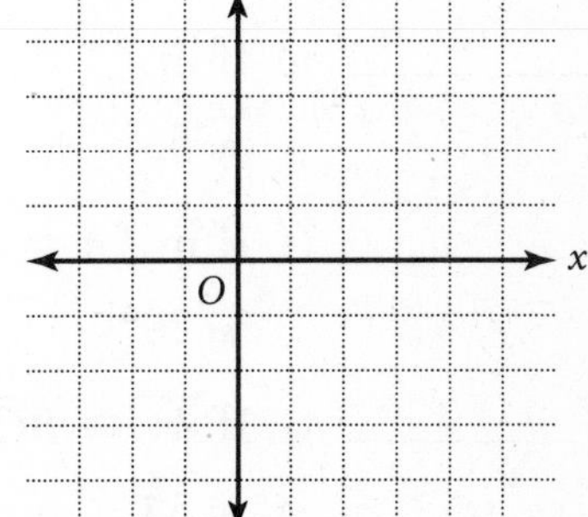

13. Plot points $E(-1, -2)$, $F(-2, 1)$, $G(4, 3)$, and $H(5, 0)$. Identify quadrilateral $EFGH$ and justify your answer.

______

Chapter Assessment

Chapter 3, Form A, page 1

Write the letter that best answers the question or completes the statement.

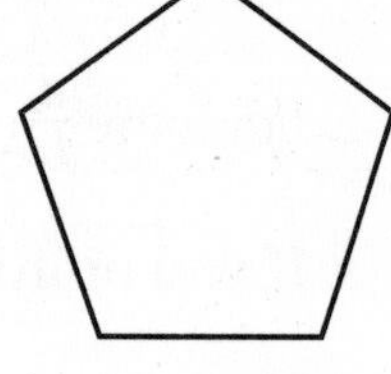

______ 1. How many lines of symmetry does the figure at right have?

a. 0 b. 1 c. 5 d. 10

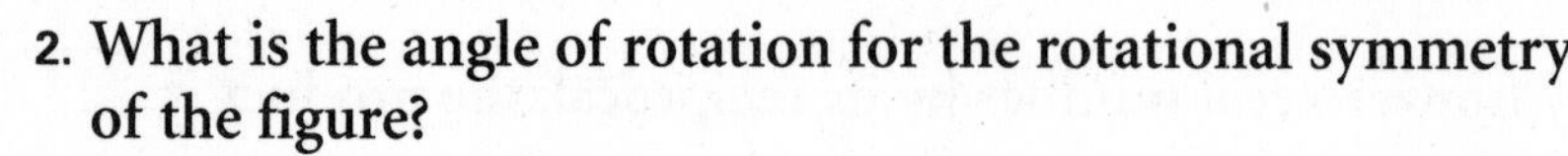

______ 2. What is the angle of rotation for the rotational symmetry of the figure?

a. 0° b. 50° c. 72° d. 180°

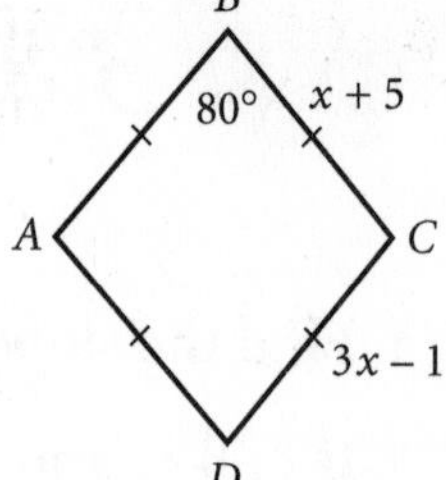

______ 3. What type of quadrilateral is *ABCD*?

a. square b. rhombus
c. rectangle d. trapezoid

______ 4. What is the length of side $\overline{AB}$?

a. 3 b. 6 c. 8 d. 9

______ 5. What is the measure of $\angle A$?

a. 80° b. 90° c. 100° d. 180°

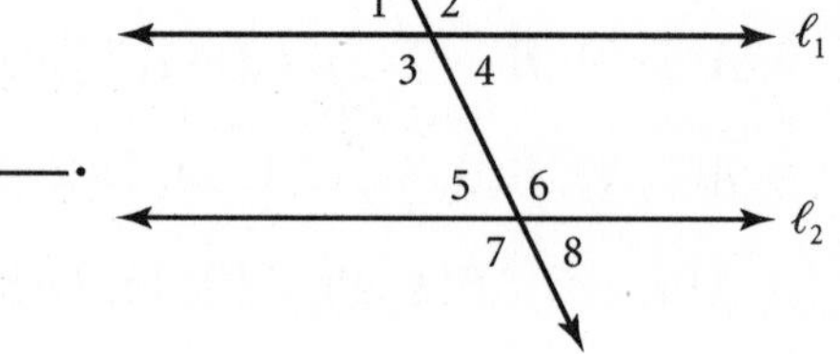

______ 6. What type of angles are $\angle 3$ and $\angle 6$?

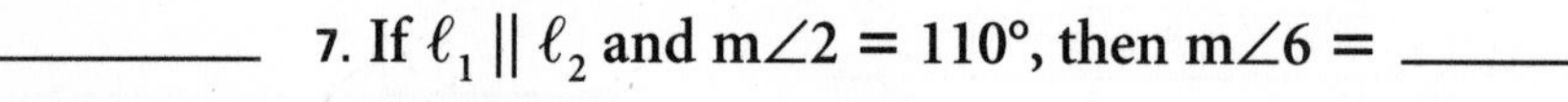

a. alternate interior b. alternate exterior
c. consecutive interior d. corresponding

______ 7. If $\ell_1 \parallel \ell_2$ and $m\angle 2 = 110°$, then $m\angle 6 =$ ______.

a. 35° b. 55°
c. 70° d. 110°

______ 8. If $\ell_1 \parallel \ell_2$ and $m\angle 5 = 75°$, then $m\angle 3 =$ ______.

a. 15° b. 75° c. 90° d. 105°

______ 9. If $m\angle 5 = 55°$ and $m\angle 4 = 35°$, then ℓ_1 and ℓ_2 ______.

a. are perpendicular b. are parallel
c. intersect at an acute angle d. intersect at an obtuse angle

______ 10. If $m\angle A = 65°$ and $m\angle BCD = 125°$, then $m\angle B =$ ______.

a. 55° b. 60° c. 65° d. 185°

______ 11. If $m\angle A = 60°$ and $m\angle B = 80°$, then $m\angle BCD =$ ______.

a. 20° b. 40° c. 60° d. 140°

B
A C D

______ 12. If $\overline{AC} \cong \overline{BC}$ and $m\angle BCD = 108°$, then $m\angle A =$ ______.

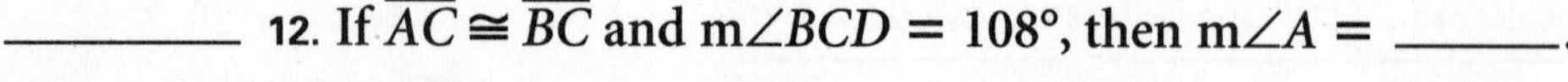

a. 54° b. 72° c. 36° d. 90°

Chapter Assessment

Chapter 3, Form A, page 2

______ 13. In $\triangle EFG$, $m\angle E = 85°$ and $m\angle F = 25°$. What is $m\angle G$?

a. 60° b. 70° c. 110° d. 180°

______ 14. What is the sum of the measures of the interior angles of a hexagon?

a. 180° b. 360° c. 540° d. 720°

______ 15. What is the measure of an interior angle of a regular pentagon?

a. 60° b. 72° c. 108° d. 120°

______ 16. If the measure of an exterior angle of a regular polygon is 45°, how many sides does the polygon have?

a. 5 b. 6 c. 8 d. 10

______ 17. If the measure of an interior angle of a regular polygon is 140°, how many sides does the polygon have?

a. 10 b. 9 c. 8 d. 5

______ 18. What is $m\angle G$ in quadrilateral $DEFG$?

a. 35° b. 70°
c. 71° d. 77°

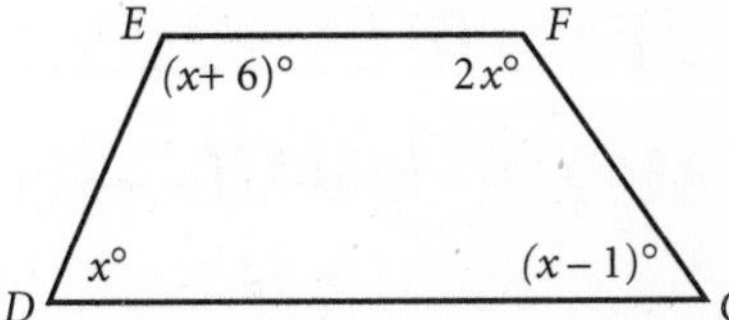

______ 19. If $HJ = 26$, then $KL =$ ______.

a. 13 b. 26
c. 30 d. 52

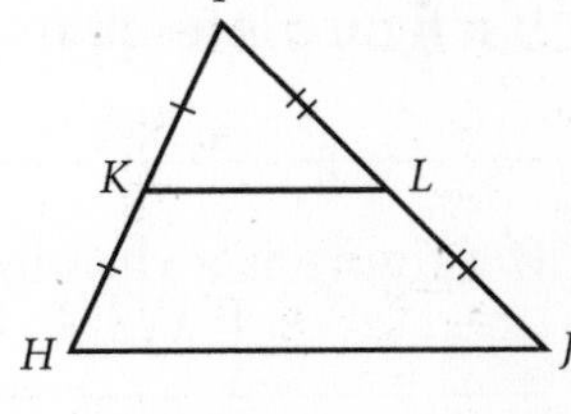

______ 20. If $KL = 15$, then $HJ =$ ______.

a. 7.5 b. 15
c. 30 d. 45

______ 21. If $HJ = 3x - 1$ and $KL = x + 1$, then $HJ =$ ______.

a. 3 b. 4 c. 8 d. 10

$\overline{AB}$ has endpoints $A(-3, -2)$ and $B(-2, 1)$, and $\overline{CD}$ has endpoints $C(2, 1)$ and $D(1, -2)$.

______ 22. What is the slope of $\overline{AB}$?

a. $-\frac{1}{3}$ b. -3 c. $\frac{1}{3}$ d. 3

______ 23. What is the relationship between $\overline{AB}$ and $\overline{CD}$?

a. $\overline{AB} \parallel \overline{CD}$ b. $\overline{AB} \perp \overline{CD}$ c. $AB = CD$ d. $AB = \frac{1}{2}CD$

NAME ____________________ CLASS __________ DATE __________

Chapter Assessment

Chapter 3, Form B, page 1

Determine whether each figure has reflectional symmetry, rotational symmetry, both, or neither.

1. 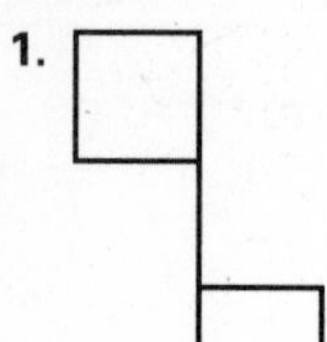

2. 

3. 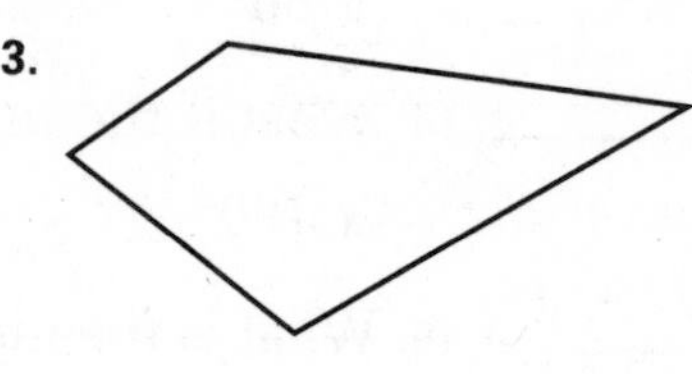

________ ________ ________

***ABCD* is a square. Determine whether each transformation below is the result of a rotation about *x*, a reflection across one of the axes, or both. The order of the letters does not matter.**

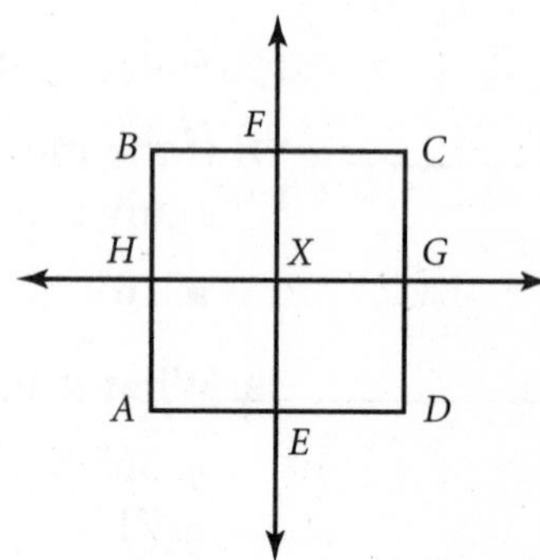

4. $\overline{AB} \rightarrow \overline{CD}$ ________
5. $\overline{AB} \rightarrow \overline{AD}$ ________
6. $\overline{BF} \rightarrow \overline{FC}$ ________
7. $\angle B \rightarrow \angle D$ ________
8. $\angle GXE \rightarrow \angle EXH$ ________
9. $\angle EXH \rightarrow \angle FXG$ ________

Decide whether each statement is true or false, and explain your answer.

10. If a figure is a square, then it is a rectangle.

____

11. If a figure is a rhombus, then it is a square.

____

***JKLM* is a parallelogram. Find each measure if m∠*KLM* = 70°, m∠*JMK* = 50°, *JM* = 10, and *JO* = 8.**

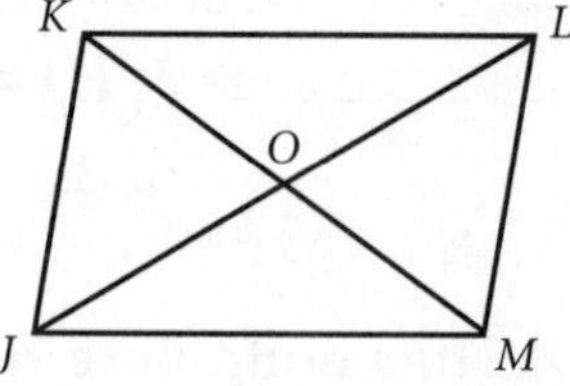

12. m∠*KJM* ________
13. *KL* ________
14. *JL* ________
15. m∠*MKL* ________
16. *LO* ________
17. m∠*JKL* ________

If $\ell_1 \parallel \ell_2$, m∠4 = $(2x + 50)°$, and m∠6 = $(x - 20)°$, find each angle measure.

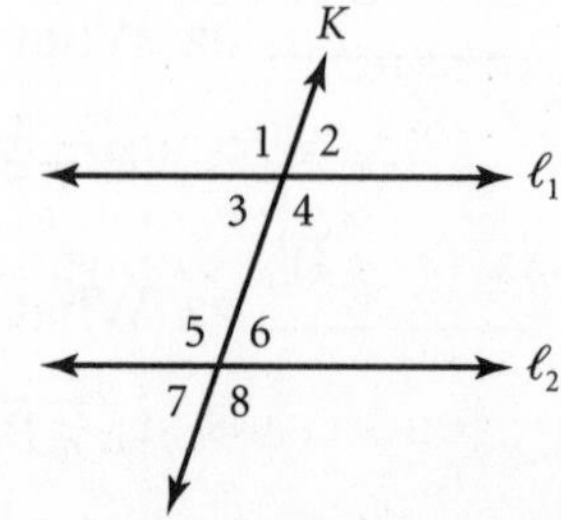

18. m∠4 ________
19. m∠6 ________
20. m∠5 ________
21. m∠2 ________
22. m∠7 ________
23. m∠8 ________

Chapter Assessment

Chapter 3, Form B, page 2

Use the figure at right to complete each statement.

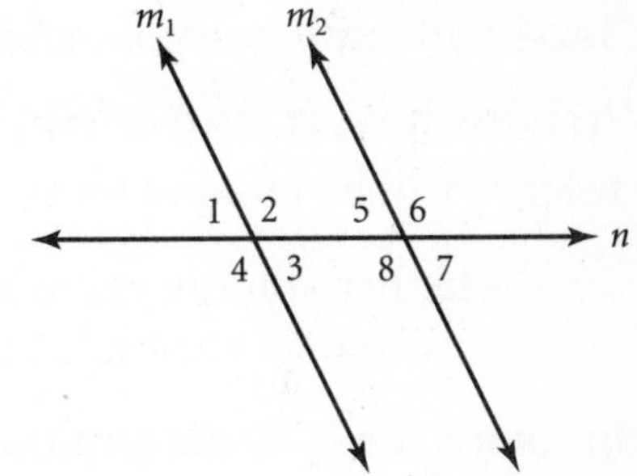

24. If $\angle 3 \cong \angle 5$, then $m_1 \parallel m_2$ because

______________________.

25. If $m\angle 2 + m\angle 5 = 180°$, then $m_1 \parallel m_2$ because

______________________.

26. If $\angle 4 \cong \angle 8$, then $m_1 \parallel m_2$, because

______________________.

Given: $\triangle ABC$ is isosceles; $\overline{EG} \parallel \overline{HJ} \parallel \overline{DC}$; $\overline{FG}$ is the midsegment of $\triangle ABC$; and H is the midsegment of trapezoid $EGCD$. Find each measure.

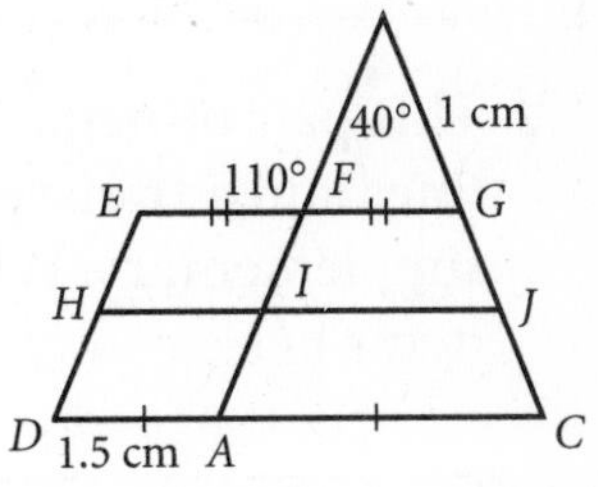

27. $m\angle C$ ______ 28. $m\angle BAC$ ______ 29. BC ______

30. FG ______ 31. $m\angle BIJ$ ______ 32. $m\angle D$ ______

33. HJ ______ 34. $m\angle DAB$ ______ 35. $m\angle DEF$ ______

36. Find the sum of the measures of the interior angles of an octagon. ______

37. Find the measure of an interior angle of a regular nonagon. ______

38. Find the sum of the measures of the exterior angles of a heptagon. ______

39. Find the measure of an exterior angle of a regular decagon. ______

N, O, and P are the midpoints of the sides of $\triangle KLM$.

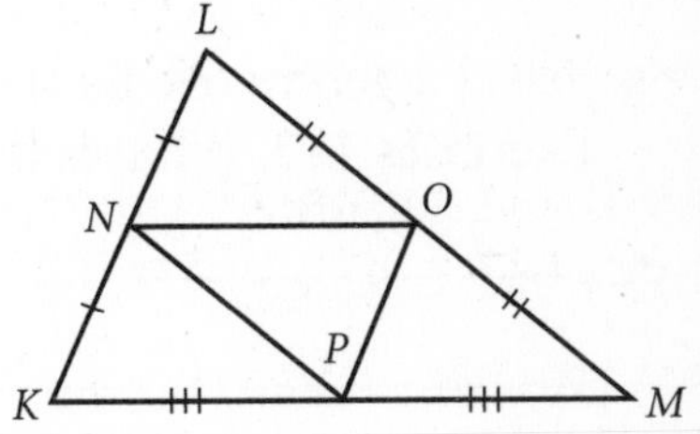

40. If $NO = 2x$ and $KM = x + 15$, find NO. ______

41. If $NO = 7$, $OP = 10$, and $NP = 12$, find the perimeter of $\triangle KLM$. ______

42. Find the slope and midpoint of the segment with endpoints at $(-1, 4)$ and $(3, -2)$. ______

43. Find the slope and midpoint of the segment with endpoints at $(1, -4)$ and $(-3, 2)$. ______

NAME ____________________ CLASS ____________ DATE ____________

Alternative Assessment

Polygons and Angles, Chapter 3, Form A

TASK: Identify polygons and angles from descriptions.

HOW YOU WILL BE SCORED: As you work through the task, your teacher will be looking for the following:

- whether you can correctly identify the item
- how well you are able to describe an item

In Exercises 1–4, determine the item or items that are being described.

1. This item has four sides. Its opposite sides are parallel. All of its angles are 90°. What could this item be?

 ____

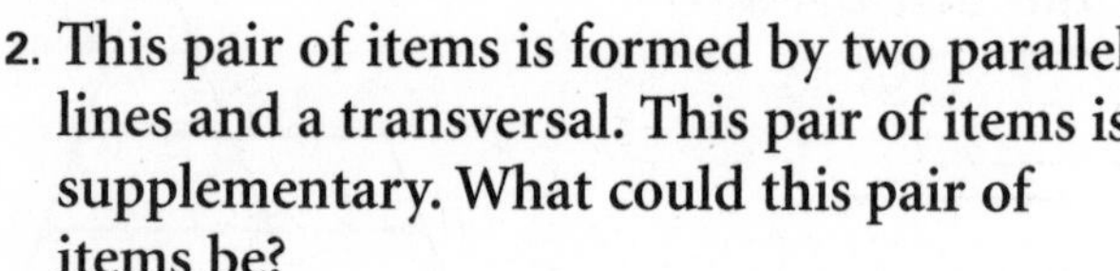

2. This pair of items is formed by two parallel lines and a transversal. This pair of items is supplementary. What could this pair of items be?

 ____

3. This item has four congruent sides and two different interior angle measures. What could this item be?

 ____

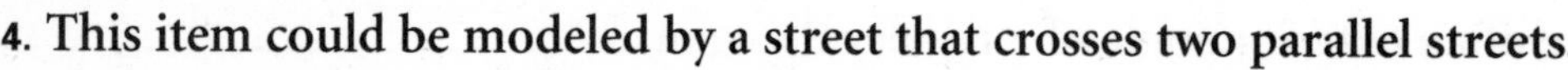

4. This item could be modeled by a street that crosses two parallel streets.

 What is this item? ____________________

5. Select a geometric figure and write a description similar to those in Exercises 1–3. What is the figure?

 ____

 ____

6. Line ℓ is perpendicular to two other lines, line m and line n, in a plane. What is the relationship between lines m and n? Why?

 ____

 ____

 ____

SELF-ASSESSMENT: Describe the polygons defined in Chapter 3 of your text.

NAME ______________________ CLASS ____________ DATE ____________

Alternative Assessment

More About Polygons, Chapter 3, Form B

TASK: Identify and mathematically describe given polygons.

HOW YOU WILL BE SCORED: As you work through the task, your teacher will be looking for the following:

- whether you can correctly identify the polygon
- how complete and accurate your mathematical description is

For each of the following exercises, read the given information and write the name of the polygon described. Then describe the figure mathematically, using information such as angle measures. Sketch the polygon in the space provided.

1. 6 sides, regular

____

____

____

2. 3 sides, two congruent sides, one right angle

____

____

3. 4 interior angles, only one pair of parallel sides

____

____

4. 8 congruent sides

____

____

SELF-ASSESSMENT: What facts and formulas did you use to answer the exercises? What other facts and formulas about polygons do you know that were not used to answer the exercises?

NAME ____________ CLASS ____________ DATE ____________

Quick Warm-Up: Assessing Prior Knowledge

4.1 Congruent Polygons

Consider the figures at right.

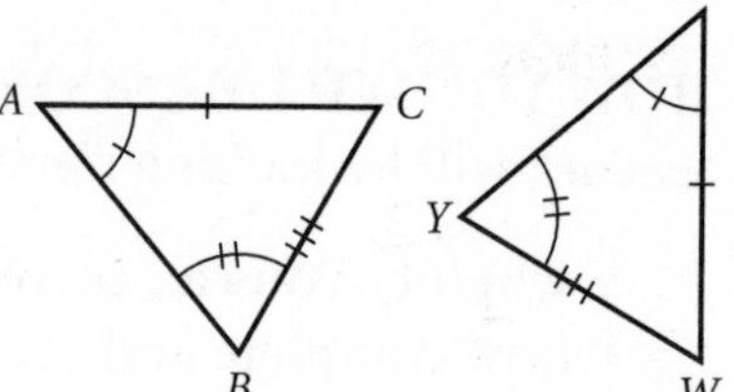

1. Name an angle congruent to $\angle Y$. ____________
2. Name a segment congruent to $\overline{BC}$. ____________
3. If $\angle A$ is 50° and $\angle Y$ is 70°, what is $\angle C$? (Hint: Use the Triangle Sum Theorem.) ____________

Lesson Quiz

4.1 Congruent Polygons

Determine whether the following pairs of figures are congruent. Explain your reasoning.

1.

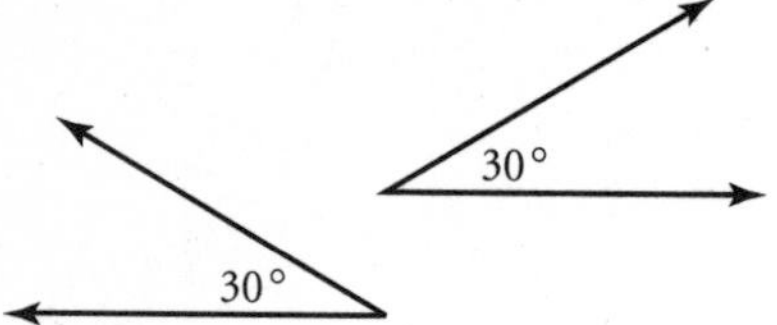

____

2.

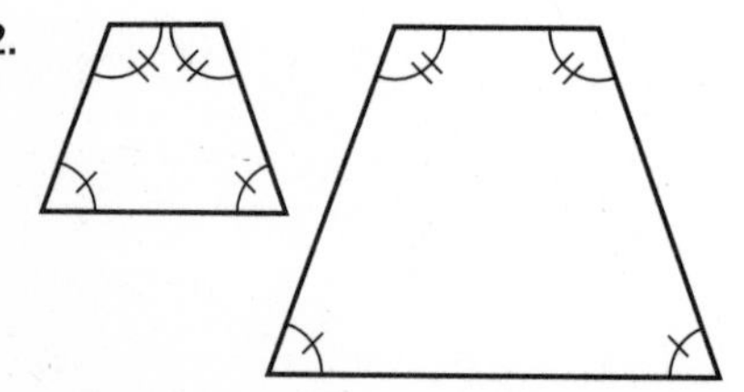

____

3. $PQ = 2.5$ feet ____________

$YZ = 30$ inches ____________

Complete the congruence statements for quadrilaterals $ABCD \cong PQRS$.

4. $\overline{AB} \cong$ ____________
5. $\angle A \cong$ ____________
6. ____________ $\cong \overline{QR}$
7. ____________ $\cong \angle Q$
8. $\overline{DC} \cong$ ____________
9. $\angle C \cong$ ____________
10. ____________ $\cong \overline{PS}$
11. ____________ $\cong \angle S$

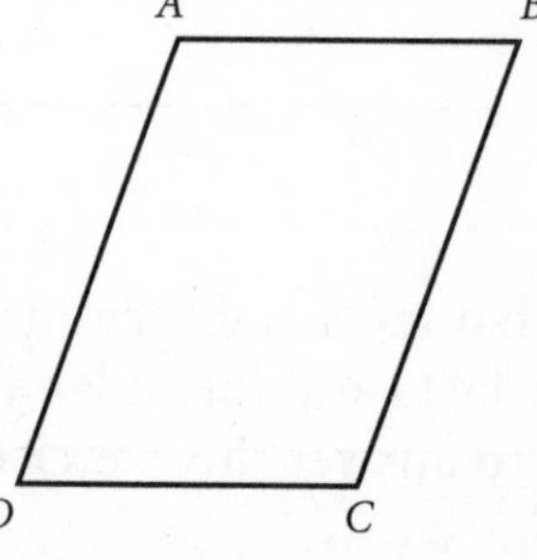

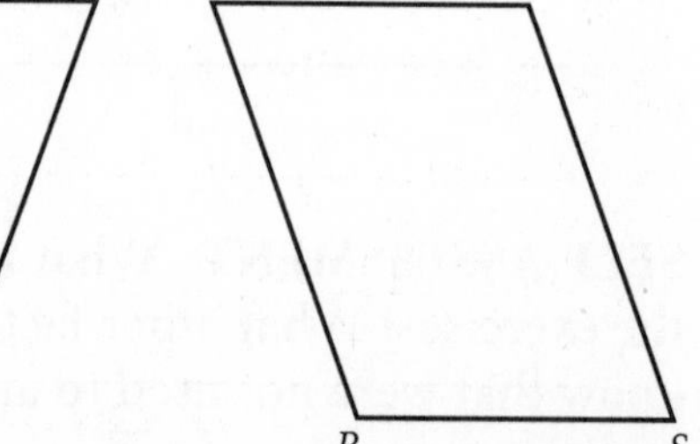

NAME ________________ CLASS ________ DATE ________

Quick Warm-Up: Assessing Prior Knowledge

4.2 Triangle Congruence

Triangles *QRS* and *FDE* are congruent. Write all pairs of corresponding parts.

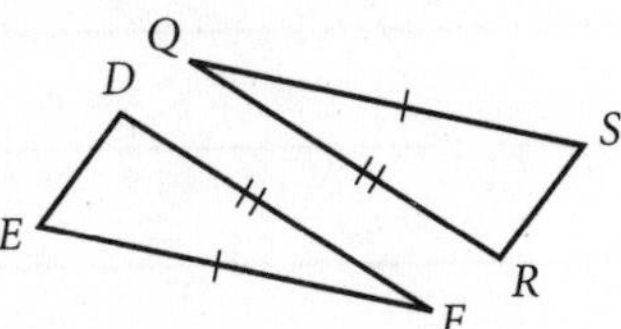

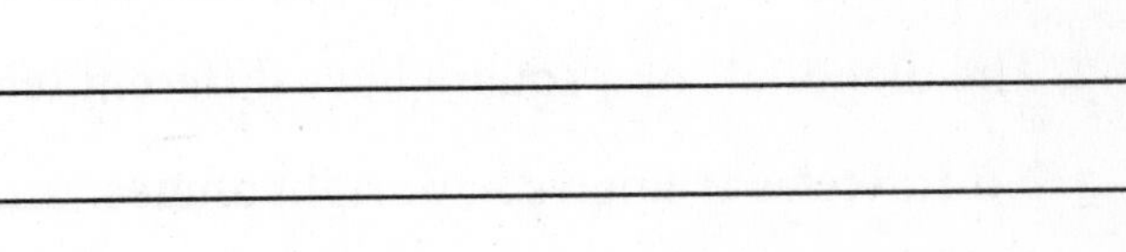

Lesson Quiz

4.2 Triangle Congruence

Decide whether each pair of triangles can be proven congruent. If so, write an appropriate congruence statement and tell which postulate you used. If not, explain why.

1. A, B, C, 60°, 20°; D, E, F, 20°, 60°

2.

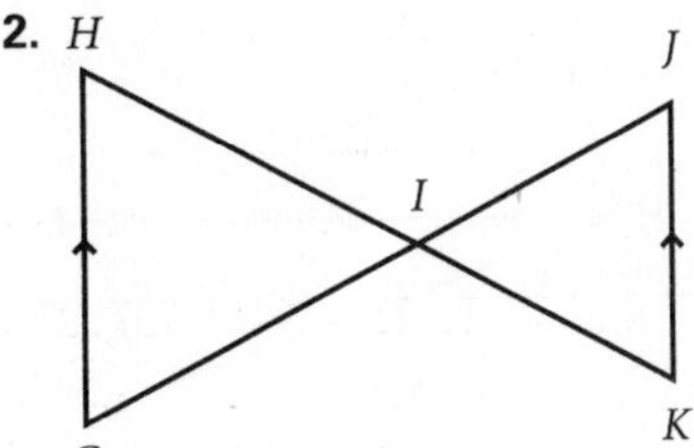

3.

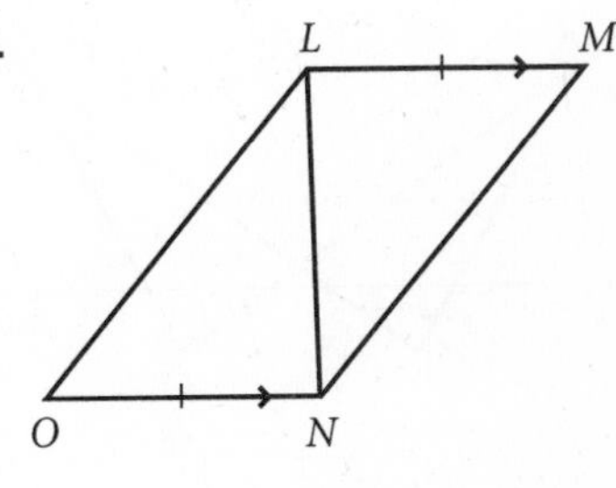

Use the diagram at right to answer Exercises 4–6.

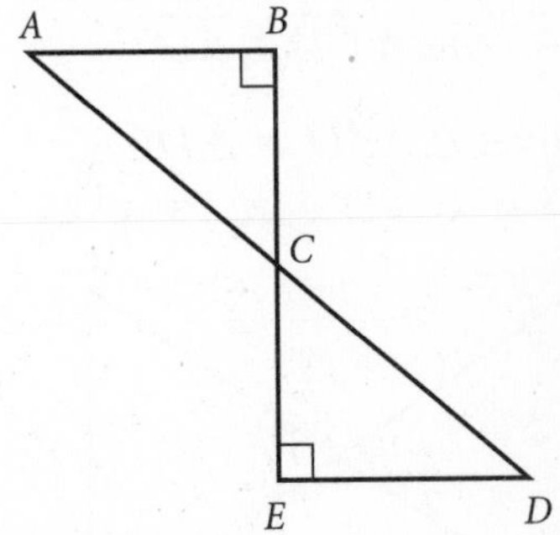

4. If C is the midpoint of $\overline{BE}$, then what two segments are congruent? ________________

5. If $\overline{BE}$ and $\overline{AD}$ intersect at C, what two angles must be congruent and why?

____

6. Name two other congruent angles, and explain why.

____

NAME ______________________ CLASS ____________ DATE ____________

Quick Warm-Up: Assessing Prior Knowledge

4.3 *Analyzing Triangle Congruence*

Can each statement be disproved with a counterexample? If yes, give the counterexample.

1. All rhombuses are squares. ____________
2. The diagonals of a square have different lengths. ____________
3. No isosceles triangles have right angles. ____________
4. An equilateral triangle cannot have right angles. ____________

Lesson Quiz

4.3 *Analyzing Triangle Congruence*

Decide whether each pair of triangles is congruent. If so, write an appropriate congruence statement and the theorem or postulate that justifies it. If not, explain why.

1.

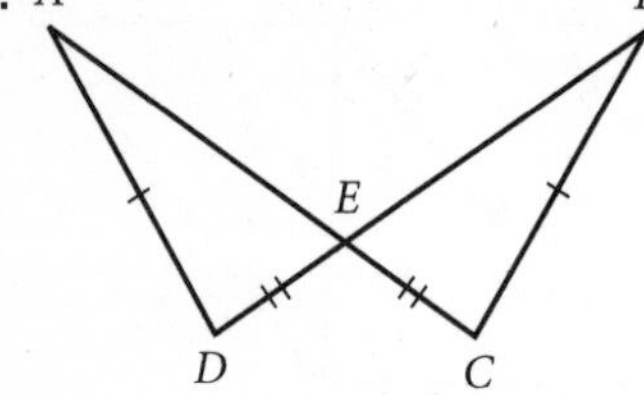

2.

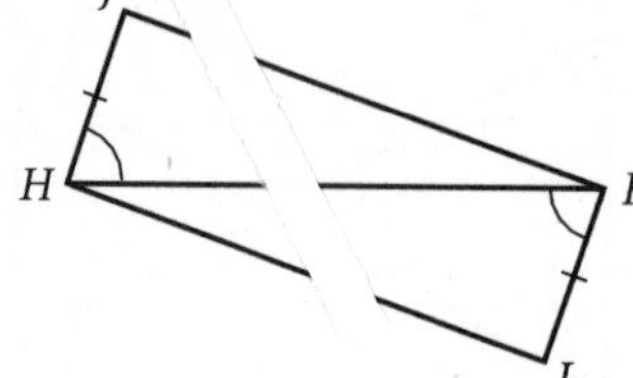

3.

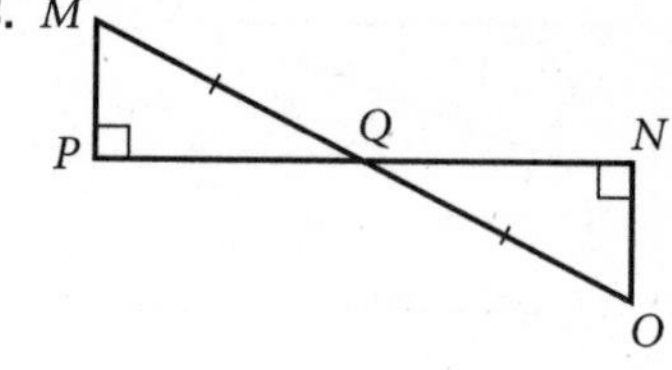

____________ ____________ ____________

Write the reason for each step in the proof.

Given: $\overline{BD} \cong \overline{AC}$, $\overline{BA} \perp \overline{AD}$, and $\overline{CD} \perp \overline{AD}$

Prove: $\triangle ABD \cong \triangle DCA$

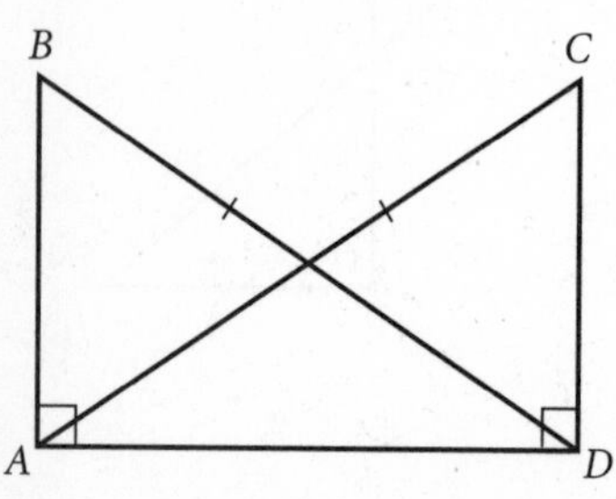

Statements	Reasons
$\overline{BD} \cong \overline{AC}$	4. ____________
$\overline{BA} \perp \overline{AD}$ and $\overline{CD} \perp \overline{AD}$	5. ____________
$m\angle BAD = 90°$, $m\angle CDA = 90°$	6. ____________
$\triangle ABD$ and $\triangle DCA$ are right triangles.	7. ____________
$\overline{AD} \cong \overline{AD}$	8. ____________
$\triangle ABD \cong \triangle DCA$	9. ____________

NAME ______________________ CLASS ____________ DATE ____________

Quick Warm-Up: Assessing Prior Knowledge

4.4 Using Triangle Congruence

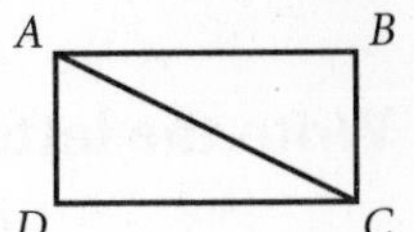

Given: Quadrilateral *ABCD* is a rectangle.
Prove: Triangles *ABC* and *CDA* are congruent by ASA.

____

____

____

Lesson Quiz

4.4 Using Triangle Congruence

1. If $\triangle ABC$ is isosceles, state two conclusions that can be drawn.

____

Write the reason for each step in the proof.

Given: $\overline{CB} \cong \overline{CD}$ and $\angle ACB \cong \angle ACD$
Prove: $\angle B \cong \angle D$

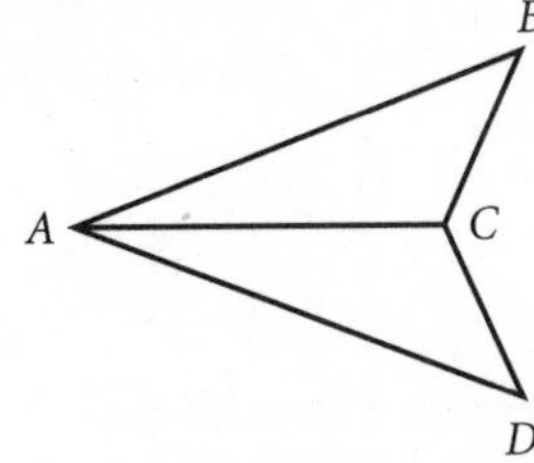

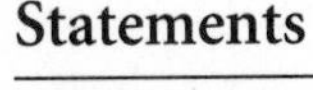

Statements	Reasons
$\overline{CB} \cong \overline{CD}$, $\angle ACB \cong \angle ACD$	2. ____________
$\overline{AC} \cong \overline{AC}$	3. ____________
$\angle ACB \cong \angle ACD$	4. ____________
$\angle B \cong \angle D$	5. ____________

6. Write a flowchart proof.
Use a separate sheet of paper if necessary.

Given: $\overline{QM} \cong \overline{QP}$ and $\overline{MN} \cong \overline{PO}$
Prove: $\angle QNP \cong \angle QOM$

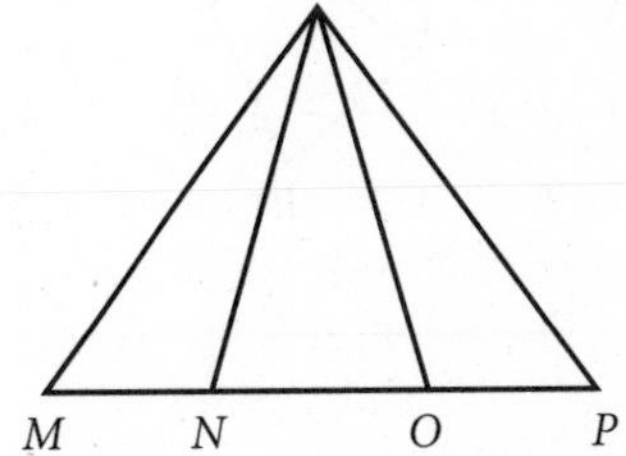

NAME ______________________ CLASS __________ DATE __________

Mid-Chapter Assessment

Chapter 4 (Lessons 4.1–4.4)

Write the letter that best answers the question or completes the statement.

______ 1. If $\triangle ABC \cong \triangle DEF$, which of the following is NOT true?

a. $\angle ABC \cong \angle DEF$ b. $\angle BCA \cong \angle EFD$

c. $\overline{AC} \cong \overline{DF}$ d. $\overline{AB} \cong \overline{EF}$

Use the diagram to answer Exercises 2–4.

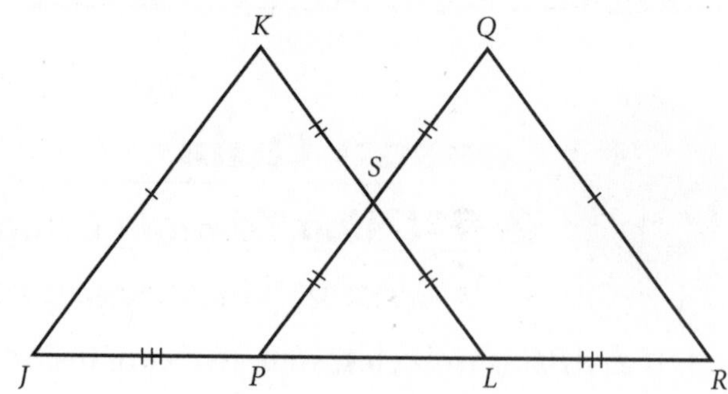

______ 2. $\triangle JKL \cong \triangle RQP$ by

a. ASA b. SSS

c. SAS d. HL

______ 3. $\angle KLJ \cong \angle QPR$ by

a. definition of vertical angles b. Isosceles Triangle theorem

c. CPCTC d. Overlapping Triangle theorem

______ 4. $\overline{JL} \cong \overline{RP}$ by

a. Transitive Property b. definition of isosceles triangle

c. Reflexive Property d. Overlapping Segment Theorem

Decide whether each pair of triangles can be proven congruent. If yes, write a congruence statement and the postulate or theorem you used.

5.

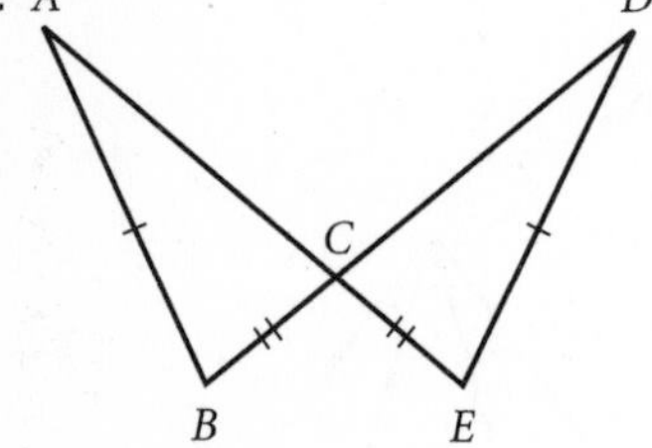

6.

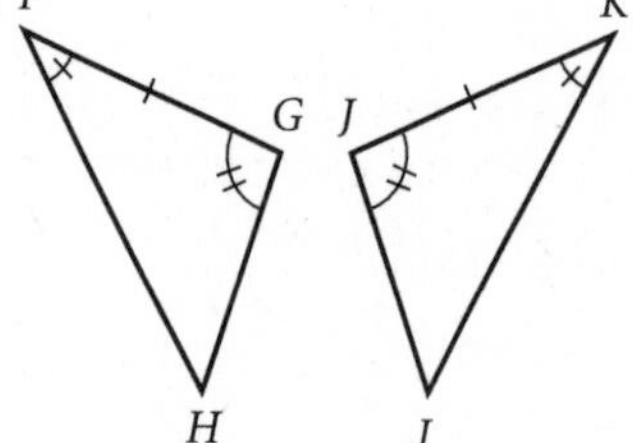

7.

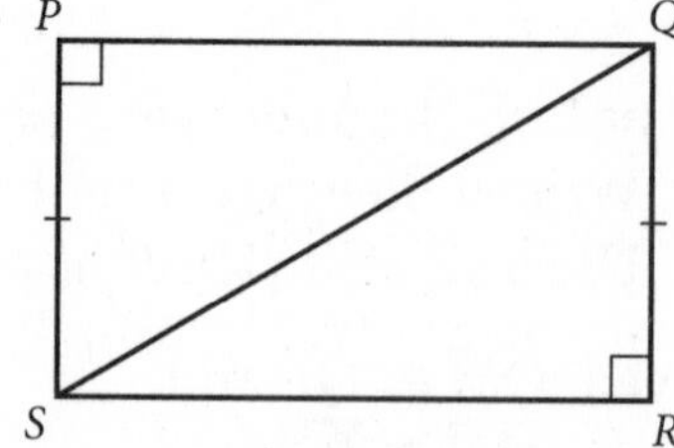

______________ ______________ ______________

Write a proof on a separate sheet of paper.

8. Given: $\overline{AB} \cong \overline{BC}$ and $\overline{AD} \cong \overline{EC}$

Prove: $\overline{AE} \cong \overline{DC}$

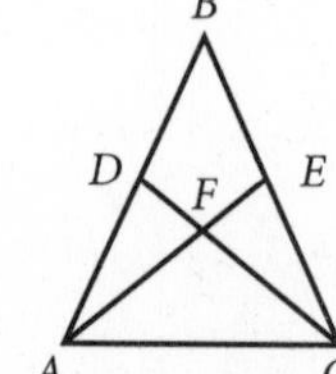

NAME ______________________ CLASS ____________ DATE ____________

Quick Warm-Up: Assessing Prior Knowledge

4.5 Proving Quadrilateral Properties

How are the quadrilaterals in each pair alike? How are they different?

1. a parallelogram and a square ______________________

2. a rhombus and a square ______________________

Lesson Quiz

4.5 Proving Quadrilateral Properties

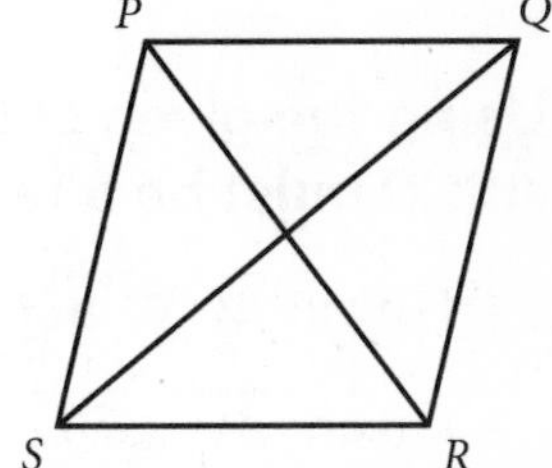

1. *PQRS* is a parallelogram. Explain why $\angle SPR \cong \angle QRP$.

____

2. How many different pairs of congruent triangles are formed when both diagonals of a parallelogram are drawn? ______________________

How many pairs of congruent triangles are formed when both diagonals of the following quadrilaterals are drawn?

3. rhombus ______________________

4. isosceles trapezoid ______________________

Complete the following proof.

Given: square *ABCD* with diagonals $\overline{AC}$ and $\overline{BD}$

Prove: $\angle DEC$ and $\angle BEC$ are right angles.

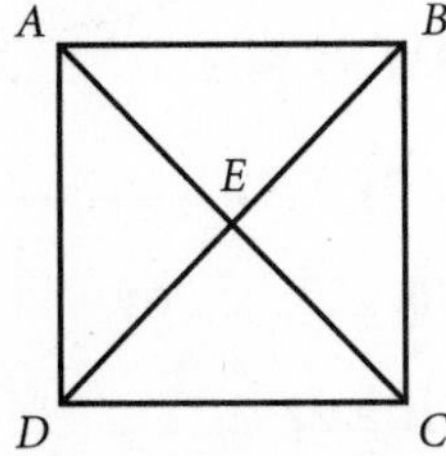

Statements	Reasons
ABCD is a square.	5. ______________
$\overline{DC} \cong \overline{BC}$	6. ______________
$\overline{DE} \cong \overline{BE}$	7. ______________
$\overline{EC} \cong \overline{EC}$	8. ______________
$\triangle DEC \cong \triangle BEC$	9. ______________
$\angle DEC \cong \angle BEC$	10. ______________
$\angle DEC$ and $\angle BEC$ are right angles.	11. ______________

NAME ______________________________ CLASS ____________ DATE ____________

Quick Warm-Up: Assessing Prior Knowledge

4.6 Conditions for Special Quadrilaterals

Draw quadrilateral *WXYZ*, and list all pairs that satisfy each description.

1. opposite sides ______________________________
2. opposite angles ______________________________
3. adjacent sides ______________________________
4. consecutive angles ______________________________

Lesson Quiz

4.6 Conditions for Special Quadrilaterals

Using the given conditions, decide whether quadrilateral *ABCD* must be a rectangle, a rhombus, or a square.

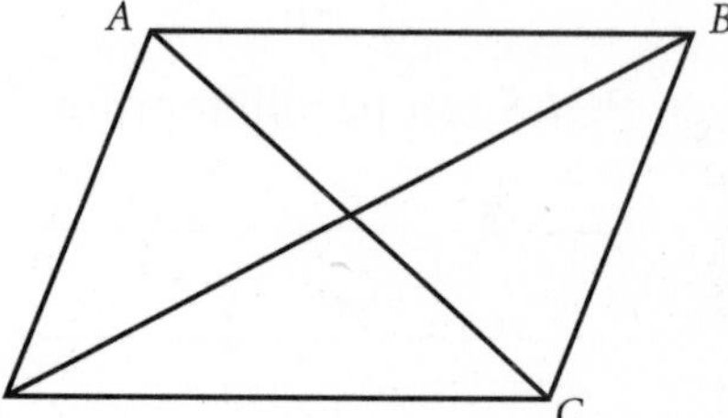

1. Given: $\overline{AC} \cong \overline{BD}$ ______________________
2. Given: $\overline{AC} \perp \overline{BD}$ ______________________
3. Given: $\overline{AC} \cong \overline{BD}$, $\overline{AC} \perp \overline{BD}$ ______________________

***GHIJ* is a parallelogram. For each statement, write *true* or *false*.**

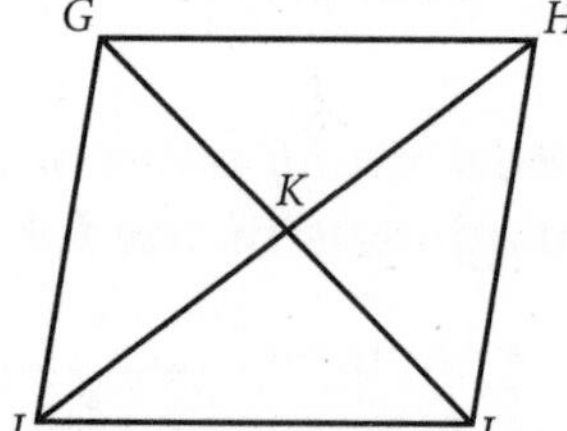

4. If $\overline{GI} \perp \overline{HJ}$, then *GHIJ* is a rhombus. ____________
5. If $\overline{GI} \cong \overline{HJ}$, then *GHIJ* is a rectangle. ____________
6. If $\overline{GI} \perp \overline{HJ}$ and $\overline{GI} \cong \overline{HJ}$, then *GHIJ* is a square. ____________

Complete the paragraph proof.

Given: *KLMN* is a parallelogram and $\overline{KM} \perp \overline{LN}$.
Prove: *KLMN* is a rhombus.

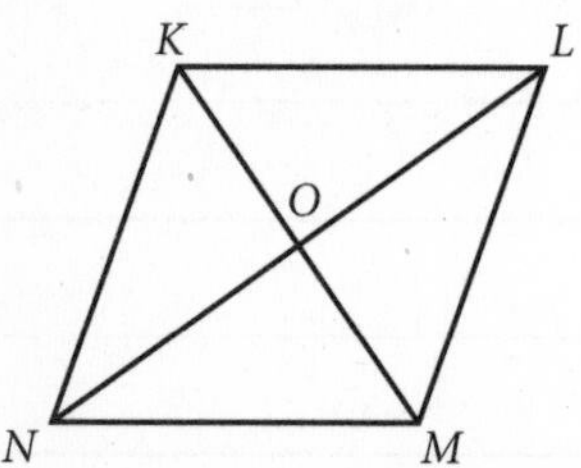

KLMN is a parallelogram, so **7.** ________ bisect each other, $\overline{NO} \cong$ **8.** ______, and $\overline{KO} \cong$ **9.** ______. $\overline{KM} \perp \overline{LN}$, so $\angle LOM \cong \angle MON \cong$ **10.** ________ $\cong$ **11.** ________. These are all **12.** ________ angles measuring **13.** ______. $\Delta LOM \cong$ **14.** ________ $\cong$ **15.** ________ $\cong$ **16.** ________ by **17.** ________. $\overline{KL} \cong$ **18.** ______ $\cong$ **19.** ______ $\cong$ **20.** ______ by **21.** ________. Thus, *KLMN* is a **22.** ________.

NAME ______________________ CLASS ______________ DATE ______________

Quick Warm-Up: Assessing Prior Knowledge

4.7 Compass and Straightedge Constructions

Use a compass and straightedge to construct the following:

1. a line segment
2. a circle with the same segment from Exercise 1 as the radius
3. a line segment and its perpendicular bisector

Lesson Quiz

4.7 Compass and Straightedge Constructions

1. Construct $\overline{CD} \cong \overline{AB}$.

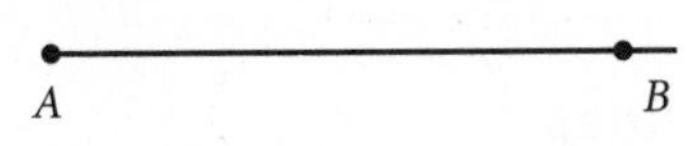

2. Construct the angle bisector of $\angle E$.

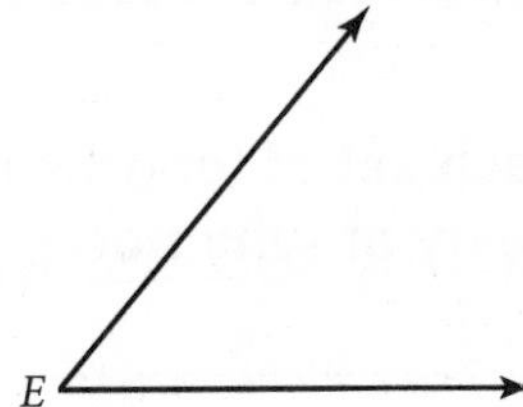

3. Construct an isosceles triangle with the base congruent to $\overline{FG}$ and the legs congruent to $\overline{HI}$.

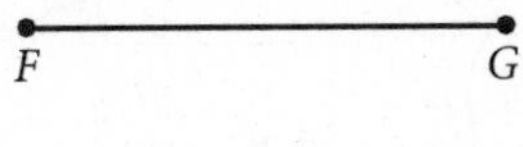

4. Describe how you would construct a square.

____

5. What justifies the construction of congruent line segments?

____

Quick Warm-Up: Assessing Prior Knowledge
4.8 Constructing Transformations

Match each transformation with the information needed to perform it.

Transformation	Needed Information
______ 1. rotation	a. the distance moved, both horizontally and vertically
______ 2. dilation	b. the center and the number of degrees
______ 3. translation	c. a "mirror" line
______ 4. reflection	d. the center and a scale factor

Lesson Quiz
4.8 Constructing Transformations

1. If $PQ = 15.28$, $PR = 7.2$, and $RQ = 8.08$, determine whether points P, Q, and R are collinear. ______________________

Determine whether each set of lengths could represent the sides of a triangle. Explain why or why not.

2. 3, 4, 5 ______________________
3. 2, 5, 9 ______________________
4. 4, 4, 8 ______________________
5. 3, 3, 5 ______________________

The diagram shows a translation. Find the image of each of the following:

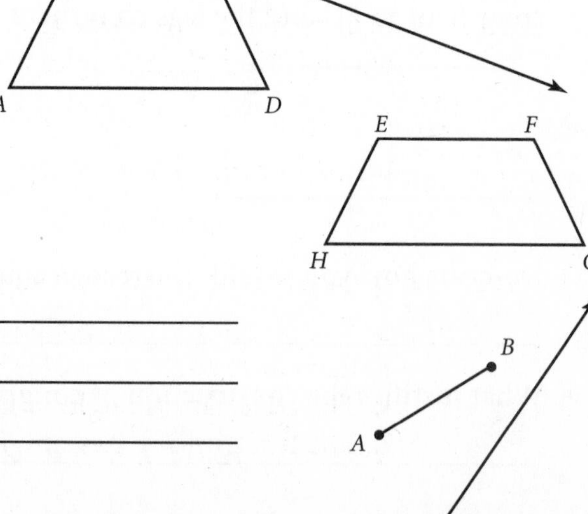

6. $\angle B$ ______________
7. $\overline{CD}$ ______________
8. $\angle D$ ______________
9. $\overline{AB}$ ______________
10. $\angle C$ ______________
11. $\overline{AD}$ ______________

12. Explain how to reflect $\overline{AB}$ across line m.

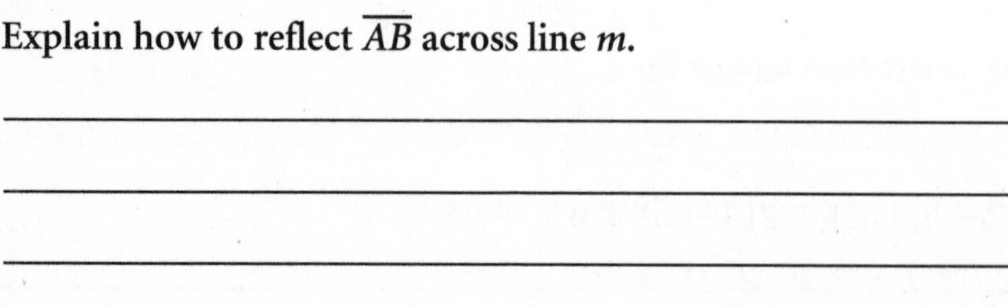

Chapter Assessment

Chapter 4, Form A, page 1

Write the letter that best answers the question or completes the statement.

Use the figure at right for Exercises 1 and 2.

______ 1. If $\triangle AOB \cong \triangle COD$, then $\angle OAB \cong$ ______.

a. $\angle ABO$ b. $\angle OCD$

c. $\angle DOC$ d. $\angle CDO$

______ 2. If $\triangle AOB \cong \triangle COD$, then $\overline{AO} \cong$ ______.

a. $\overline{DO}$ b. $\overline{OB}$

c. $\overline{BA}$ d. $\overline{CO}$

Use trapezoid *ABCD* for Exercises 3–5.

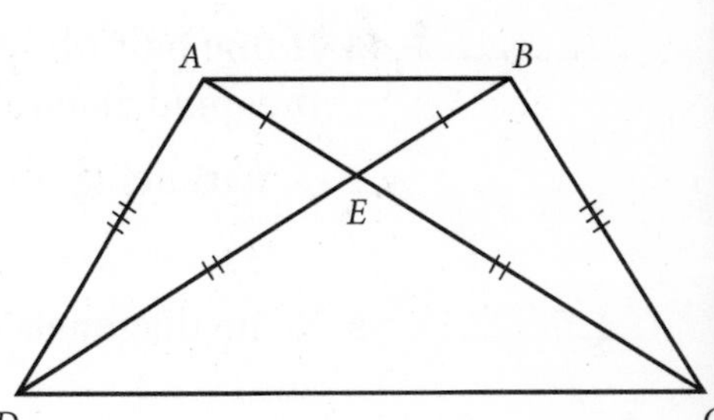

______ 3. Which congruence statement is correct?

a. $\triangle AEB \cong \triangle DEC$ b. $\triangle AED \cong \triangle CEB$

c. $\triangle AED \cong \triangle BEC$ d. $\triangle ABC \cong \triangle BCD$

______ 4. $\overline{AB} \cong \overline{AB}$ by

a. Overlapping Segments Postulate

b. Reflexive Property

c. CPCTC

d. Isosceles Triangle Theorem

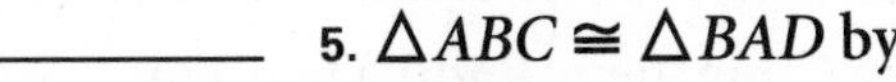

______ 5. $\triangle ABC \cong \triangle BAD$ by

a. SSS b. ASA c. SAS d. HL

______ 6. If three ______ of one triangle are congruent to three ______ of another triangle, then the triangles are congruent.

a. angles b. sides c. angle bisectors d. altitudes

______ 7. Which postulate *cannot* be used to prove that two triangles are congruent?

a. SSS b. SAS c. ASA d. SSA

______ 8. If $\triangle FGH$ is an isosceles triangle, then which of the following is always true?

a. The bisector of the vertex angle is the perpendicular bisector of the base.

b. Each base angle measures 50°.

c. The consecutive angles are congruent.

d. The bisector of a base angle is the median of the side to which it is drawn.

Chapter Assessment

Chapter 4, Form A, page 2

Use the figures at right for Exercises 9–11.

________ 9. If m$\angle IJK = 72°$, then m$\angle JIL =$ ________.

a. 108° b. 72°

c. 54° d. 24°

________ 10. If $IM = 8$ and $LM = 6$, what is MK?

a. 6 b. 14

c. 2 d. 5

11. If m$\angle JLM = 65°$, then m$\angle LJM =$ ________.

a. 65° b. 130°

c. 115° d. 50°

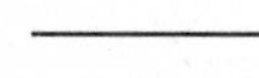

12. If one pair of opposite sides of a quadrilateral is parallel and congruent, then the quadrilateral is a ________.

a. parallelogram b. kite c. trapezoid d. cannot be determined

13. If the diagonals of a parallelogram are congruent, then the parallelogram is a ________.

a. rhombus b. rectangle c. kite d. cannot be determined

14. To construct a segment congruent to a given segment, use ________.

a. the Reflexive Property b. the definition of congruent segments

c. CPCTC d. the Congruent Radii Theorem

15. Which segments are congruent?

a. $\overline{AB}$ and $\overline{CD}$ with $A(3, 0)$, $B(8, 0)$, $C(0, -2)$, and $D(0, 3)$
b. $\overline{AB}$ and $\overline{CD}$ with $A(3, 0)$, $B(8, 0)$, $C(3, 0)$, and $D(0, 8)$
c. $\overline{AB}$ and $\overline{CD}$ with $A(3, 0)$, $B(8, 0)$, $C(3, 0)$, and $D(-8, 0)$
d. $\overline{AB}$ and $\overline{CD}$ with $A(3, 0)$, $B(8, 0)$, $C(-3, 0)$, and $D(8, 0)$

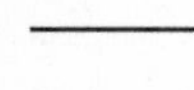

16. What is true about $\triangle ABC$?

a. $x + 4 > 5$ b. $x - 5 > 4$

c. $4 + 5 < x$ d. $5 - 4 > x$

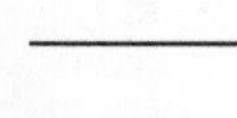

17. What is the smallest possible value for x?

a. 1 b. 3

c. just larger than 1 d. just less than 9

NAME ______________________ CLASS ______________ DATE ______________

Chapter Assessment

Chapter 4, Form B, page 1

For Exercises 1 and 2, decide whether or not the polygons are congruent. Justify your conclusion.

1. 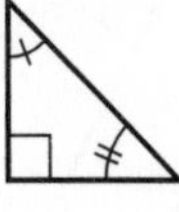 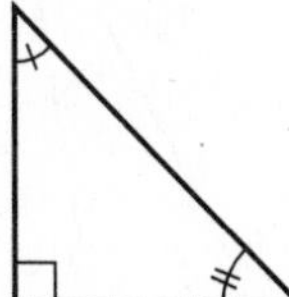

2. 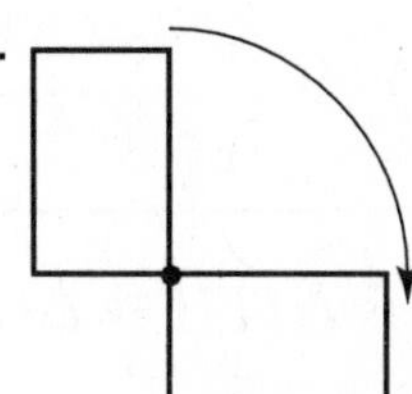

____

____

____

Supply the reasons for each step in the proof below.

Given: $\triangle ABC$ is isosceles and $\overline{BD} \perp \overline{AC}$.

Prove: $\overline{AD} \cong \overline{CD}$

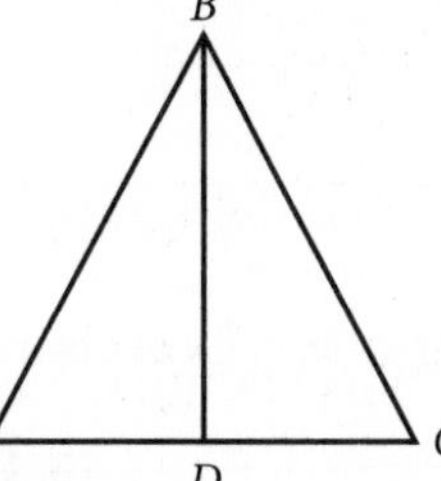

Statements	Reasons
$\triangle ABC$ is isosceles and $\overline{BD} \perp \overline{AC}$.	**3.** ______
$\overline{AB} \cong \overline{BC}$	**4.** ______
$\overline{BD} \cong \overline{BD}$	**5.** ______
$\angle ADB$ and $\angle BDC$ are right angles.	**6.** ______
$\triangle ABD \cong \triangle CBD$	**7.** ______
$\overline{AD} \cong \overline{CD}$	**8.** ______

According to the Polygon Congruence Postulate, Pentagon *KLMNO* is congruent to pentagon *PQRST* if and only if

9. ______________________ and

10. ______________________.

NAME ______________________ CLASS __________ DATE __________

Chapter Assessment

Chapter 4, Form B, page 2

For Exercises 11–13, use the figure at right.

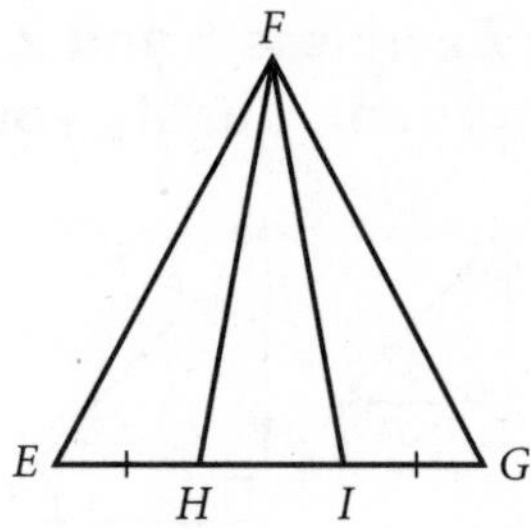

11. How many different triangles are in the figure? __________

12. Is $\overline{EI} \cong \overline{HG}$? Why or why not? __________

____

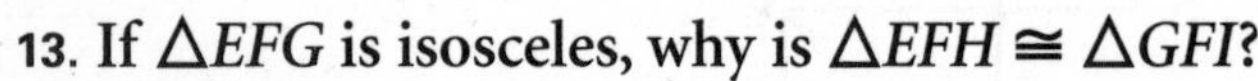

13. If $\triangle EFG$ is isosceles, why is $\triangle EFH \cong \triangle GFI$?

____

14. Given: Parallelogram *JKLM* is a rectangle.

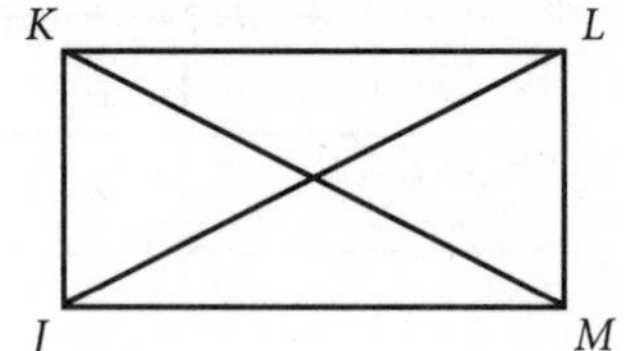

Write a paragraph proof showing that the diagonals are congruent.

____

____

____

____

Use a compass and a straightedge for Exercises 15 and 16.

15. Copy $\overline{XY}$.

16. Construct the angle bisector of the given angle.

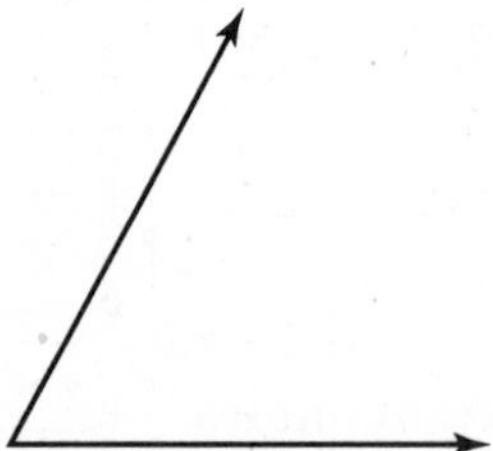

State whether each set of lengths will form a triangle.

17. $AB = 6$, $BC = 7$, and $CA = 9$ __________

18. $AB = 6$, $BC = 15$, and $CA = 6$ __________

19. $AB = \frac{2}{3}$, $BC = \frac{3}{5}$, and $CA = \frac{1}{2}$ __________

NAME ______________________ CLASS ____________ DATE ____________

Alternative Assessment

Exploring Congruency, Chapter 4, Form A

TASK: Prove that parts of a figure are congruent.

HOW YOU WILL BE SCORED: As you work through the task, your teacher will be looking for the following:

- whether you can apply the congruence theorems to the figure shown
- how well you describe the different possibilities from the figure and given information

Study the figure and the given information.

P is the midpoint of $\overline{OM}$ and $\overline{LN}$.
LMNO is a rectangle.

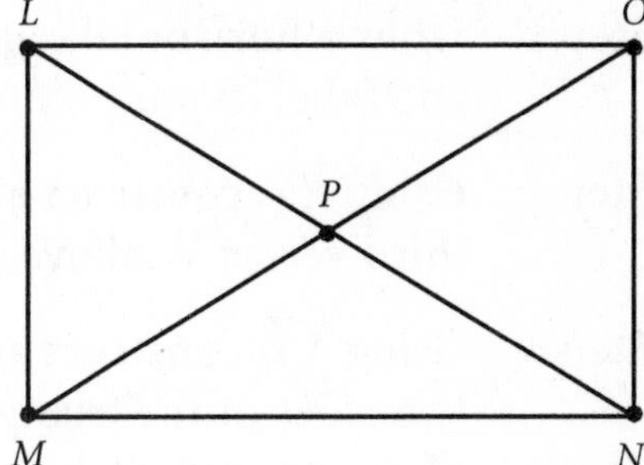

1. Describe one way in which you can prove that two triangles in the figure are congruent.

____

____

2. There are several ways in which triangles can be proven congruent from the figure and given information. Make a list of all of the congruent pairs and how they can be proven congruent.

____

____

3. Could all of the triangle pairs be proven congruent in at least one way with only the information that *LMNO* is a rectangle? Why or why not?

____

____

4. Could all of the triangle pairs be proven congruent in at least one way with only the information that *P* is the midpoint of $\overline{OM}$ and $\overline{LN}$? Why or why not?

____

____

5. On your own paper, choose one of the triangle pairs and use the given information to write a two-column proof showing that the two triangles are congruent.

SELF-ASSESSMENT: Design a diagram in which triangles can be proven congruent in many different ways. Make a list of all congruent triangle pairs and how they can be proven congruent.

NAME ______________________ CLASS ______________ DATE ______________

Alternative Assessment

Constructions, Chapter 4, Form B

TASK: Construct a figure and prove items congruent.

HOW YOU WILL BE SCORED: As you work through the task, your teacher will be looking for the following:

- whether you can complete the described construction steps
- how well you can apply the congruence theorems to the figure you create

Follow the construction steps to create a figure in the space provided. You will need a compass and a straightedge to complete the construction. After you have completed the steps below, use your construction to answer the questions.

Step 1 Draw a horizontal segment approximately 1.5 to 2 inches long and label its endpoints X and Y.

Step 2 Using $\overline{XY}$, construct an equilateral triangle on $\overline{XY}$ and label the third vertex W above $\overline{XY}$.

Step 3 Using $\overline{XY}$, construct an isosceles triangle on $\overline{XY}$ with the other two sides of the triangle longer than $\overline{XY}$. Label the third vertex of this triangle Z, below $\overline{XY}$.

Step 4 Using the straightedge, draw $\overline{WZ}$. Label the intersection of $\overline{XY}$ and $\overline{WZ}$ as point V.

1. List all of the pairs of triangles in your figure that can be proven congruent. Next to each pair, list the way(s) in which that pair can be proven congruent. ______________________

____

2. Using this figure, prove that $\angle XVZ \cong \angle YVZ$.

Given: ______________________

SELF-ASSESSMENT: Create a list of construction steps that can be used to prove items congruent.

Quick Warm-Up: Assessing Prior Knowledge

5.1 Perimeter and Area

Draw a rectangle with a length of 8 units and a width of 6 units.

1. What is the perimeter of this rectangle? ______
2. What is the area of this rectangle? ______
3. Draw two other rectangles with the same perimeter.
4. Draw two other rectangles with the same area.

Lesson Quiz

5.1 Perimeter and Area

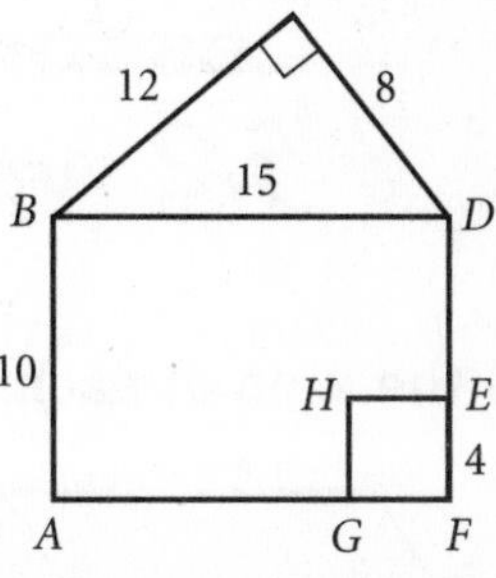

Refer to the diagram at right. Find the perimeter of each figure.

1. rectangle *ABDF* ______
2. triangle *BCD* ______
3. square *EFGH* ______
4. pentagon *ABCDF* ______
5. hexagon *ABDEHG* ______
6. heptagon *ABCDEHG* ______

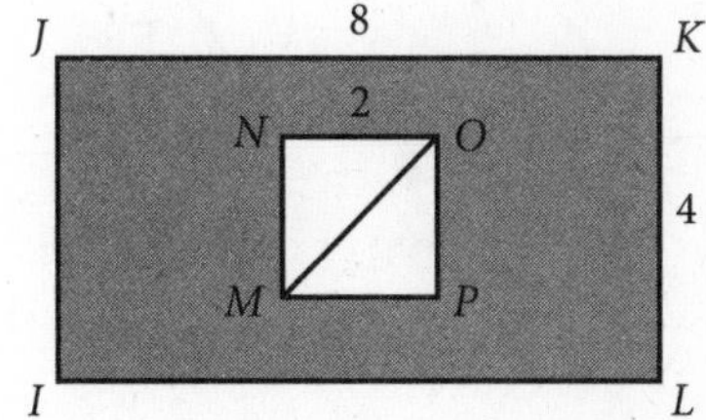

Find the area of each figure in square units.

7. rectangle *IJKL* ______
8. square *MNOP* ______
9. triangle *MNO* ______
10. shaded region ______

Find the perimeter and area of a rectangle with vertices at the given points.

11. (0, 0), (−3, 0), (−3, 4), and (0, 4)

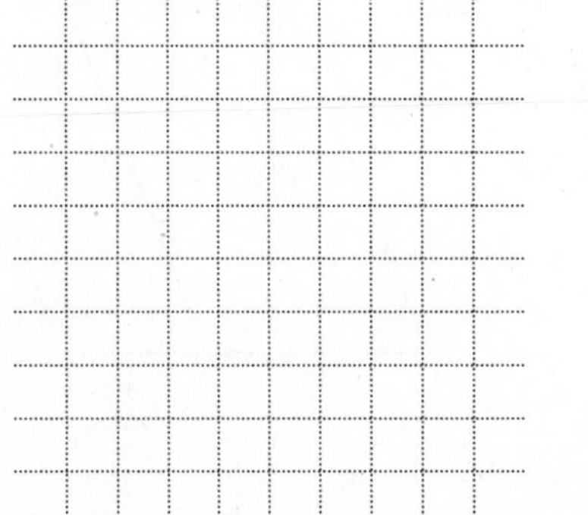

______

______

12. (−3, 2), (−3, −5), (5, 2), and (5, −5)

______

______

NAME ______________________ CLASS ____________ DATE ____________

Quick Warm-Up: Assessing Prior Knowledge

5.2 Areas of Triangles, Parallelograms, and Trapezoids

Define each figure by describing key properties of the sides and angles.

1. right triangle ______________________

2. acute triangle ______________________

3. parallelogram ______________________

Lesson Quiz

5.2 Areas of Triangles, Parallelograms, and Trapezoids

Find the area of each figure.

1. 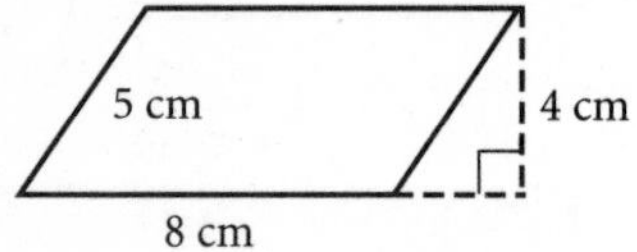

2. 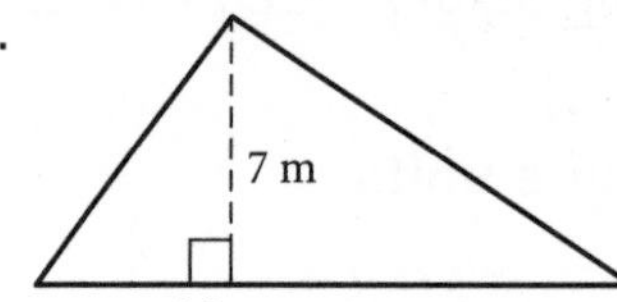

3. 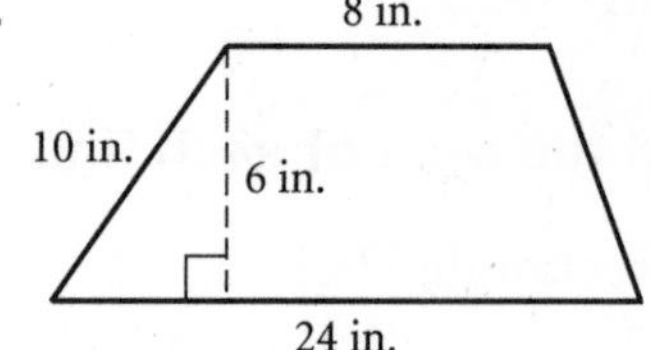

4. Is it possible for two rectangles to have the same area and not be congruent? If so, give an example.

____

Use the diagram at right for Exercises 5–8.

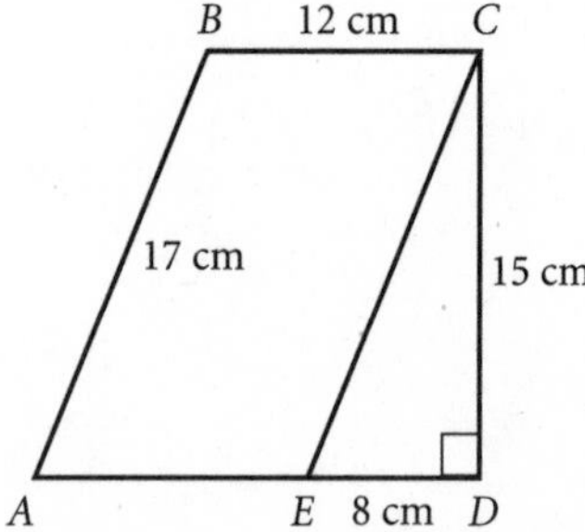

5. Find the area of parallelogram *ABCE*. ______________

6. Find the area of $\triangle CDE$. ______________

7. Find the area of trapezoid *ABCD*. ______________

8. The area of a triangle is 48 square centimeters. Find the height of the triangle if its base is 6 centimeteres.

____

Quick Warm-Up: Assessing Prior Knowledge

5.3 Circumferences and Areas of Circles

Evaluate. Round answers to the nearest hundredth. You may use a calculator if you wish.

1. 5^2 ______________
2. 6.7^2 ______________
3. $\sqrt{36}$ ______________
4. $\sqrt{48}$ ______________
5. π ______________
6. 2π ______________
7. $3\pi + 1.3$ ______________
8. $3^2 \cdot \pi$ ______________
9. $(3\pi)^2$ ______________

10. Draw a circle in the space at right. Label the center, radius, and diameter.

Lesson Quiz

5.3 Circumferences and Areas of Circles

Find the circumference of each circle. Round your answers to the nearest tenth.

1. $r = 12$ ______________
2. $d = 5$ ______________

Find the area of each circle. Round your answers to the nearest tenth.

3. $r = 8$ ______________
4. $d = 9$ ______________

For Exercises 5 and 6, find the area of the shaded region. Round your answers to the nearest hundredth.

5.

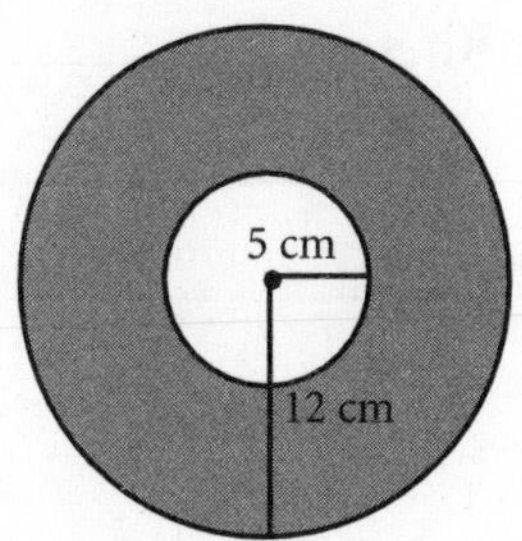

____

6.

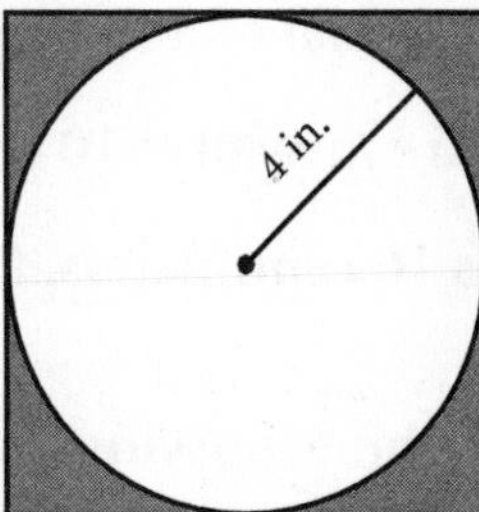

____

7. A circular garden has an area of 616 square feet. What is the diameter of the garden? ______________

NAME ______________________ CLASS ____________ DATE ____________

Quick Warm-Up: Assessing Prior Knowledge

5.4 The Pythagorean Theorem

Evaluate. Round your answers to the nearest hundredth.

1. $\sqrt{72}$ ______________
2. $\sqrt{10}$ ______________
3. $\sqrt{120}$ ______________
4. $\sqrt{6^2 + 4^2}$ ______________
5. $\sqrt{14^2 - 5^2}$ ______________

Lesson Quiz

5.4 The Pythagorean Theorem

In $\triangle ABC$, $\angle C$ is a right angle. Find the length c of the hypotenuse to the nearest tenth.

1. $a = 6, b = 8, c =$ ______________
2. $a = 8, b = 15, c =$ ______________
3. $a = 5, b = 10, c =$ ______________
4. $a = 9, b = 14, c =$ ______________

Find the missing length in each triangle. Give exact answers.

5. 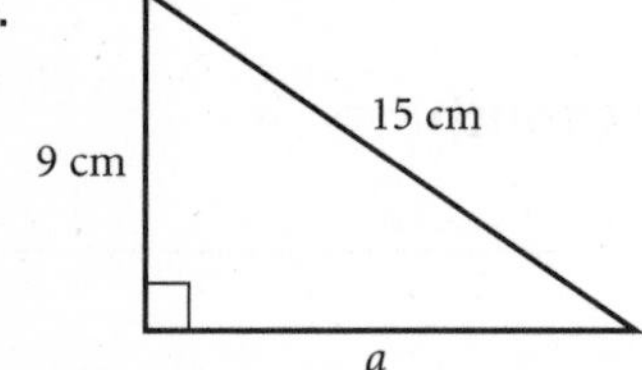

____

6. 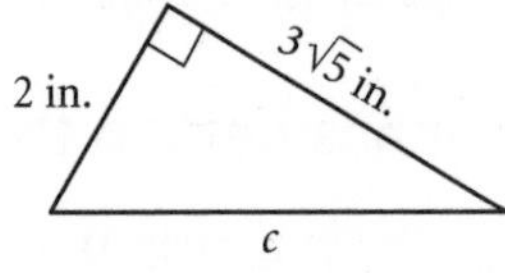

____

7. 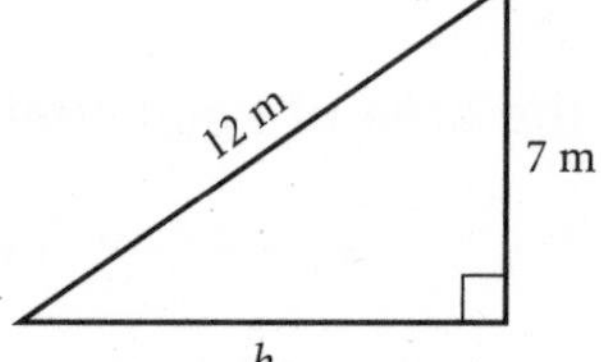

____

8. Find the length of a diagonal of a square whose side measures 8 centimeters. ______________
9. The diagonal of a square measures 10 centimeters. Find its area. ______________
10. The area of a square is 16 centimeters. Find the length of a diagonal. ______________

Decide whether each of the following triples could be the side lengths of a triangle. If so, determine whether the triangle is right, obtuse, or acute.

11. 8, 12, 16 ______________
12. 3, 7, 10 ______________
13. 10, 24, 26 ______________
14. 6, 6, 11 ______________
15. 16, 20, 38 ______________
16. 15, 15, 15 ______________

NAME ______________________ CLASS ____________ DATE ____________

Mid-Chapter Assessment

Chapter 5 (Lessons 5.1–5.4)

Write the letter that best answers the question or completes the statement.

3 2 2 8 9

______ 1. What is the perimeter of the figure at right?

a. 24 units b. 34 units

c. 38 units d. 64 units

2. What is the area of the figure?

a. 72 square units b. 64 square units

c. 40 square units d. 42 square units

3. The circumference of the inscribed circle at right is about ______.

a. 9.4 units b. 18.8 units

c. 28.3 units d. 37.7 units

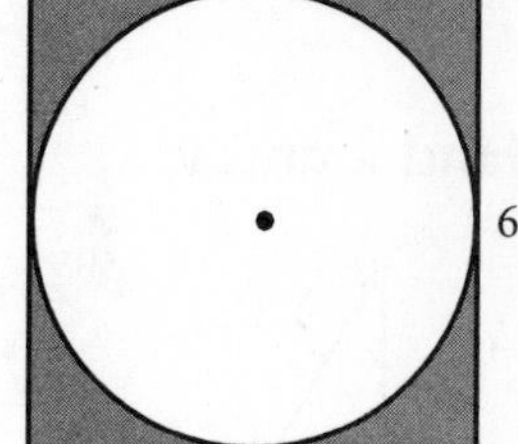

______ 4. The area of the shaded region is about ______.

a. 7.7 square units b. 18.8 square units

c. 17.2 square units d. 28.3 square units

Find the perimeter or circumference of each figure.

5.

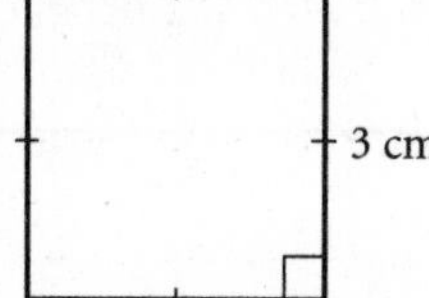

6.

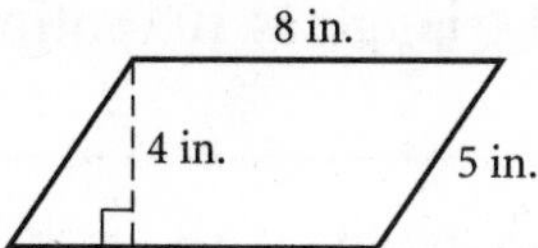

7.

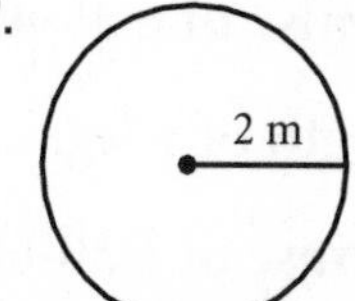

______ ______ ______

Find the area of each figure.

8.

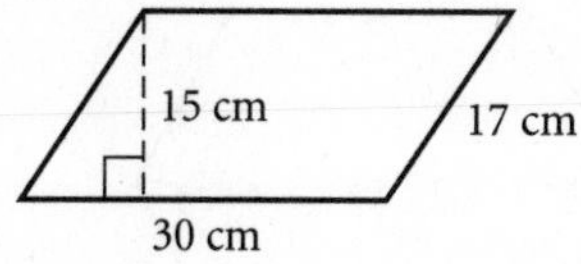

9.

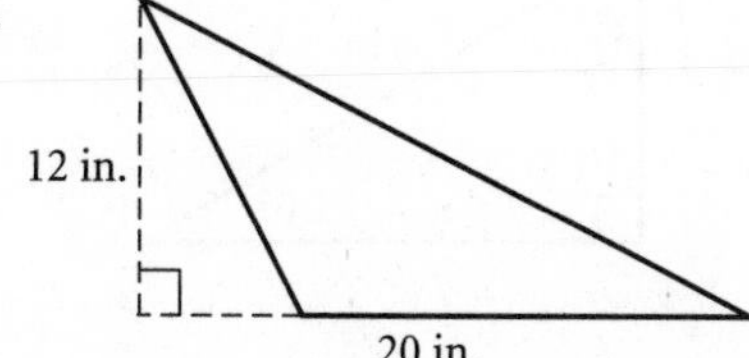

10.

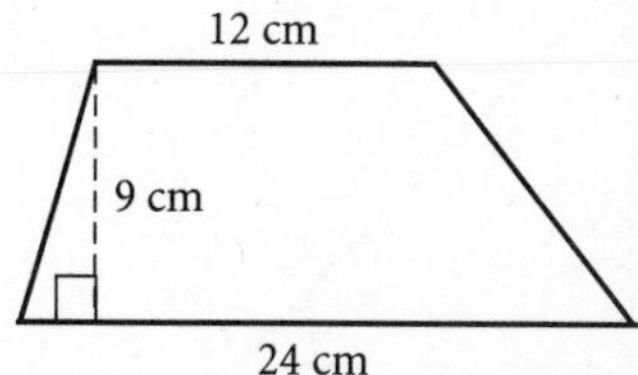

______ ______ ______

11. At $9 per square foot, determine the cost of carpeting a 12-foot by 8-foot rectangular room. Include sales tax of 7%. ______

Quick Warm-Up: Assessing Prior Knowledge

5.5 Special Triangles and Areas of Regular Polygons

Simplify each radical expression by giving the values for *a* and *b*.

1. $\sqrt{32} = \sqrt{a} \cdot \sqrt{2} = b\sqrt{2}$ ____________________
2. $\sqrt{54} = \sqrt{a} \cdot \sqrt{6} = b\sqrt{6}$ ____________________
3. $\sqrt{45} = \sqrt{a} \cdot \sqrt{5} = b\sqrt{5}$ ____________________

Lesson Quiz

5.5 Special Triangles and Areas of Regular Polygons

Find *x* and *y*.

1. 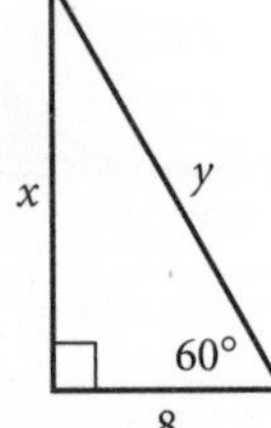

2. 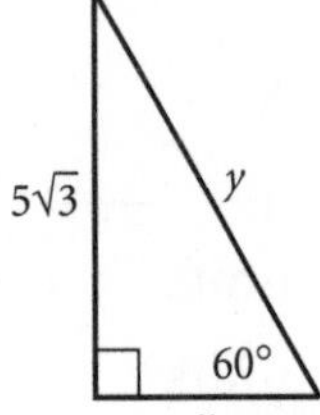

3. 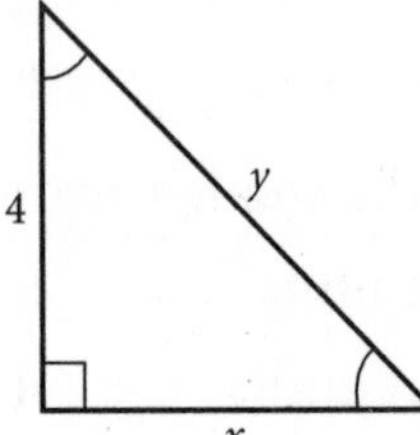

4. 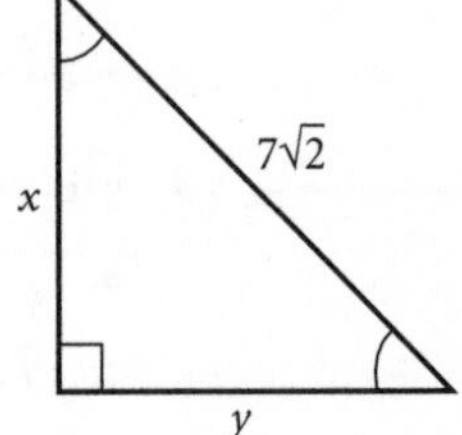

____________ ____________ ____________ ____________

5. The hypotenuse of a 45-45-90 right triangle is 10 centimeters long.
   Find the length of a leg. ____________________
6. The hypotenuse of a 30-60-90 right triangle is 12 centimeters long.
   Find the length of the shorter leg. ____________________

Find the area of each figure. In Exercise 9, the figure is a regular hexagon.

7. 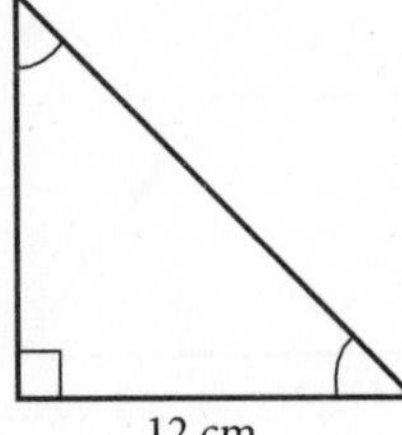

8. 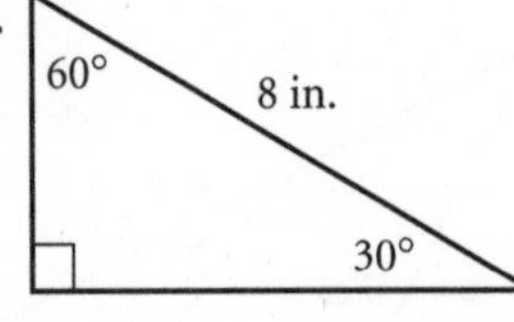

9. 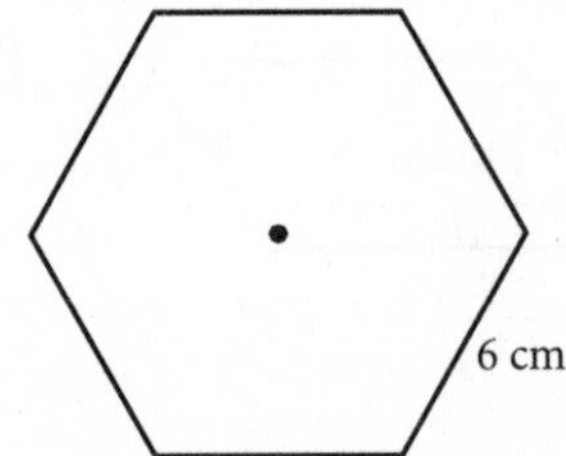

____________ ____________ ____________

10. The altitude of an equilateral triangle is $6\sqrt{3}$ centimeters. Find the area. ____________

NAME ______________________ CLASS ______________ DATE ______________

Quick Warm-Up: Assessing Prior Knowledge

5.6 The Distance Formula and the Method of Quadrature

Let (x_1, y_1) be (2, 1) and (x_2, y_2) be (6, 4). Find the value of each expression.

1. $|x_2 - x_1|$ ______________
2. $|x_2 - x_1|^2 + |y_2 - y_1|^2$ ______________
3. $\sqrt{(x_2 - x_1)^2 + (y_2 - y_1)^2}$ ______________

Lesson Quiz

5.6 The Distance Formula and the Method of Quadrature

Find the distance between each pair of points. Round your answers to the nearest hundredth.

1. (5, 3) and (2, 7) ______________
2. (−2, 4) and (10, −1) ______________
3. (−2, 3) and (−5, 4) ______________
4. (−3, 3) and (1, −1) ______________

For Exercises 5–10, refer to the coordinate plane below.

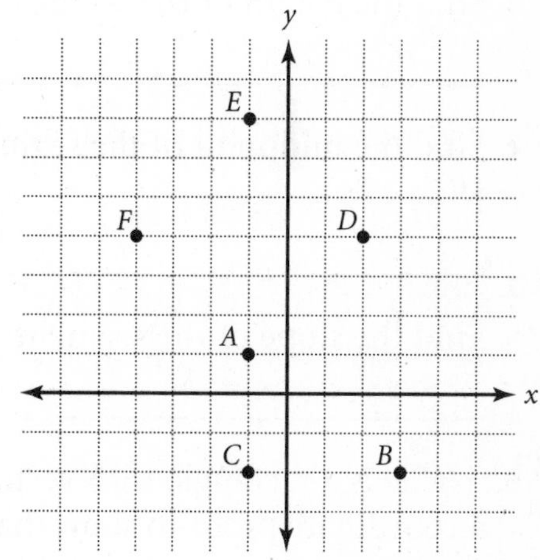

5. Find the distance between points *A* and *B*. ______________

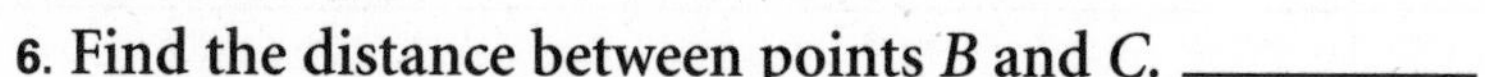

6. Find the distance between points *B* and *C*. ______________
7. Is the triangle formed by points *A*, *B*, and *C* a right triangle? Why or why not?

____

____

8. Is the quadrilateral formed by points *A*, *D*, *E*, and *F* a rhombus? Why or why not?

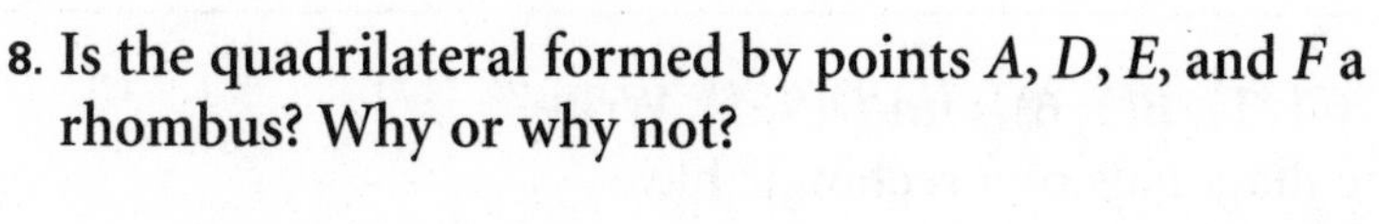

____

____

9. Find the coordinates of the midpoint of $\overline{BD}$. ______________
10. Find the area of $\triangle ABC$. ______________
11. Use the method of quadrature to estimate the area between $y = x^2 + 1$ and the *x*-axis for $0 \leq x \leq 2$. Use the left-hand rule. ______________

Quick Warm-Up: Assessing Prior Knowledge

5.7 Proofs Using Coordinate Geometry

1. Find the slope of the line containing $(-2, 4)$ and $(-1, 3)$. ______________
2. Find the length of the segment with endpoints $(4, 7)$ and $(-2, -1)$. ______________
3. What is the image of $(-2, 4)$ under a reflection across the line $y = x$? ______________

Lesson Quiz

5.7 Proofs Using Coordinate Geometry

1. *ABCD* is a square with vertices $A(0, 0)$ and $B(0, 4)$. Find the coordinates of vertices *C* and *D*.

____

2. *ABC* is an isosceles triangle with $\overline{AB} \cong \overline{BC}$. Two vertices are $A(0, 0)$ and $B(3, 2)$. What are the coordinates of vertex *C*?

____

3. Find the length of the segment with endpoints $A(0, 0)$ and $B(3, 4)$.

____

4. Find the midpoint of the segment with endpoints $C(2a, 2b)$ and $D(2m, 2n)$.

____

5. Find the slope of the segment with endpoints $S(s, 0)$ and $T(0, t)$.

____

6. *ABCD* is a rectangle with vertices $A(1, 1)$, $B(1, 6)$, and $D(9, 1)$. Write a coordinate proof to show that the diagonals of a rectangle have the same measure.

____

____

____

____

NAME ______________________ CLASS __________ DATE __________

Quick Warm-Up: Assessing Prior Knowledge
5.8 Geometric Probability

Use graph paper for the following exercises:

1. Draw a rectangle in which the side lengths are whole numbers.
2. Inside the rectangle, draw a circle with a whole-number radius.
3. Find the ratio of the area of the circle to the area of the rectangle. ______________

Lesson Quiz
5.8 Geometric Probability

Convert each probability to percent.

1. 0.2 __________ 2. $\frac{3}{5}$ __________ 3. $\frac{5}{8}$ __________

Convert each percent to a fractional probability.

4. $37\frac{1}{2}\%$ __________ 5. 75% __________ 6. 20% __________

For Exercises 7–9, suppose that a point *P* is selected at random from the points shown on this number line.

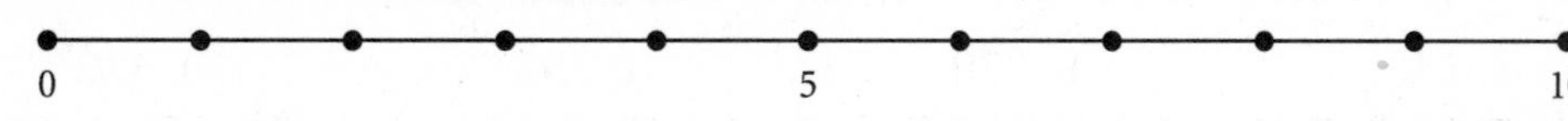

7. What is the probability that $0 \leq P \leq 5$? __________
8. What is the probability that $1 \leq P \leq 9$? __________
9. What is the probability that $7 \leq P \leq 8$? __________

Find the probability that a dart thrown at random will land in the shaded area.

10. 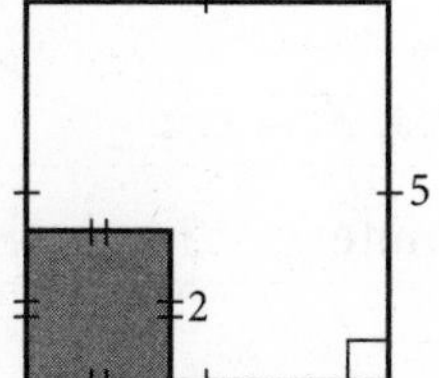

11. 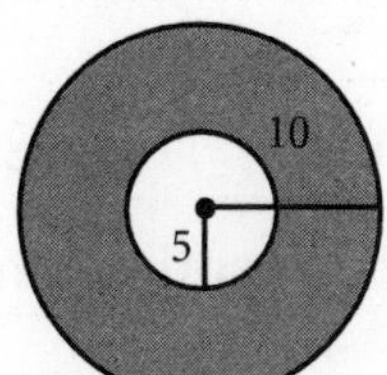

12. 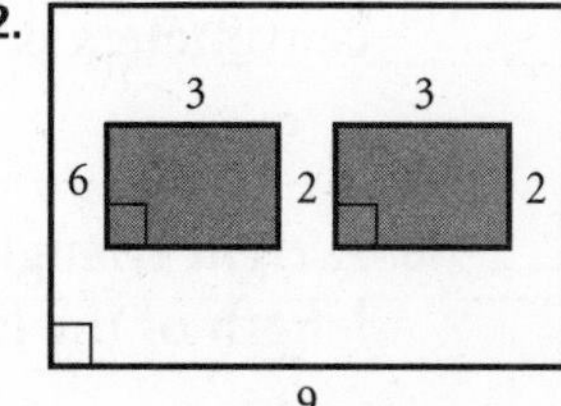

__________ __________ __________

NAME ______________________ CLASS ______________ DATE ______________

Chapter Assessment

Chapter 5, Form A, page 1

Write the letter that best answers the question or completes the statement.

______ 1. For a fixed perimeter, what shape has the maximum area?

a. parallelogram b. trapezoid c. square d. circle

______ 2. The perimeter of a rectangle is 54 centimeters and the base length is twice the height. What is the area?

a. 108 sq cm b. 154 sq cm
c. 162 sq cm d. 324 sq cm

______ 3. The perimeter of a rectangle is 40 centimeters. If the width is 8 centimeters, what is the length?

a. 12 cm b. 16 cm c. 32 cm d. 48 cm

______ 4. What is the area, in square units, of parallelogram *ABDE*?

a. 120 b. 136
c. 210 d. 240

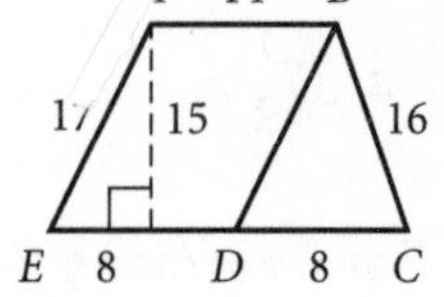

______ 5. What is the area, in square units, of triangle *BCD*?

a. 60 b. 128
c. 120 d. cannot be determined

______ 6. What is the area, in square units, of trapezoid *ABCE*?

a. 240 b. 270 c. 352 d. 420

______ 7. The perimeter of *ABCE* is ________.

a. 63 b. 69 c. 75 d. cannot be determined

______ 8. What is the measure of the base of a triangle with a height of 20 centimeters and an area of 100 centimeters?

a. 5 cm b. 6 cm c. 8 cm d. 10 cm

______ 9. To the nearest tenth, the circumference of a circle with a radius of 3 centimeters is

a. 9.4 cm b. 18.8 cm c. 28.3 cm d. 31.4 cm

______ 10. A right triangle has lengths of 15 centimeters and 20 centimeters. What is the length of the hypotenuse?

a. 18 cm b. 25 cm c. 35 cm d. 50 cm

Chapter Assessment

Chapter 5, Form A, page 2

______ 11. What is the area, in square units, of the shaded region?

a. 34.4 b. 108

c. 339.3 d. 37.7

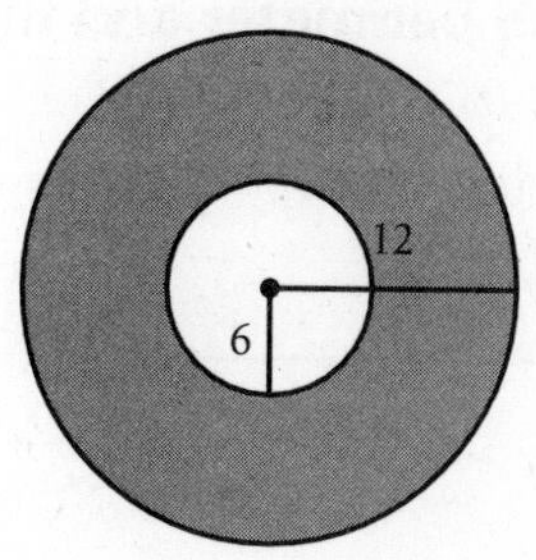

______ 12. If $AB = 10$, what is BC?

a. 5 b. $5\sqrt{3}$

c. $10\sqrt{3}$ d. 20

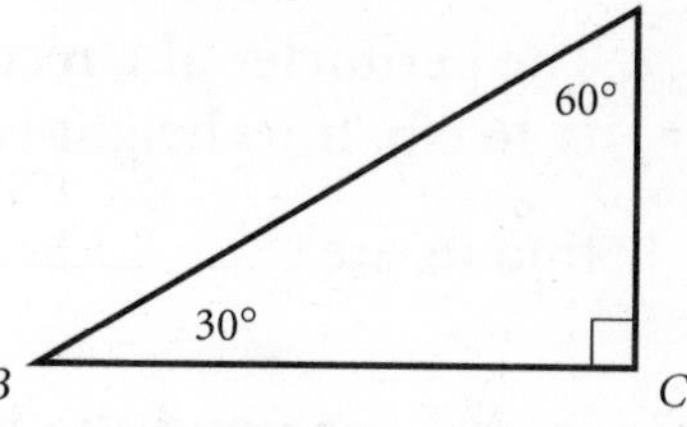

______ 13. If $BC = 9$, what is AC?

a. $3\sqrt{3}$ b. 4.5 c. $9\sqrt{3}$ d. 18

______ 14. What is the area, in square units, of the figure at right?

a. 4 b. 8

c. 16 d. 32

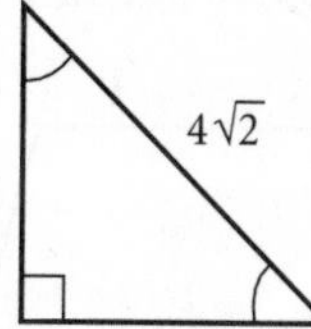

______ 15. The distance between $(5, -2)$ and $(-3, 4)$ is ______.

a. 8 b. 10 c. 64 d. 100

______ 16. The midpoint of the line segment from $(-3, 4)$ and $(-1, -6)$ is ______.

a. $(-1, 5)$ b. $(-1, -2)$ c. $(-2, -1)$ d. $(-2, 1)$

______ 17. The triangle formed by H, I, and J is ______.

a. scalene b. isosceles

c. right d. equilateral

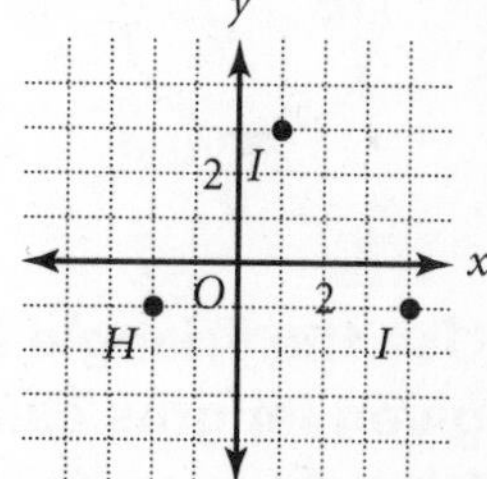

______ 18. What is the probability that a dart thrown at random will land in a circle at right?

a. 52% b. 60%

c. 79% d. 85%

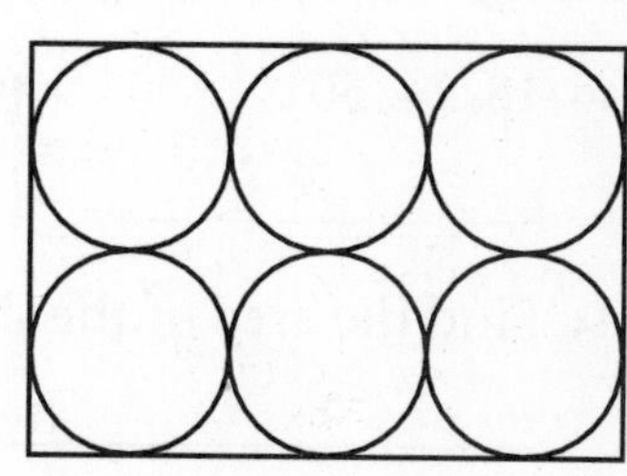

______ 19. What is the probability that the dart will *not* land in a circle?

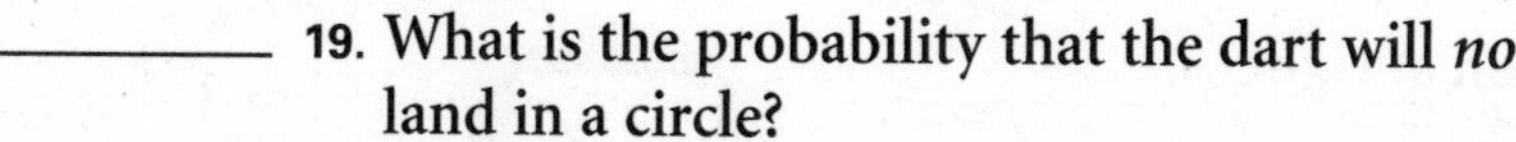

a. 48% b. 40%

c. 21% d. 15%

NAME ______________________ CLASS ____________ DATE ____________

Chapter Assessment

Chapter 5, Form B, page 1

Find the perimeter and the area of each figure.

1.

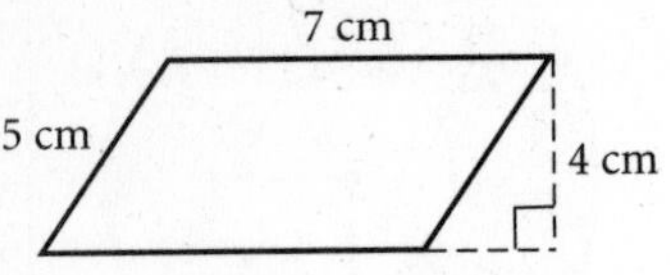

____

2.

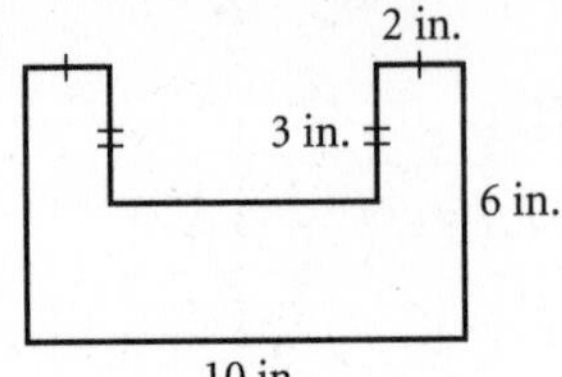

____

3.

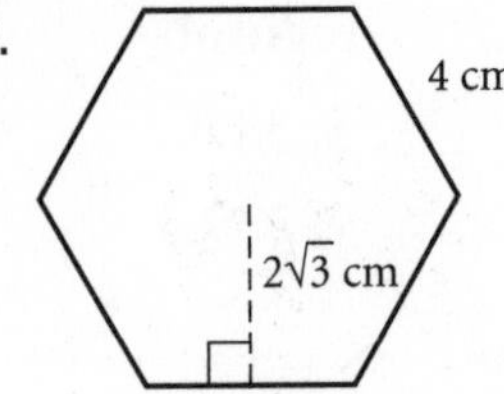

____

4. The perimeter of a rectangle is 46 cm. If its height is 8 cm, find its area. ______________________

5. The perimeter of a rectangle is 48 cm. If its base is three times its height, find its dimensions. ______________________

Given the area and the information shown in the diagram, find *x*.

6. A = 28.14 square units ______________________

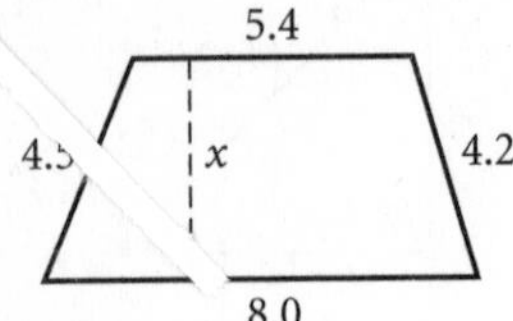

7. A = 19.58 square inches ______________________

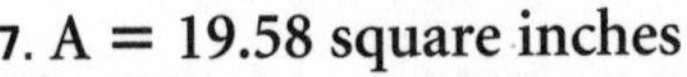

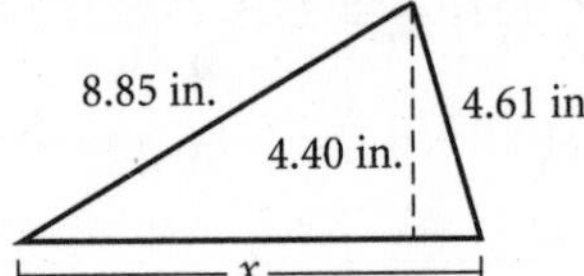

Find the circumference and area of each circle.

8.

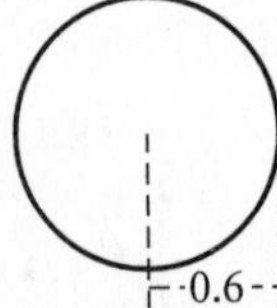

____

____

9.

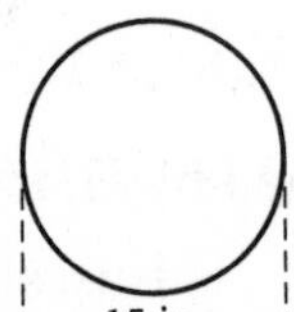

____

____

Use the Triangle Inequality Theorem to determine whether the given lengths can be lengths of a triangle. If so, tell whether the triangle is acute, right, or obtuse.

10. 18, 24, 30 ______________________

11. 12, 15, 25 ______________________

12. 10, 13, 24 ______________________

13. 15, 18, 21 ______________________

14. Find the area of the shaded region at right.

____

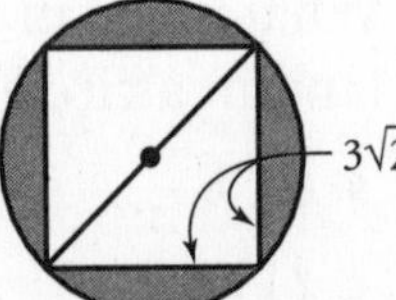

Chapter Assessment

Chapter 5, Form B, page 2

Find the value of each variable.

15.

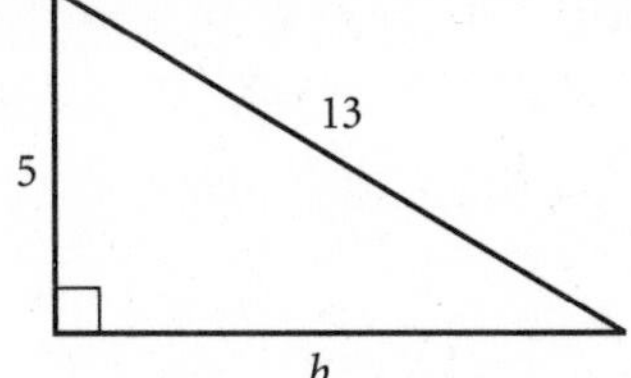

16.

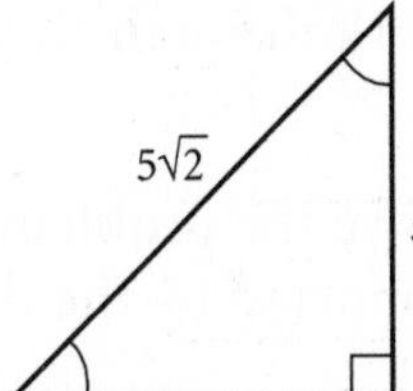

17.

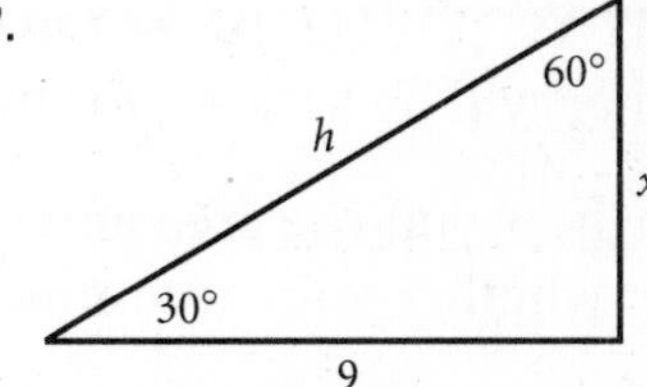

______________ ______________ ______________

18. A tricycle wheel has a six-inch diameter. About how many revolutions must the wheel make to travel 6 feet? ______________

19. An isosceles trapezoid has nonparallel sides 15 inches long. One parallel base is twice as long as the other. The perimeter of the trapezoid is 60 inches. Find the area of the trapezoid. ______________

For Exercises 20–22, refer to the coordinate plane at right.

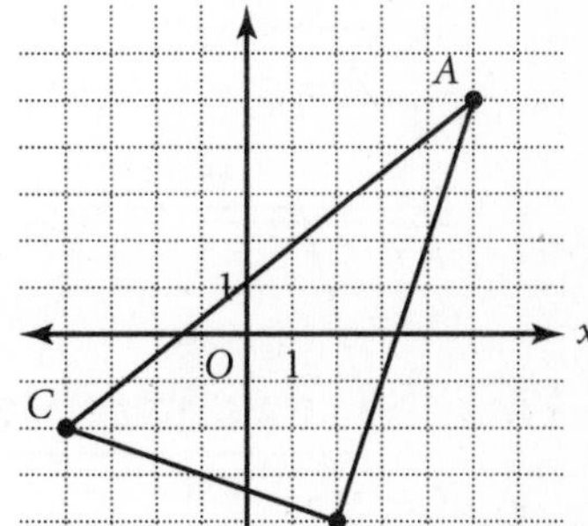

20. Use the distance formula to find AB, BC, and AC.

____

21. Use the converse of the Pythagorean Theorem to show that $\triangle ABC$ is a right triangle. ______________

22. Find the area of $\triangle ABC$. ______________

In Exercises 23 and 24, find the perimeter and area of each regular polygon.

23.

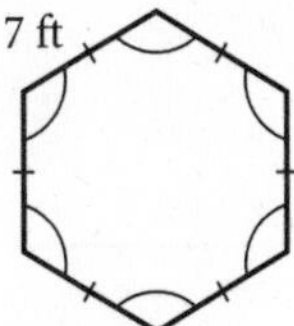

24.

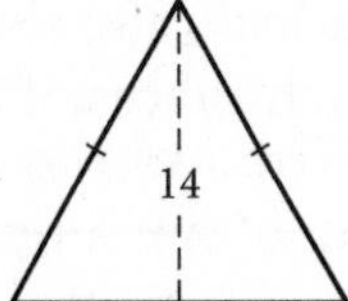

______________ ______________

25. The diagram at right shows two squares inside a rectangle. Estimate the probability that a dart randomly thrown at the rectangle will land inside one of the squares. ______________

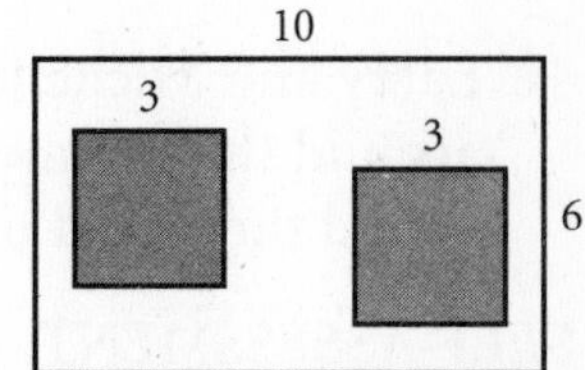

NAME ______________________ CLASS ____________ DATE ____________

Alternative Assessment

Retail Space, Chapter 5, Form A

TASK: Determine which of several possible store sites best meets the needs described.

HOW YOU WILL BE SCORED: As you work through the task, your teacher will be looking for the following:

- how you use area and perimeter to solve the problem
- whether your recommendation is supported by the data

Larry and Michelle own and operate a small, successful store. They would like to move into a space in a large, new shopping mall being built in an area of the city experiencing a lot of growth. The available spaces that they are considering are different shapes and sizes, as shown in the figures. The cost of each space is given in dollars per square foot per year. They must decide on a space so that they can reserve it and begin to plan the layout of the store.

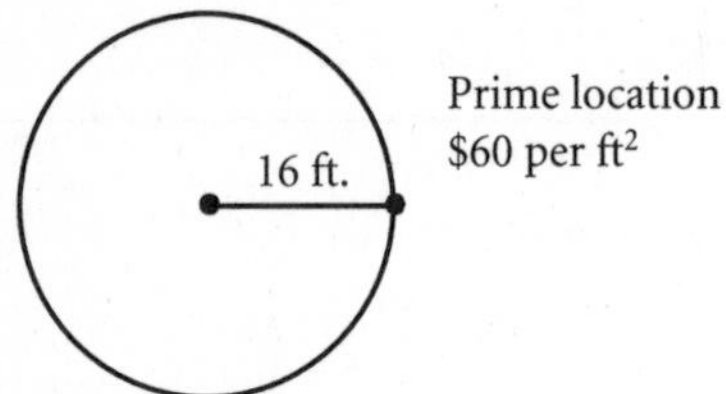

Prime location
$60 per ft^2

30 ft.
29 ft.
28.5 ft.
35 ft.

Good location
$45 per ft^2

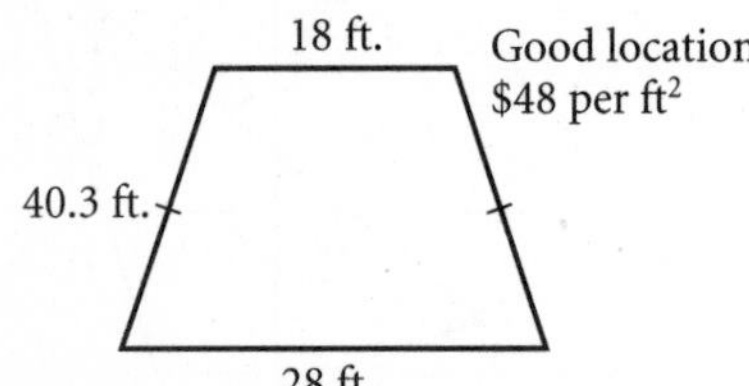

Good location
$48 per ft^2

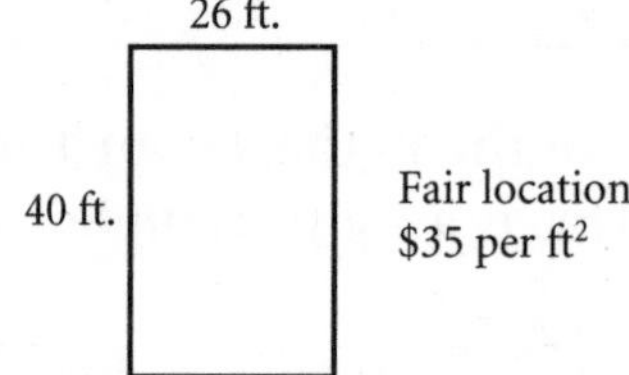

Fair location
$35 per ft^2

1. Which is the least expensive of the available spaces? ____________

2. Which space has the largest area? ____________

3. Larry and Michelle need at least 800 square feet at a price of less than $48,000 per year. Which spaces, if any, meet these needs?

   ____

4. Which space would you recommend that they lease? Why?

   ____

5. Larry and Michelle would like to install a strip of neon lighting around the perimeter of the store's ceiling. How much neon lighting would they need if they took the space you recommended? ____________

SELF-ASSESSMENT: Make a list of all area or perimeter formulas you used in this task. Describe any other formulas that could have been used.

Alternative Assessment

Exploring Triangles, Chapter 5, Form B

TASK: Describe what can be determined about a given trapezoid.

HOW YOU WILL BE SCORED: As you work through the task, your teacher will be looking for the following:

- how many different ways you can describe a trapezoid
- whether each of your descriptions can be supported by calculations, formulas, or other explanations

Study the figure below. Make a list of all the things that can be shown to be mathematically true about the figure. For example, "A horizontal line can divide the figure into two trapezoids whose areas can be added to find the area of the figure. Use the formula $\frac{1}{2}(b_1 + b_2)h$." You may want to label the vertices so that you can express the relationships symbolically.

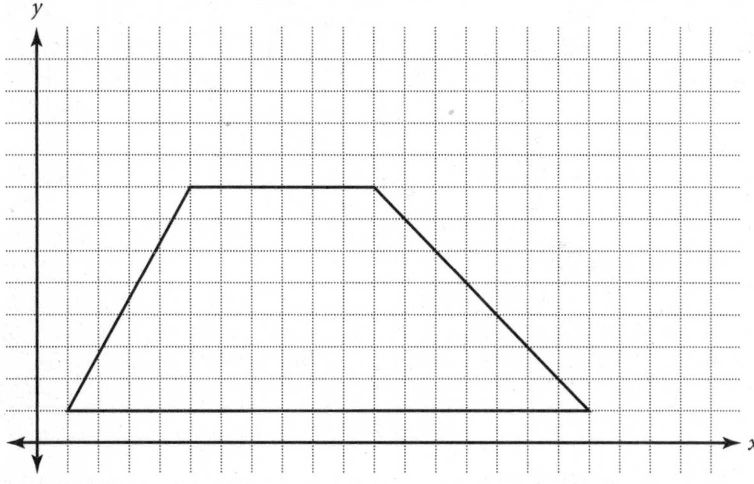

Description

____

____

____

____

____

____

____

SELF-ASSESSMENT: Draw a figure using several different shapes. What mathematical determinations can you make about the figure?

NAME ______________________ CLASS ____________ DATE ____________

Quick Warm-Up: Assessing Prior Knowledge

6.1 Solid Shapes

1. How many faces does a cube have? ______________

2. How many faces of a cube are ordinarily hidden from view? ______

3. Find the area of a square with side lengths of 1 cm. ______________

Lesson Quiz

6.1 Solid Shapes

Refer to the isometric drawing of a solid shown at right. Assume that no cubes are hidden from view.

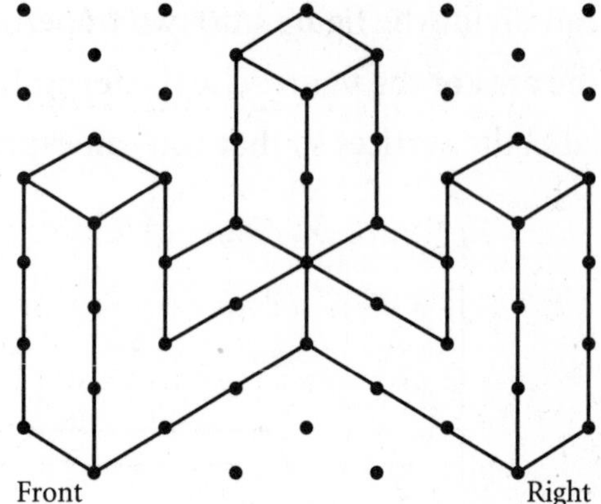

1. Give the volume in cubic units. ______________

2. Give the surface area in square units. ______________

In Exercises 3–8, draw the six orthographic projections for the solid.

3. 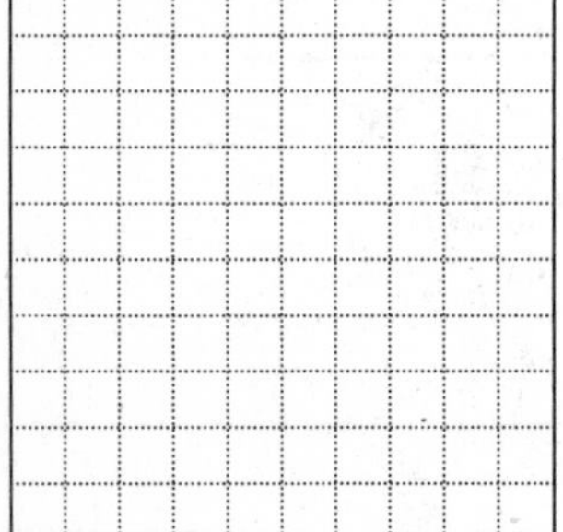

4. 

5. 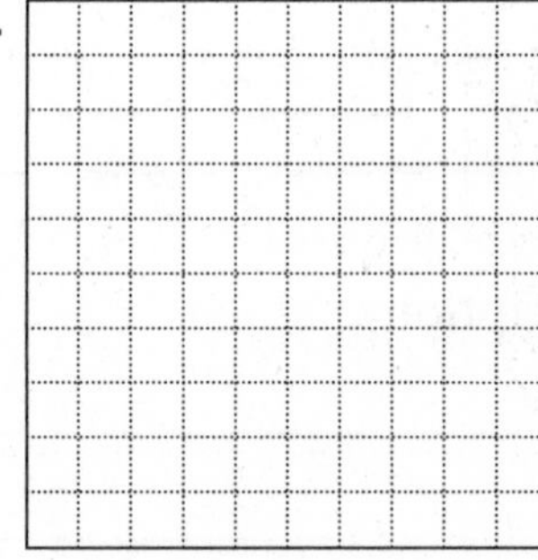

6. 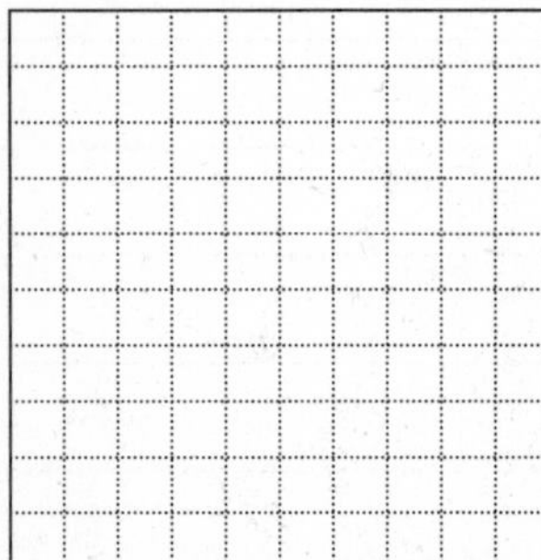

7. 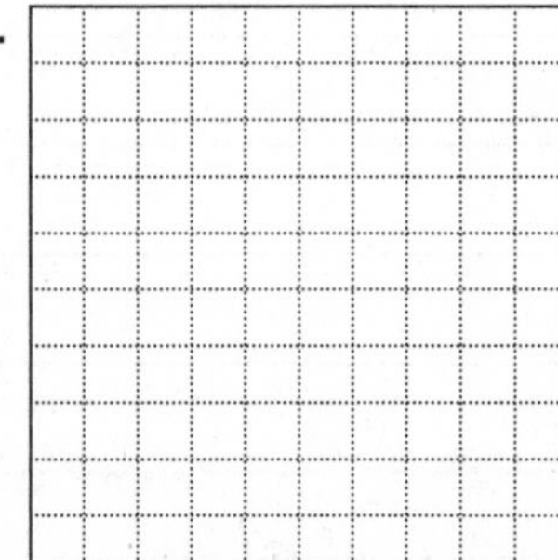

8. 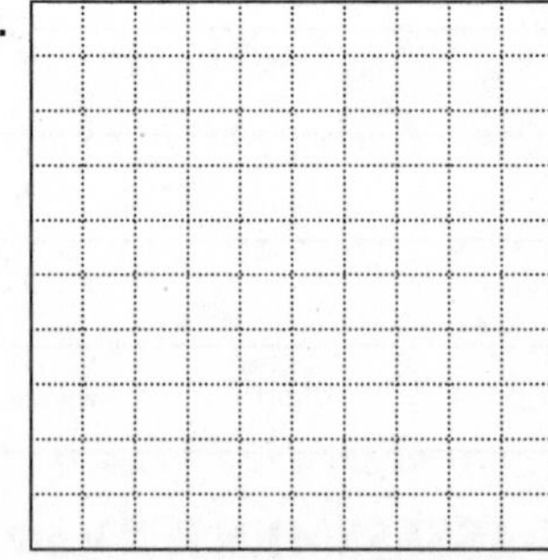

NAME ______________________ CLASS ____________ DATE ____________

Quick Warm-Up: Assessing Prior Knowledge

6.2 Spatial Relationships

Classify each statement below as true or false.

________ 1. Two lines in three-dimensional space that do not intersect are parallel.

________ 2. A cube has 6 faces.

________ 3. A cube has 12 edges.

________ 4. Parallel lines are coplanar.

Lesson Quiz

6.2 Spatial Relationships

Use the figure at right for Exercises 1–5.

1. Name a plane parallel to the plane containing *ABCD*. ______________
2. Name all planes perpendicular to the plane containing *ABCD*.

   ____
3. Name three segments parallel to $\overline{BC}$. ______________
4. Name all planes perpendicular to $\overline{BC}$. ______________
5. Name all segments perpendicular to the plane containing *ABCD*. ______________

Use the figure at right for Exercises 6–11.

6. Are points *I*, *P*, *K*, and *N* coplanar? If so, identify the plane.

   ____
7. Are points *J*, *K*, and *M* collinear? If so, identify the line.

   ____
8. Name two segments skew to $\overline{LO}$. ______________
9. Name two planes parallel to $\overline{IL}$. ______________
10. Name the edge of the dihedral angle formed by *JKNM* and *IJKL*.

    ____
11. Name the faces of the dihedral angle whose edge is $\overline{OP}$.

    ____

NAME ______________________ CLASS ____________ DATE ____________

Quick Warm-Up: Assessing Prior Knowledge
6.3 Prisms

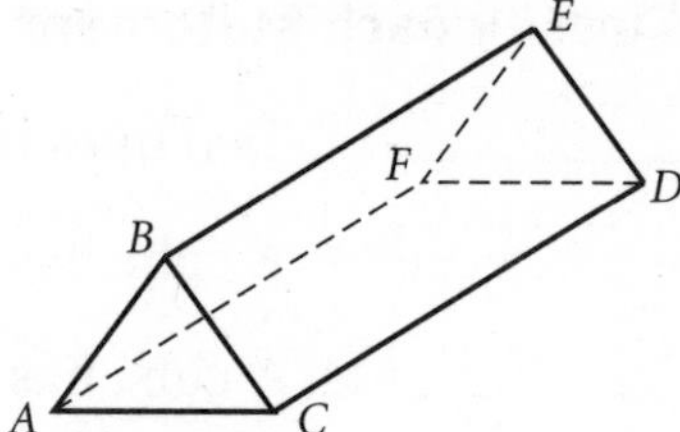

1. Name the planes in the figure at right.

____

2. Which planes are parallel?

____

Lesson Quiz
6.3 Prisms

Classify each prism.

1. 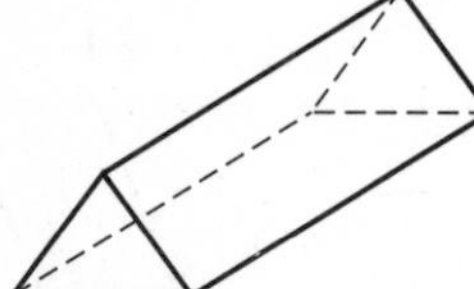

____

2. 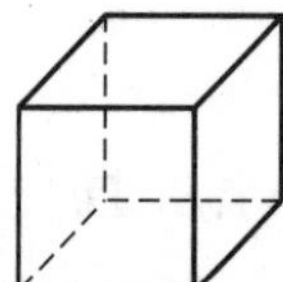

____

3. 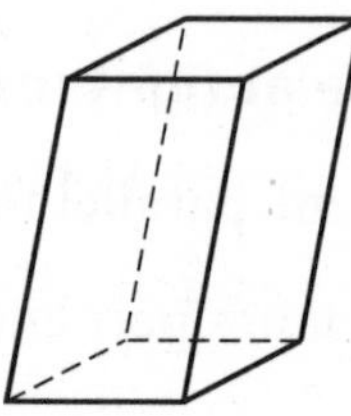

____

Use the figure at the right for Exercises 4–7.

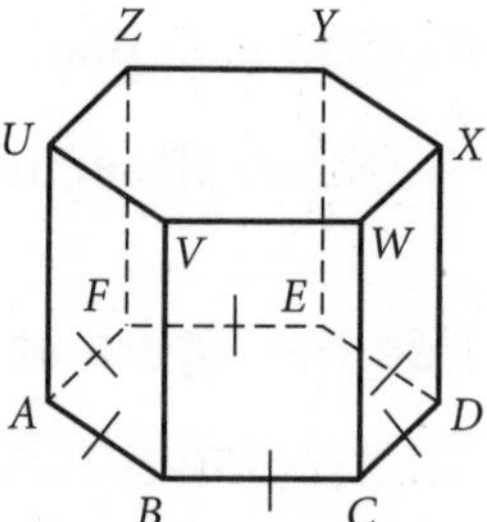

4. Name all segments congruent to $\overline{UA}$. ______________
5. Which face is congruent to *ABCDEF*? ______________
6. Name two congruent lateral faces. ______________
7. If each angle in *ABCDEF* measures 120°, what type of polygons are *ABCDEF* and *UVWXYZ*? ______________

Find the length of the diagonal for each rectangular prism.

8. 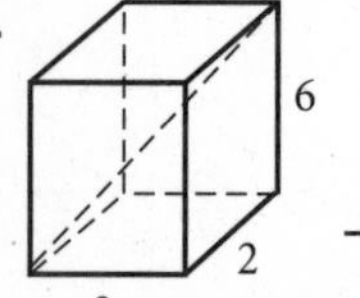

____

9. 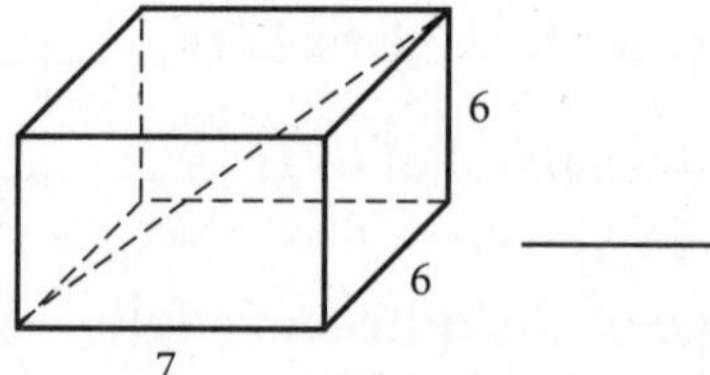

____

10. The base of a prism is a 10-sided polygon. How many faces, vertices, and edges does the prism have? ______________

NAME ______________________ CLASS ____________ DATE ____________

Mid-Chapter Assessment

Chapter 6 (Lessons 6.1–6.3)

Write the letter that best answers the question or completes the statement.

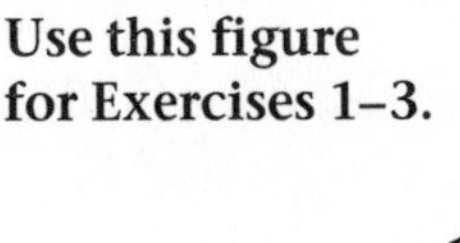

Use this figure for Exercises 1–3.

______ 1. Which of the following is the top view?

a. 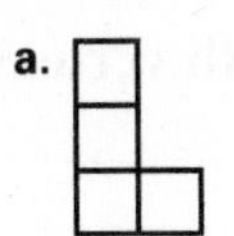

b. 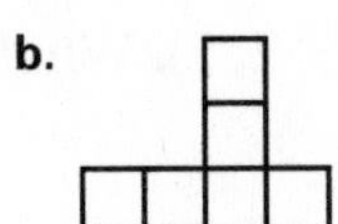

c.

d.

______ 2. What is the surface area if each cube edge is 1 unit in length?

a. 30 square units b. 31 square units c. 15 square units d. 25 square units

______ 3. What is the volume if each cube edge is 1 unit in length?

a. 6 cubic units b. 7 cubic units c. 8 cubic units d. 9 cubic units

Use this figure for Exercises 4–6.

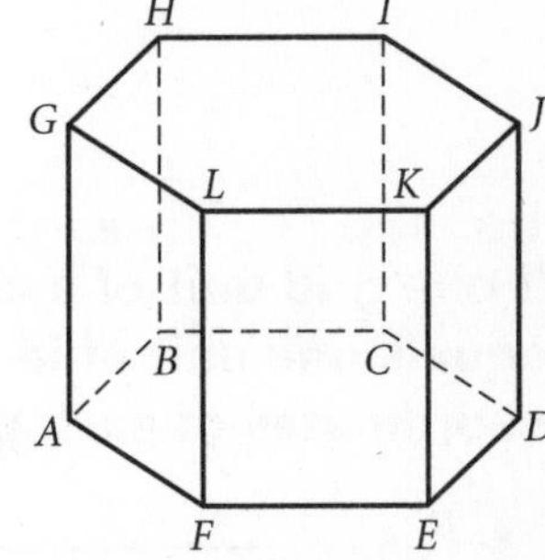

______ 4. The figure at right is classified as ______.

a. a cube b. a hexagonal prism

c. an oblique prism d. a hexagonal pyramid

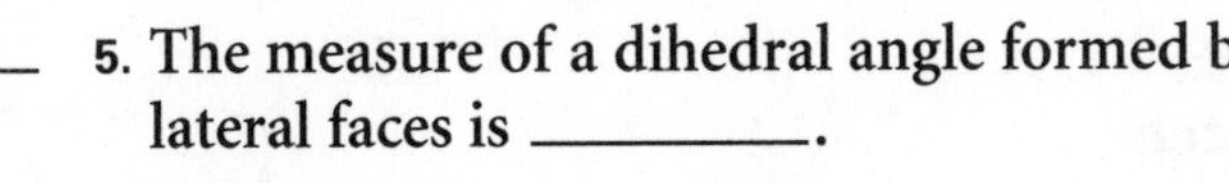

______ 5. The measure of a dihedral angle formed by the lateral faces is ______.

a. 60° b. 90°

c. 120° d. 18°

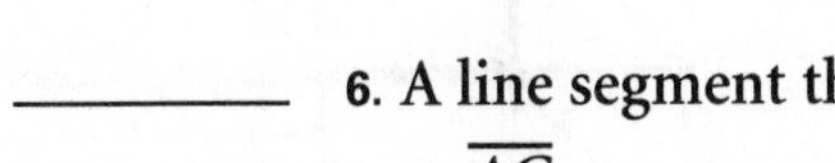

______ 6. A line segment that is skew to $\overline{IC}$ is ______.

a. $\overline{AG}$ b. $\overline{EK}$ c. $\overline{IJ}$ d. $\overline{FE}$

Find the length of the diagonal of each right rectangular prism.

7.

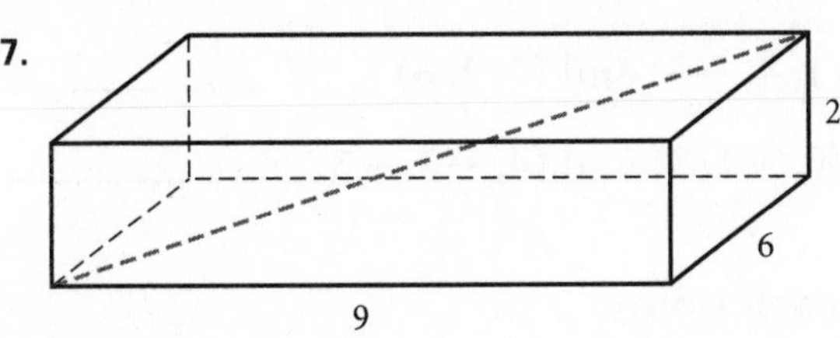

______

8.

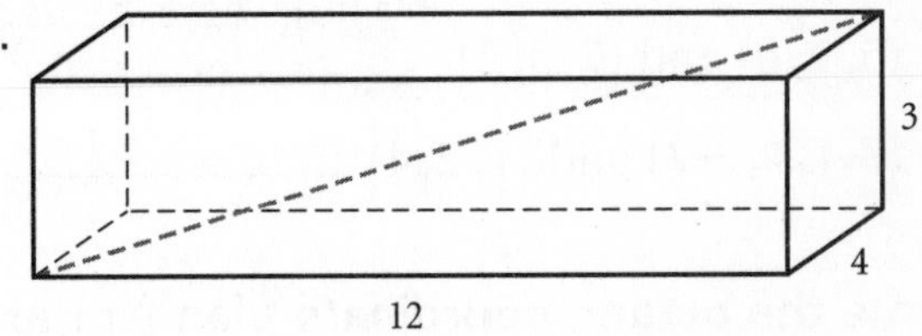

______

9. $l = 12, w = 9, h = 8$ ______

10. cube with edge = 6 ______

NAME ______________________ CLASS ____________ DATE ____________

Quick Warm-Up: Assessing Prior Knowledge

6.4 Coordinates in Three Dimensions

1. Find the midpoint of the segment joining points (3, −5) and (8, −2). ____________

2. Name the axis containing the point (−4, 0). ____________

3. A square's vertices are (5, 0), (−2, 0), and (−2, 7). What is its fourth vertex? ____________

Lesson Quiz

6.4 Coordinates in Three Dimensions

Plot each point on the grid provided.

1. $A(3, 0, 5)$

2. $B(-1, 4, 0)$

3. $C(2, 3, 6)$

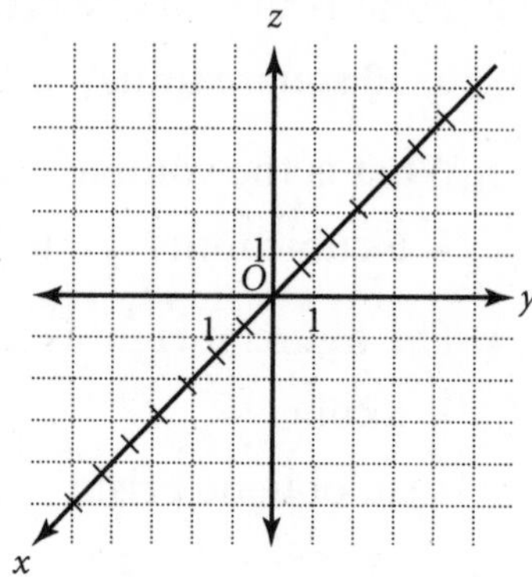

If one grid unit of a coordinate system equals one unit of length, determine the coordinates of each point below.

4. B ____________ 5. F ____________

6. G ____________ 7. H ____________

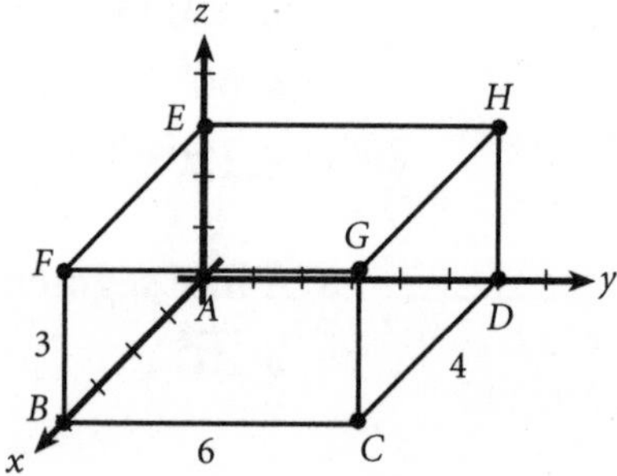

Find the length of the segment joining each pair of points. Then find the midpoint.

8. (5, 2, 0) and (7, 3, 2) ____________

9. (1, 4, −3) and (7, 2, 6) ____________

10. (−1, 5, −2) and (1, 2, 4) ____________

11. (0, −1, 3) and (4, −3, −3) ____________

Name the octant, coordinate plane, or axis for each point.

12. both points in Exercise 8 ____________

13. both points in Exercise 11 ____________

NAME ______________________ CLASS __________ DATE __________

Quick Warm-Up: Assessing Prior Knowledge
6.5 Lines and Planes in Space

1. Determine the x- and y-intercepts of the line defined by $2x + 3y = 6$. __________
2. Is the point $(-2, 4)$ on the graph of $2x + 3y = 6$? __________
3. Find the equation of the horizontal line that passes through the point $(0, -3)$. __________

Lesson Quiz
6.5 Lines and Planes in Space

Use intercepts to sketch the graph of the plane represented by each equation.

1. $4x + 2y + 3z = 12$

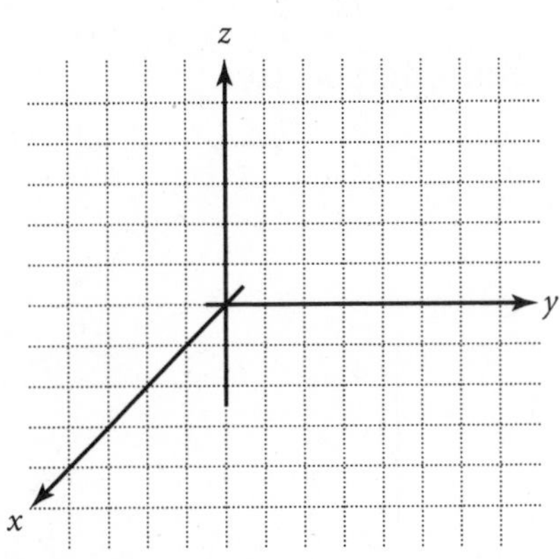

2. $9x + 3y + 6z = 18$

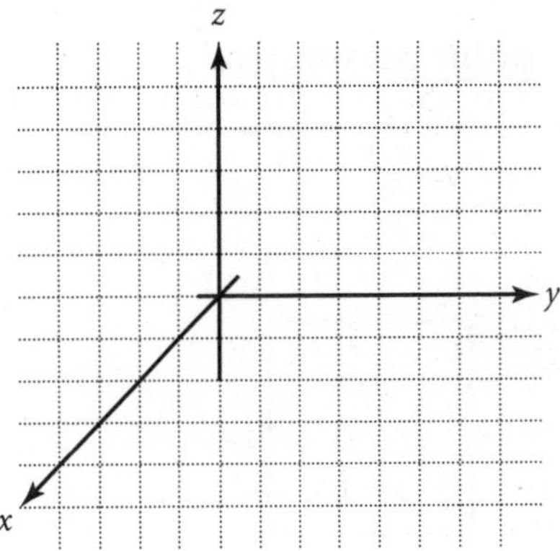

Use the grid provided to sketch the specified line in the plane or in space.

3. $x = t + 1$
$y = 2t$

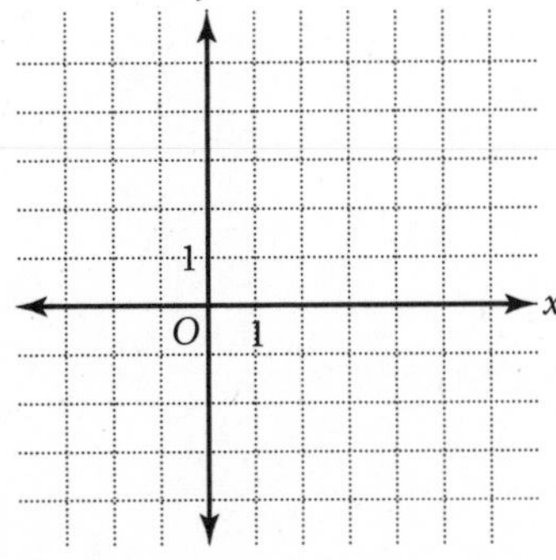

4. $x = 2t - 1$
$y = 3t$
$z = t + 2$

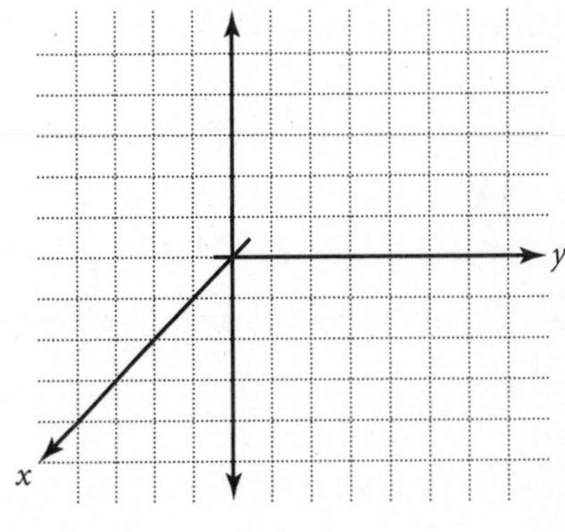

NAME ______________________________ CLASS ______________ DATE ______________

Quick Warm-Up: Assessing Prior Knowledge

6.6 *Perspective Drawing*

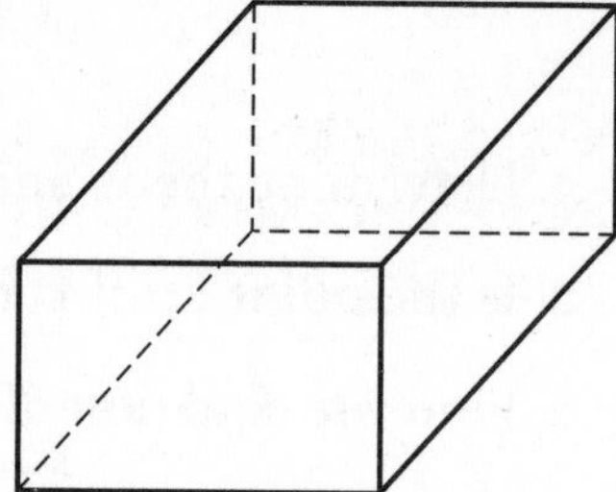

1. Draw a rectangular prism. ______________________________

2. Explain how to make a two-dimensional drawing appear three-dimensional.

____

____

Lesson Quiz

6.6 *Perspective Drawing*

In Exercises 1 and 2, use the space provided to sketch each perspective drawing.

1. cube: one-point perspective

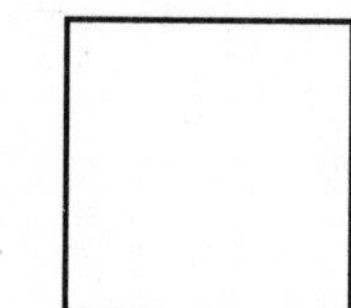

2. cube: two-point perspective

3. For the solid shown below in two-point perspective, locate the vanishing points and horizon.

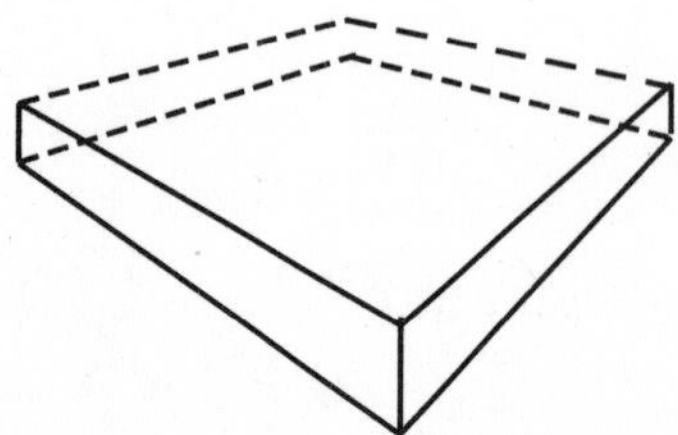

NAME ______________________ CLASS ____________ DATE ____________

Chapter Assessment

Chapter 6, Form A, page 1

Write the letter that best answers the question or completes the statement. Use the given figure for Exercises 1–4. Assume that no cubes are hidden from view.

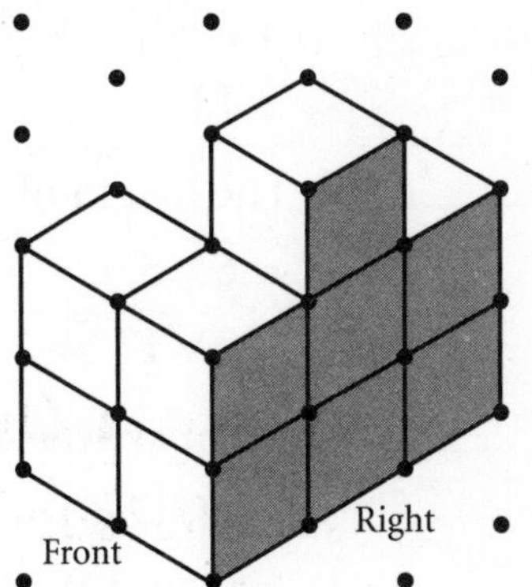

_______ 1. Which of the following is the left side view?

a. 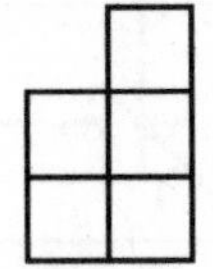

b. 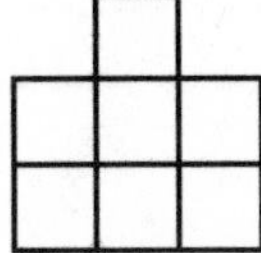

c. 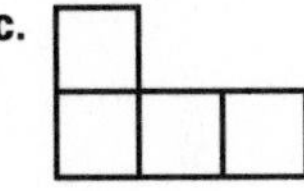

d. 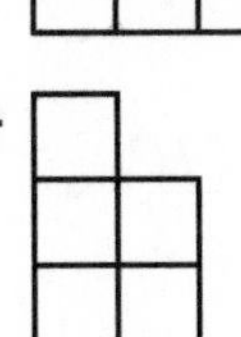

_______ 2. What is the surface area if each cube edge is 1 unit in length?

a. 17 square units b. 25 square units
c. 32 square units d. 36 square units

_______ 3. What is the volume if each cube edge is 1 unit in length?

a. 9 cubic units b. 10 cubic units
c. 27 cubic units d. 34 cubic units

_______ 4. What is the measure of the dihedral angle formed by the left face and the front face?

a. 30° b. 60° c. 90° d. 120°

For Exercises 5–8, refer to the prism at right.

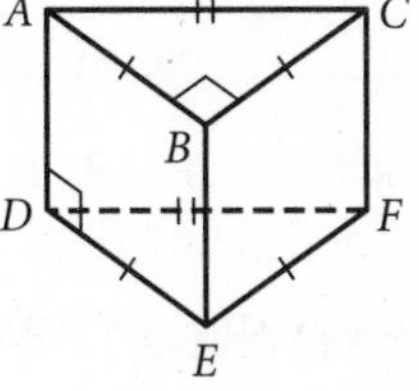

_______ 5. Which is the classification of the figure?

a. triangular prism b. oblique prism
c. rectangular prism d. pentagonal prism

_______ 6. What is the measure of the dihedral angle formed by the intersection of planes *ADFC* and *BEFC*?

a. 30° b. 45° c. 60° d. 90°

_______ 7. Which plane is parallel to plane *ABC*?

a. *ADF* b. *BEF* c. *ADE* d. *DEF*

_______ 8. Which segment is is perpendicular to $\overline{EB}$?

a. $\overline{FC}$ b. $\overline{DF}$ c. $\overline{BC}$ d. $\overline{AC}$

Chapter Assessment

Chapter 6, Form A, page 2

For Exercises 9–15, use the rectangular solid at right.

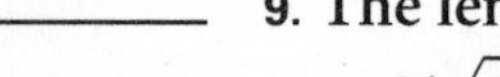

______ **9.** The length of $\overline{GI}$ in the xy-plane is ______.

a. $3\sqrt{5}$ **b.** $3\sqrt{17}$
c. 13 **d.** $6\sqrt{30}$

______ **10.** The length of diagonal $\overline{GM}$ is ______.

a. $6\sqrt{5}$ **b.** $4\sqrt{10}$
c. $2\sqrt{13}$ **d.** $3\sqrt{21}$

______ **11.** What are the coordinates of point H?

a. (4, 12, 0) **b.** (12, 4, 0)
c. (4, 0, 12) **d.** (0, 4, 12)

______ **12.** What are the coordinates of point L?

a. (4, 6, 12) **b.** (12, 6, 4) **c.** (4, 12, 6) **d.** (6, 12, 4)

______ **13.** The midpoint of $\overline{LM}$ is ______.

a. (2, 12, 6) **b.** (2, 12, 0) **c.** (2, 6, 3) **d.** $(2, 2\sqrt{3}, 6)$

______ **14.** Which line is skew to $\overleftrightarrow{GH}$?

a. $\overleftrightarrow{KL}$ **b.** $\overleftrightarrow{NM}$ **c.** $\overleftrightarrow{LH}$ **d.** $\overleftrightarrow{NJ}$

______ **15.** Name the location of point K.

a. xz-plane **b.** first octant **c.** yz-plane **d.** top-front-left octant

______ **16.** Find the distance between $(-1, 5, 2)$ and $(3, 1, -4)$ in space.

a. 9 **b.** $2\sqrt{17}$ **c.** $2\sqrt{22}$ **d.** $2\sqrt{14}$

For Exercises 17 and 18, use the figure at right.

______ **17.** An equation of the plane is ______.

a. $3x + 4y + 2z = 12$ **b.** $4x + 3y + 6z = 12$
c. $3x + 2y + 4z = 12$ **d.** $4x + 2y + 3z = 12$

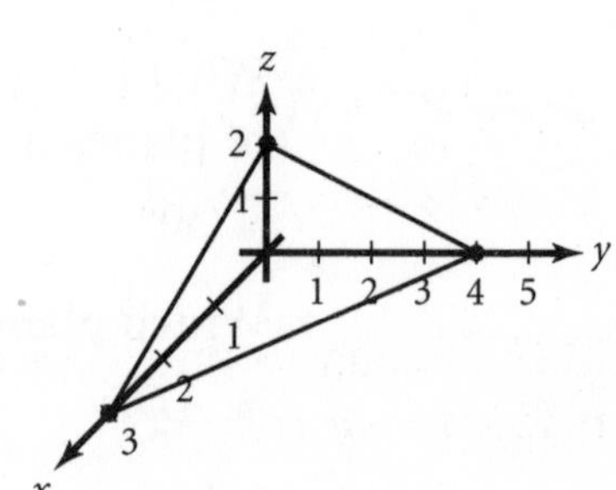

______ **18.** An equation of the trace defined by $2x + 5y - 3z = 8$ is ______.

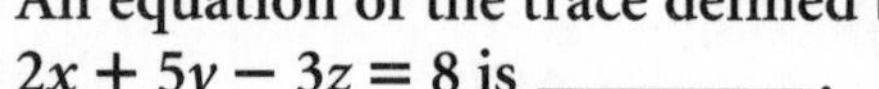

a. $2x + 5y = 8$ **b.** $2x - 3z = 8$
c. $5y - 3z = 8$ **d.** $2x + 5y = 0$

NAME ______________________ CLASS ______________ DATE ______________

Chapter Assessment

Chapter 6, Form B, page 1

In Exercises 1–5, refer to the isometric drawing at right. Each cube has edges measuring 1 unit in length. Assume that no cubes are hidden from view.

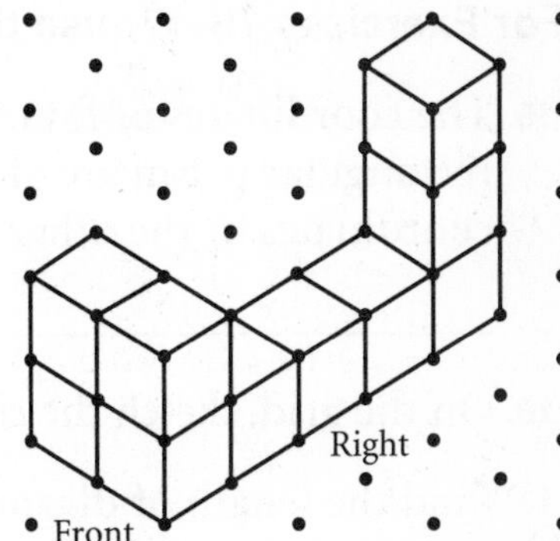

1. How many cubes make up the solid? Find the volume of the solid. ______________

2. How many faces does the solid have? ______________

3. Find the surface area of the solid. ______________

In Exercises 4 and 5, sketch each orthographic projection.

4. the front view

5. the left view

In Exercises 6–10, refer to the prism at right.

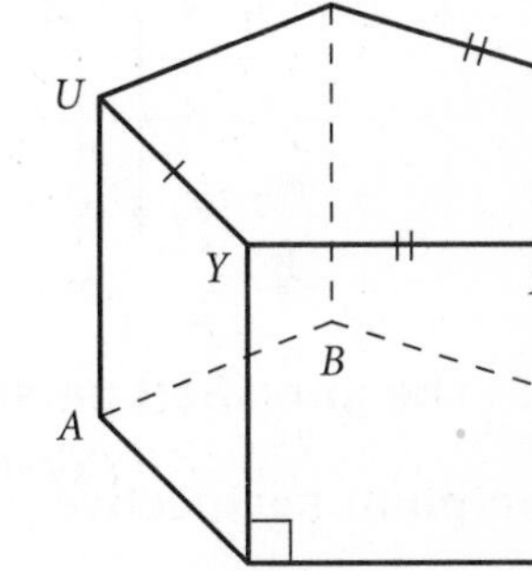

6. Classify the prism. ______________________________

7. Name the congruent lateral faces.

____

8. What is the measure of the dihedral angle formed by the planes containing a base and a lateral face? ______________

9. Classify quadrilateral $XWCD$. ______________________________

10. Name all segments congruent to $\overline{VB}$. ______________________________

Given the values of *l*, *w*, and *h*, find the length of diagonal $\overline{AB}$.

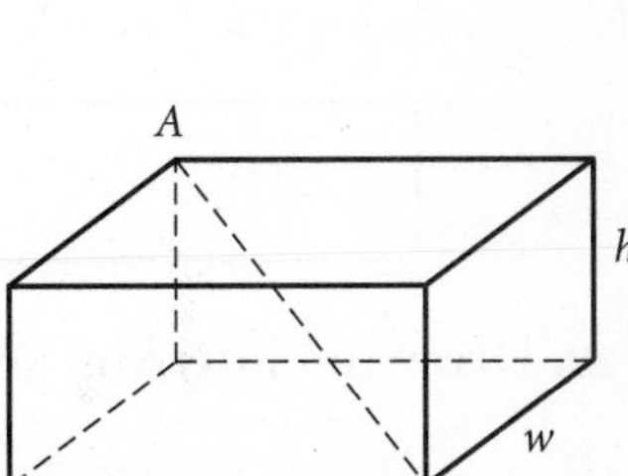

11. $l = 9, w = 2, h = 6$ ______________

12. $l = 12, w = 8, h = 9$ ______________

In Exercises 13 and 14, find *PQ* and the coordinates of the midpoint of $\overline{PQ}$.

13. $P(-3, 5, -4)$ and $Q(6, 4, -7)$

____

14. $P(0, 3, -1)$ and $Q(10, 0, 4)$

____

Chapter Assessment

Chapter 6, Form B, page 2

For Exercises 15–17, use the grid at the right.

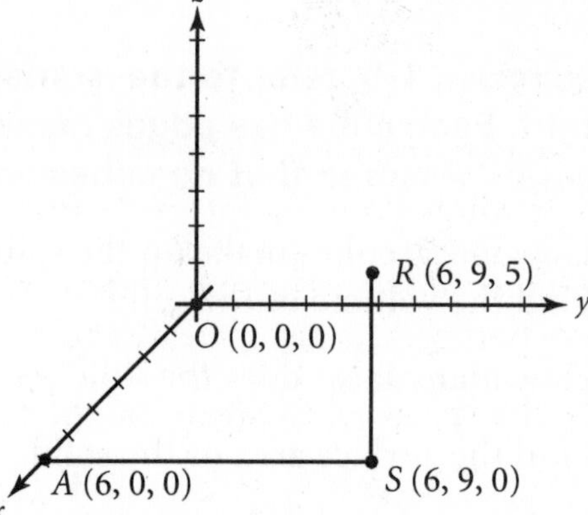

15. The coordinates of four vertices of a right rectangular prism are given. Find the coordinates of the other four vertices.

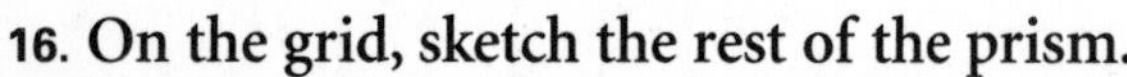

16. On the grid, sketch the rest of the prism.

17. Find the length of diagonal $\overline{OR}$. ____________

In Exercises 18 and 19, use the grid provided to sketch the specified graph.

18. the line in the xy-plane defined by $x = 2t - 3$ and $y = -t - 3$

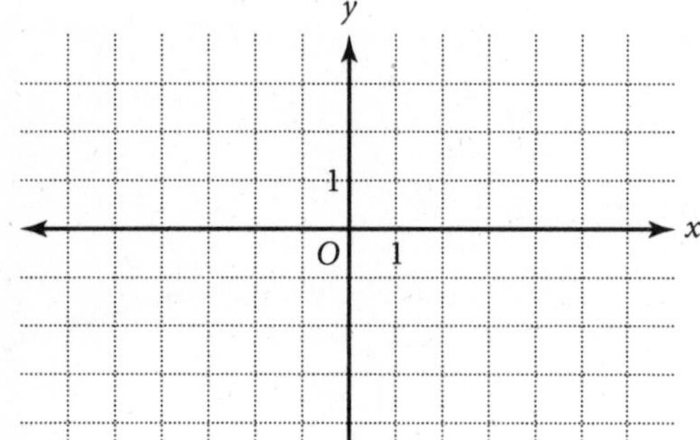

19. the plane in three-dimensional space defined by $6x + 2y + 3z = 12$

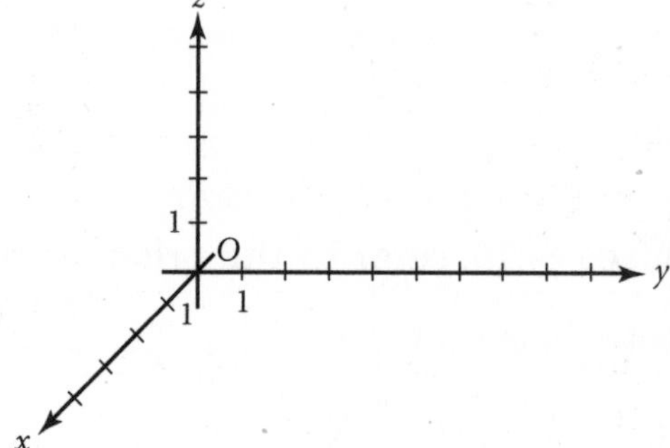

Sketch the specified perspective drawing of a right rectangular solid.

20. one-point perspective

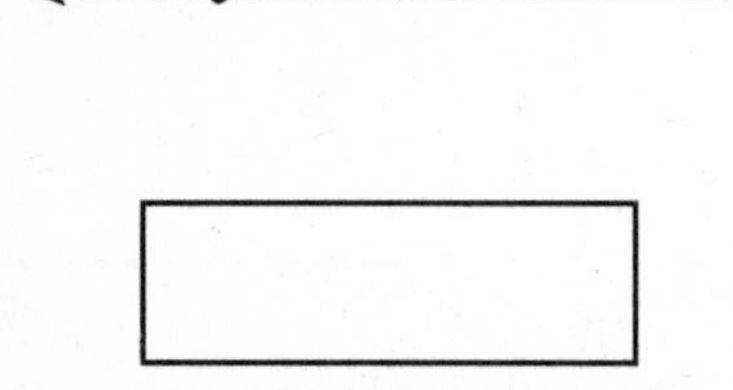

21. two-point perspective

22. Draw the horizon, and locate the two vanishing points.

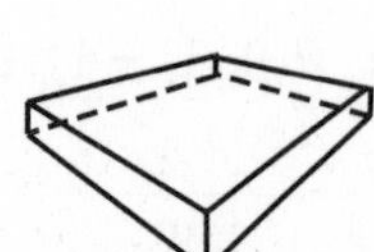

NAME ______________________ CLASS ____________ DATE ____________

Alternative Assessment

Prisms, Chapter 6, Form A

TASK: Draw and identify properties of prisms.

HOW YOU WILL BE SCORED: As you work through the task, your teacher will be looking for the following:

- how well you are able to draw and label a prism
- whether you can identify the features and properties of prisms

1. In the space below, draw a right rectangular prism. Label the vertices.

2. Name the bases of your prism. What type of quadrilateral are they?

____

3. List pairs of congruent lateral faces. What types of quadrilateral are they?

____

4. Name three pairs of parallel planes.

____

5. Name three pairs of perpendicular planes.

____

6. Which angles measure 90°? Explain why.

____

7. Assuming that your right rectangular prism has length l, width w, and height h, explain how you can find the length of its diagonal.

____

SELF-ASSESSMENT: Draw an oblique rectangular prism with the same length, width, and height as your right rectangular prism. Explain how it differs from the right rectangular prism.

NAME ______________________ CLASS ____________ DATE ____________

Alternative Assessment

Three-Dimensional Systems, Chapter 6, Form B

TASK: Locate points, lines, and planes in a three-dimensional coordinate system and explore the relationships among the figures.

HOW YOU WILL BE SCORED: As you work through the task, your teacher will be looking for the following:

- whether you can locate points, lines, and planes in a three-dimensional system
- how well you understand the relationships among three-dimensional figures

1. Locate and label these points in the three-dimensional coordinate system at right.

   $A(5, 0, 0)$ $B(0, 0, 10)$ $C(0, 12, 0)$ $D(6, 8, 2)$

2. Connect the points. Name the figure formed.

   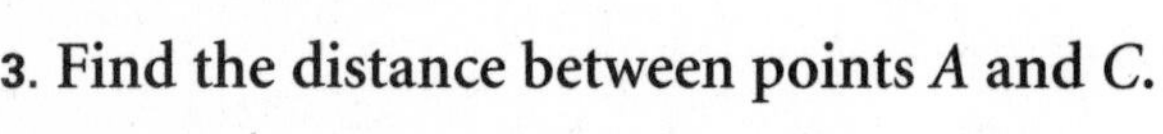

3. Find the distance between points A and C.

   ____

4. Find the distance between points B and D.

   ____

5. Find the distance between points A and C.

   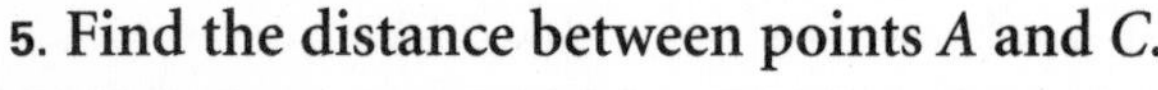

6. Sketch the graph of the plane defined by the equation $12x + 5y + 6z = 60$ in the same three-dimensional system.

7. Find the equation of the trace. Sketch the trace.

   ____

8. What do you notice?

   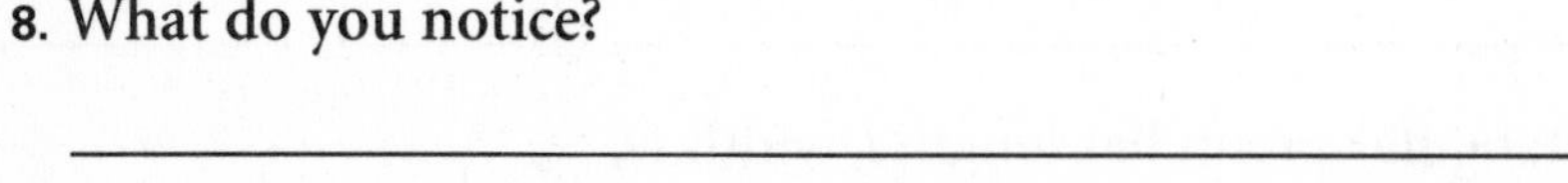

SELF-ASSESSMENT: How does the equation of a line in space compare with the equation of a line in a plane?

NAME ______________________ CLASS ____________ DATE ____________

Quick Warm-Up: Assessing Prior Knowledge

7.1 Surface Area and Volume

1. Find the surface area and volume of a cube with sides of 4 in.

____

2. Find the surface area and volume of a rectangular prism with a length of 5 in., height of 4 in., and width of 3 in.

____

3. Suppose that there are 20 students in geometry class and 8 of them are boys. What is the ratio of boys to girls? ____________________

Lesson Quiz

7.1 Surface Area and Volume

Find the surface-area-to-volume ratio for each of the following.

1. a cube with an edge length of 3 cm ____________________
2. a rectangular prism with a length of 3 in., width of 2 in., and height of 2 in. __________
3. a rectangular prism with a length of 3 m., width of 3 m., and height of 1 m. __________
4. a cube with a volume of 125 ft^3 ____________________
5. a cube with a volume of x^3 cubic units ____________________
6. Suppose that you are going to cut squares out of the corners of an 8-in. by 10-in. piece of paper to form an open box. Fill in the table to determine the possible dimensions.

Side (in.)	Length (in.)	Width (in.)	Height (in.)	Volume ($in.^3$)
1				
1.5				
2				
2.5				
3				

7. Which of the squares will maximize the volume of the box in problem 5? __________

Quick Warm-Up: Assessing Prior Knowledge

7.2 Surface Area and Volume of Prisms

1. Find the area of a triangle with a height of 6 cm and a base of 4 cm. ________
2. Find the distance between points (0, 3, −2) and (4, −1, 3) in a three-dimensional coordinate system. ________
3. Find the surface area and volume of a cube with a side length of 8 in. ________

Lesson Quiz

7.2 Surface Area and Volume of Prisms

1. In the space below, draw a net for a right pentagonal prism.

Find the surface area and volume of a right rectangular prism with the given dimensions.

2. $\ell = 8$ in., $w = 4$ in., $h = 2$ in. ________ 3. $\ell = 1.5$ m, $w = 2.4$ m, $h = 10$ m ________
4. The volume of an oblique rectangular prism is 576 ft^3. The area of its base is 48 ft^2. What is the height of the prism? ________
5. A right triangular prism has a height of 6 cm. The base has legs of 8 cm and 15 cm. Find the volume of the prism. ________
6. Find the volume of the right hexagonal prism shown at right. ________
7. Find the volume of the prism at right if its base edges are shortened to 3 feet in length. ________

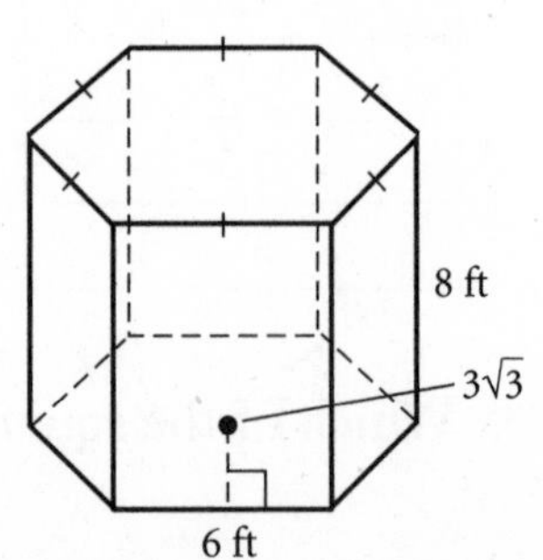

NAME ______ CLASS ______ DATE ______

Quick Warm-Up: Assessing Prior Knowledge

7.3 Surface Area and Volume of Pyramids

1. Find the volume of a right rectangular prism with a length of 12, width of 8, and height of 4. ______
2. Find the height of an isosceles triangle with sides of 8, 8, and 6. ______
3. Find the lateral area of a right triangular prism with an equilateral base length of 5 cm and a height of 10 cm. ______

Lesson Quiz

7.3 Surface Area and Volume of Pyramids

1. In the space at right, draw a net for a right regular octagonal pyramid.

Find the surface area of a regular pyramid with side length *s* and slant height ℓ. The number of sides of the base is given by *n*.

2. $n = 4, s = 10$ in., $\ell = 13$ in. ______
3. $n = 3, s = 8$ cm, $\ell = 4\sqrt{3}$ cm ______
4. The base of a regular pyramid is a square. The perimeter of the base is 20 ft, and the surface area of the pyramid is 65 ft^2. Find the slant height of the pyramid. ______

Find the volume of each pyramid.

5.

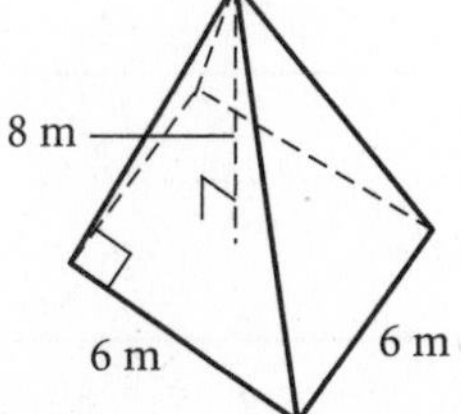

6.

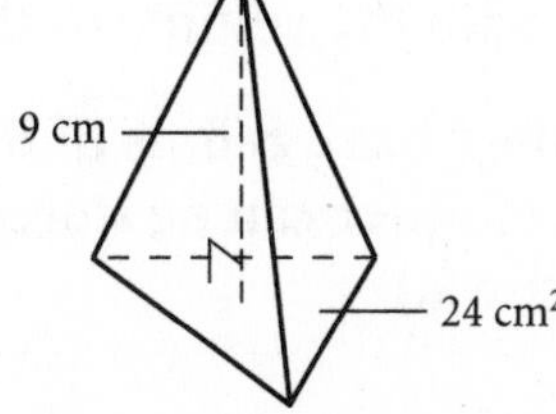

______ ______

Quick Warm-Up: Assessing Prior Knowledge
7.4 Surface Area and Volume of Cylinders

1. Find the circumference and area of a circle with a radius of 6. ______________

2. Find the surface area and volume of a right rectangular prism whose base is a square with a side length of 6 and whose height is 10. ______________

Lesson Quiz
7.4 Surface Area and Volume of Cylinders

Find the surface area of each cylinder with radius *r* and height *h*.

1. $r = 6$ cm, $h = 5$ cm ______________
2. $r = 3$ m, $h = 12$ m ______________
3. $r = \frac{1}{4}$ ft, $h = 5$ ft ______________

Find the volume of each cylinder with radius *r* and height *h*.

4. $r = 8$ in., $h = 3$ in. ______________
5. $r = 4$ m, $h = 10$ m ______________
6. $r = 6$ cm, $h = 1.5$ cm ______________

7. What is the height of a cylinder with a radius of 5 centimeters and a volume of 100π cm^3? ______________

8. What is the radius of a cylinder with a height of 6 centimeters and a volume of 54π cm^3? ______________

9. A cylindrical water tank has a diameter of 90 ft and a height of 30 ft. If 1 ft^3 of storage area holds approximately 7.48 gal of water, how many gallons of water can be stored in the tank? ______________

10. A cylindrical shipping container is 1 ft in diameter and has a length of 4 ft. A new container is made by doubling the diameter. What is the volume of the new container? How does the volume of the new container compare with the volume of the original container? ______________

11. A cylindrical container has a radius of 8 ft and a height of 10 ft. How many gallons of water can be stored in this container? (Hint: Refer back to Exercise 9.) ______________

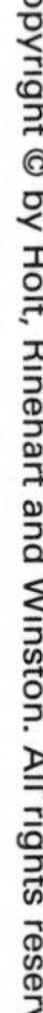

NAME ______________________ CLASS __________ DATE __________

Mid-Chapter Assessment

Chapter 7 (Lessons 7.1–7.4)

Write the letter that best answers the question or completes the statement. For Exercises 1–3, use the given net.

______ 1. The net at right is for a ______.

a. triangular prism
b. triangular pyramid
c. square prism
d. square pyramid

______ 2. The surface area of the figure is ______.

a. 240 cm^2 b. 256 cm^2
c. 400 cm^2 d. 736 cm^2

______ 3. The volume of the figure is approximately ______.

a. 80 cm^3 b. 240 cm^3 c. 1082.8 cm^3 d. 1450.7 cm^3

______ 4. The volume of a cube is 64 cm^3. What is its surface area?

a. 16 cm^2 b. 96 cm^2 c. 256 cm^2 d. 384 cm^2

______ 5. The volume of a rectangular prism is 120 cm^3 and the area of the base is 24 cm^2. What is its height?

a. 4 cm b. 5 cm c. 6 cm d. 10 cm

6. Find the ratio of surface area to volume for a cube with a volume of 64 cubic inches. ______

7. A square pyramid has a slant height of 10 m. Each side of the base is 6 m. What is the surface area of the pyramid? ______

8. A right triangular pyramid has a base that measures 3 ft by 4 ft by 5 ft. The height of the pyramid is 8 ft. What is the volume of the pyramid? ______

9. The volume of a square pyramid is 162 cm^3. The height of the pyramid is 6 cm. What is the length of each side of the base? ______

10. Find the surface area of a cylinder with a radius of 5 cm and a height of 3 cm. ______

11. Find the volume of a cylinder with a diameter of 18 in. and a height of 6 in. ______

12. Labels for a soup can are cut from a sheet of paper that is 2 ft × 3 ft. The soup can has a circumference of 8 in. and a height of 6 in. What is the maximum number of labels that can be cut from the sheet of paper? ______

Quick Warm-Up: Assessing Prior Knowledge

7.5 Surface Area and Volume of Cones

Classify each statement as true or false.

1. The area of a circle with a radius of 8 is 64π. ____________
2. The volume ratio of a pyramid to a prism is 1:3. ____________
3. A prism may be divided into three pyramids of equal volume. ____________
4. The slant height of a pyramid is used to find its volume. ____________

Lesson Quiz

7.5 Surface Area and Volume of Cones

Find the surface area of each cone.

1. 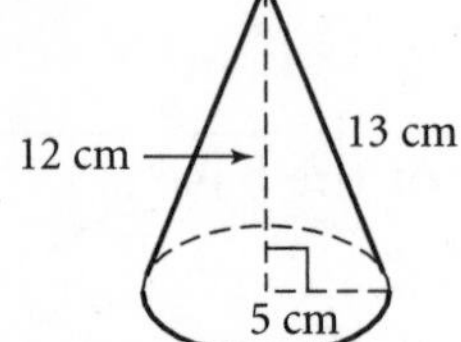

2. 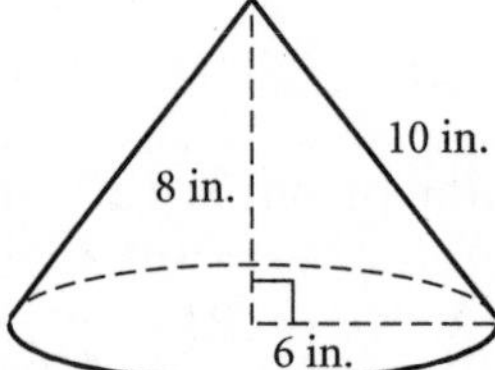

3. 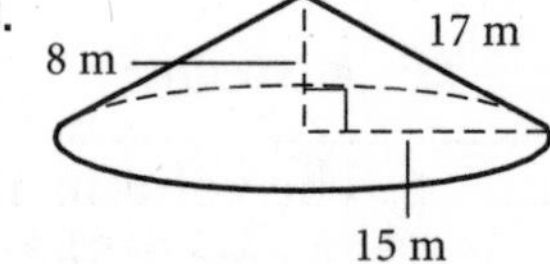

____________ ____________ ____________

Find the volume of each cone.

4. 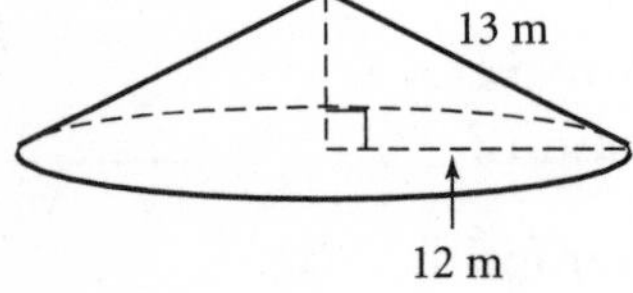

5. 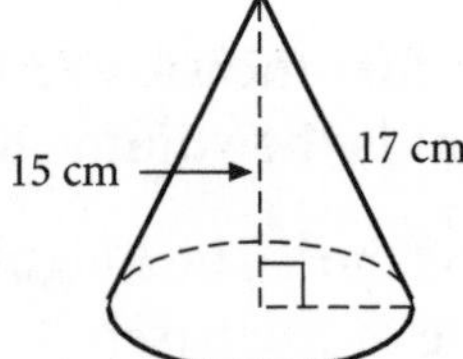

6. 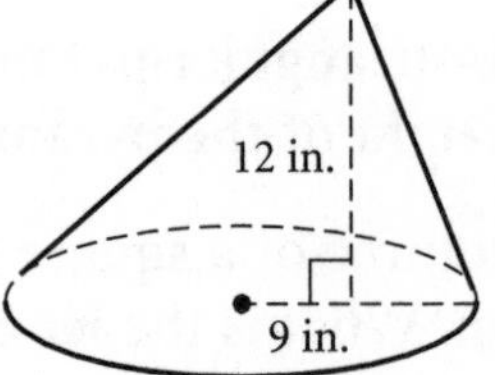

____________ ____________ ____________

7. The height of a cone is 2 cm, and its radius is 1.5 cm. What is the slant height of the cone? ____________

8. A cone has a volume of 128π cm^3, and its radius is 8 cm. What is the height of the cone? ____________

9. The same cone in Exercise 8 is narrowed to a radius of 4 cm. What is its height now? ____________

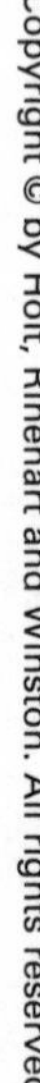

Quick Warm-Up: Assessing Prior Knowledge

7.6 Surface Area and Volume of Spheres

1. Find the area of a circle with a radius of 8 cm. ______________

2. Find the volume of a cylinder with a radius of 6 in. and a height of 8 in. ______________

3. Find the volume of a right cone with a radius of 4 cm and a height of 6 cm. ______________

Lesson Quiz

7.6 Surface Area and Volume of Spheres

1. What happens to the surface area of a sphere when the radius is halved?

____

2. A sphere of radius r is inscribed in a cylinder. Show that the surface area of the sphere is equal to the lateral area of the cylinder.

____

____

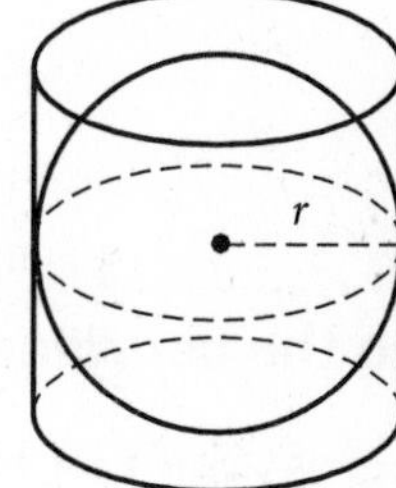

Find the surface area and volume of each sphere for the given radius *r* or diameter *d*. Round your answers to the nearest hundredth.

3. $r = 4$ in. ______________

4. $r = 20$ cm ______________

5. $d = 10$ m ______________

6. $r = 0.5$ cm ______________

7. $r = 6$ in. ______________

8. $d = 9$ ft ______________

9. A beach ball has a diameter of 15 in. How much vinyl is needed to make the ball? ______________

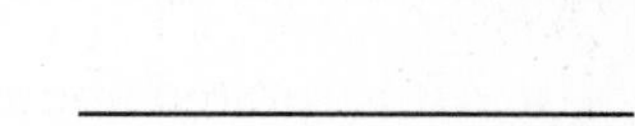

10. A ball with a diameter of 4 in. is placed in a cylinder filled to the top with water. The cylinder is 5 in. in diameter and 10 in. tall. How much water is forced out of the cylinder? ______________

Quick Warm-Up: Assessing Prior Knowledge

7.7 Three-Dimensional Symmetry

1. Find the image of $(-2, 4)$ reflected across the y-axis. ________
2. Find the image of $(3, -4)$ reflected across the x-axis. ________
3. Point $P(0, -2, 4)$ is located in which plane of a three-dimensional coordinate system? ________

Lesson Quiz

7.7 Three-Dimensional Symmetry

1. Draw a three-dimensional coordinate system, and graph $\overline{AB}$ with endpoints $A(3, -2, 1)$ and $B(-1, 3, 5)$. Draw the image $\overline{A'B'}$ of the segment by multiplying each y-coordinate by -1. ________

Write the coordinates of the image of $A(3, 5, 6)$ after each reflection.

2. across the xy-plane ________
3. across the xz-plane ________
4. across the yz-plane ________

5. Describe and give the dimensions of the figure that results when

   a. $\overline{AB}$ is rotated about the y-axis.

   ____

   b. $\overline{AB}$ is rotated about the z-axis.

   ____

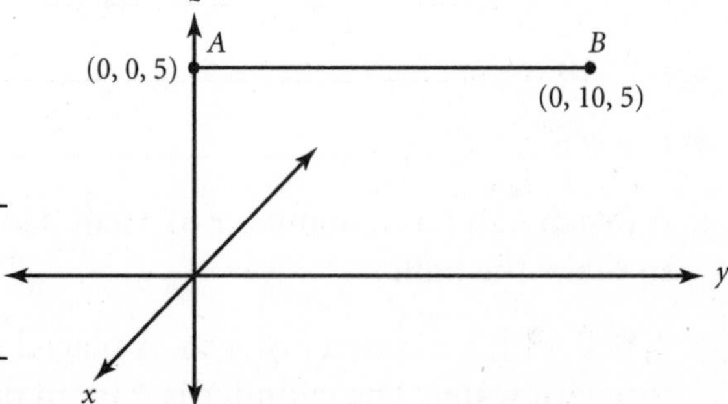

6. Find the area of each of the resulting figures.

____

NAME ________________________ CLASS ____________ DATE ____________

Chapter Assessment

Chapter 7, Form A, page 1

Write the letter that best answers the question or completes the statement.

______ 1. What is the ratio of surface area to volume for a cube with a volume of 27 cm^3?

a. 1:2 b. 1:3 c. 2:1 d. 3:1

______ 2. What is the volume of a cube with a surface area of 600 cm^2?

a. 100 cm^3 b. 216 cm^3 c. 750 cm^3 d. 1000 cm^3

______ 3. The height of an oblique rectangular prism is 4 cm. The length of the base is 3 cm. If the volume of the prism is 96 cm^3, the width of the base is ______.

a. 4 cm b. 5 cm c. 8 cm d. 12 cm

______ 4. The base of a triangular prism is a right triangle. The lengths of the legs of the base are 3 cm and 4 cm. The height of the prism is 8 cm. What is the surface area of the prism?

a. 96 cm^2 b. 108 cm^2 c. 148 cm^2 d. 480 cm^2

______ 5. The base of a triangular prism is a right triangle. The lengths of the legs of the base are 3 cm and 4 cm. The height of the prism is 8 cm. What is the volume of the prism?

a. 48 cm^3 b. 96 cm^3 c. 108 cm^3 d. 148 cm^3

______ 6. A right isosceles trapezoidal prism has a height of 6 m. The parallel sides and altitude of the trapezoid measure 8 m, 10 m, and 5 m, respectively. What is the volume of the prism?

a. 138 m^3 b. 270 m^3 c. 540 m^3 d. 2400 cm^3

______ 7. The surface area of the pyramid is ______.

a. 48 cm^2 b. 96 cm^2
c. 144 cm^2 d. 156 cm^2

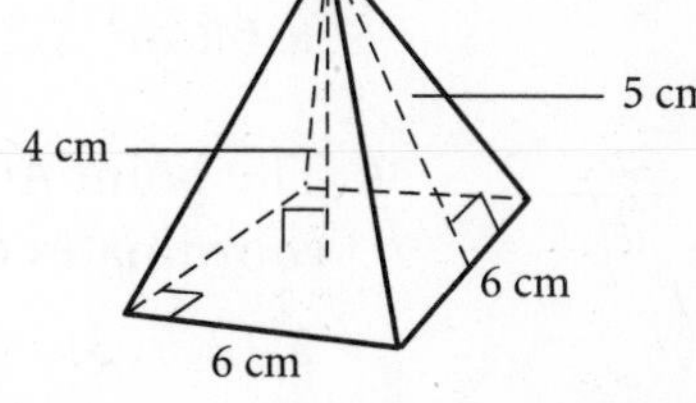

______ 8. The volume of the pyramid is ______.

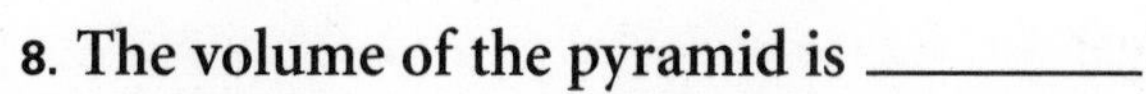

a. 48 cm^3 b. 72 cm^3
c. 96 cm^3 d. 144 cm^3

______ 9. A square pyramid has a height of 9 cm and a volume of 75 cm^3. What is the length of a side of the base?

a. 5 cm b. $8\frac{1}{3}$ cm c. 9 cm d. 25 cm

NAME ______________________ CLASS ____________ DATE ____________

Chapter Assessment

Chapter 7, Form A, page 2

______ 10. A cylinder has a radius of 6 ft and a height of 10 ft. The surface area of the cylinder is about ________.

a. 200.96 ft^2 b. 602.8 ft^2 c. 640 ft^2 d. 603.19 ft^2

______ 11. A cylindrical tank is being constructed to hold 20,000 ft^3 of water. The diameter of the tank will be 40 ft. What is the minimum height needed for the tank?

a. 14 ft b. 15 ft c. 16 ft d. 17 ft

______ 12. A semicircular solid is formed by cutting a cylinder in half lengthwise. If the diameter of the cylinder is 6 ft and its height is 10 ft, the surface area of the semicircular solid is about ________.

a. 182.5 ft^2 b. 276.7 ft^2 c. 122.52 ft^2 d. 465.1 ft^2

______ 13. The surface area of the cone is ________.

a. $\approx$ 78.5 cm^2 b. $\approx$ 204.1 cm^2
c. $\approx$ 282.7 cm^2 d. $\approx$ 267.0 cm^2

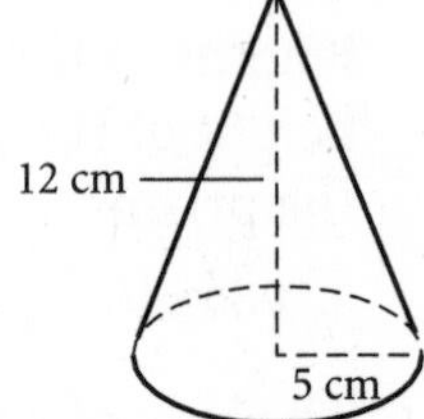

______ 14. The volume of the cone is about ________.

a. 260 cm^3 b. 314 cm^3
c. 360 cm^3 d. 780 cm^3

______ 15. The surface area of a sphere with a diameter of 12 in. is about ________.

a. 113.0 in.^{2} b. 150.7 in.^{2} c. 266.1 in.^{2} d. 452.4 in.^{2}

______ 16. The diameter of a golf ball is about 4.2 cm. The volume of the golf ball is about ________.

a. 8.8 cm^3 b. 18.5 cm^3 c. 38.8 cm^3 d. 55.4 cm^3

______ 17. The surface area of a sphere is 144π m^2. What is the volume of the sphere?

a. 6π m^3 b. 36π m^3 c. 144π m^3 d. 288π m^3

______ 18. The point $A(2, 7, -3)$ is reflected across the xy-plane. What are the coordinates of the image?

a. $(2, 7, 3)$ b. $(-2, 7, 3)$ c. $(2, -7, 3)$ d. $(-2, 7, -3)$

______ 19. Segment $\overline{AB}$ with endpoints $A(3, 0, 5)$ and $B(3, 0, 0)$ is rotated about the z-axis. The volume of the figure formed is about ________.

a. 94.2 cubic units b. 141.4 cubic units
c. 235.5 cubic units d. 423.9 cubic units

Chapter Assessment

Chapter 7, Form B, page 1

1. What is the surface-area-to-volume ratio of a rectangular prism with a length of 3 ft, width of 2 ft, and height of 4 ft? ______________

2. What is the surface area of a cube if the area of one side of the cube is 100 cm^2? ______________

3. What is the surface area of a sphere with a radius of 3 in.? ______________

4. What is the surface area of a rectangular prism with a length of 8 m, width of 4 m, and height of 5 m? ______________

5. What is the surface area of a cylinder that has a diameter of 10 ft and a height of 8 ft? ______________

Find the surface area of each figure.

6.

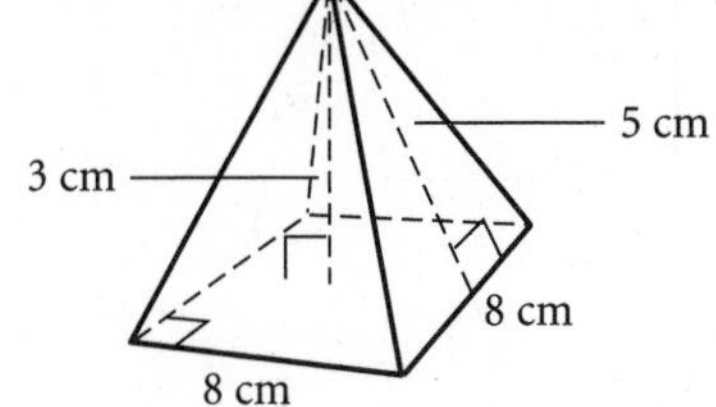

7.

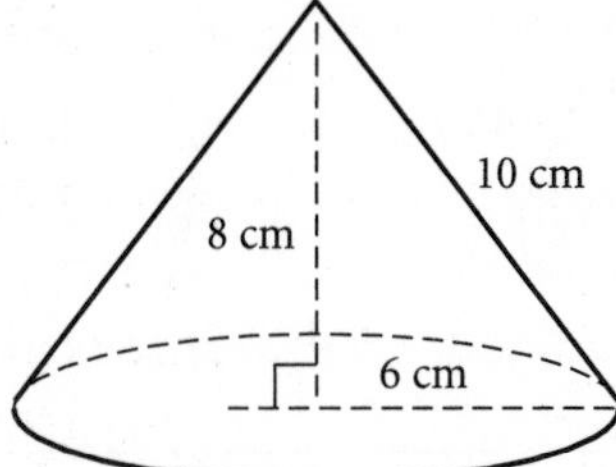

8. The surface area of a cube is 216 yd^3. What is the length of an edge? ______________

9. What is the height of a cylinder with a radius of 6 in. and a surface area of 168π in.^{2}? ______________

10. What is the radius of a sphere with a surface area of 100π m^2? ______________

11. The diameter of a baseball is about 1.4 in. How much leather is needed to cover a baseball? ______________

12. A cone has a radius of 5 units and a slant height of 10 units. A sphere has a radius of 5 units. Which figure has a greater surface area? ______________

13. A triangular pyramid has a base that is congruent to its lateral faces. The base edge is 24 cm, and the slant height is 16 cm. What is the surface area of the pyramid? ______________

14. The radius of the Earth is about 7900 miles. About 70% of the Earth's surface is water. How much of the Earth's surface is land? Round your answer to the nearest million square miles. ______________

Chapter Assessment

Chapter 7, Form B, page 2

15. What is the volume of a rectangular prism with a length of 6 m, width of 3 m, and height of 5 m? ____________

16. What is the volume of a cylinder with a radius of 8 ft and a height of 3 ft? ____________

17. A square pyramid has a slant height of 13 cm. The base edge measures 10 cm. What is the volume of the pyramid? ____________

18. What is the volume of a sphere with a diameter of 12 ft? ____________

Find the volume of each figure.

19.

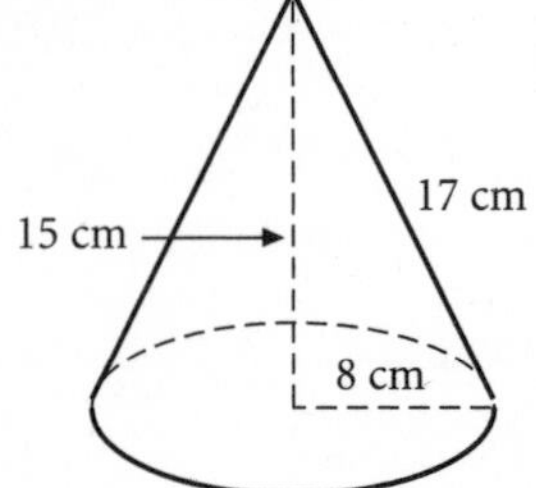

____

20.

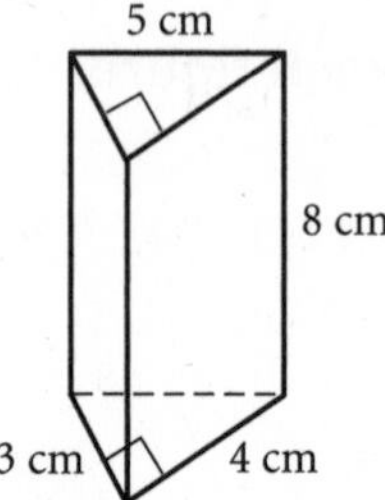

____

21. Find the volume of a cube with a surface area of 96 ft^2. ____________

22. A volcano has the shape of a cone. The diameter of the base of the volcano is about 9 miles. The height of the volcano is about 1000 ft. What is the approximate volume of the volcano to the nearest tenth of a cubic mile? (Hint: 1 mi = 5280 ft) ____________

23. The volume of a cylindrical water can is 100 $in.^3$ If the radius of the can is doubled, how many cubic inches of water will the new can hold? ____________

The coordinates of point *A* are (2, 3, −1). Write the coordinates of its image after a reflection across each plane.

24. *xy*-plane ____________ 25. *yz*-plane ____________ 26. *xz*-plane ____________

27. Segment $\overline{AB}$ with endpoints $A(0, 3, 5)$ and $B(0, 3, 0)$ is rotated about the *z*-axis. Name the figure that results, and find its volume. ____________

28. A segment with endpoints (2, 4, 0) and (7, 1, 0) is reflected across the *x*-axis. Find the area of the figure formed by connecting the points. ____________

NAME ______________________ CLASS ____________ DATE ____________

Quick Warm-Up: Assessing Prior Knowledge
8.1 Dilations and Scale Factors

1. Find the distance between the points $(-2, 1)$ and $(4, 3)$. ____________
2. What is the slope of the line given by $y = -3x + 2$? ____________
3. The points $(-2, 4)$, $(3, 4)$, and $(3, -1)$ are three vertices of a square. Find the fourth vertex of this square. ____________
4. Write the equation, in slope-intercept form, of the line containing the points $(-1, 5)$ and $(1, 3)$. ____________

Lesson Quiz
8.1 Dilations and Scale Factors

Find the image of each point transformed by the given dilation.

1. $(-3, 2)$; $D(x, y) = (2x, 2y)$ ____________
2. $(6, -4)$; $D(x, y) = (-1.5x, -1.5y)$ ____________
3. The dashed figure at the right represents the preimage of a dilation, and the solid figure represents its image. What is the scale factor of the dilation? ____________

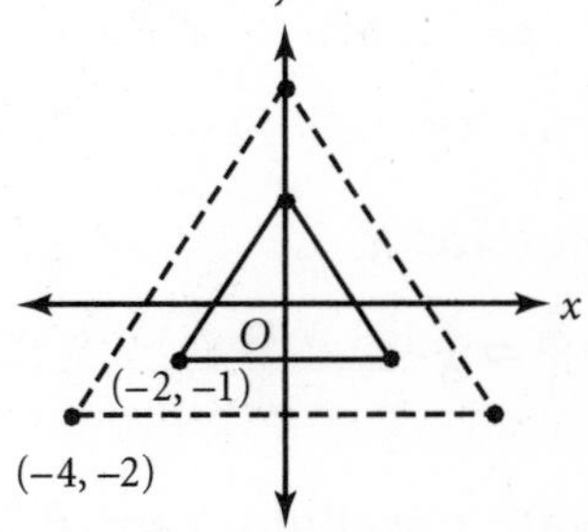

4. Plot the image of $P\left(\frac{1}{2}, -\frac{3}{4}\right)$ after a dilation with a scale factor of -4.

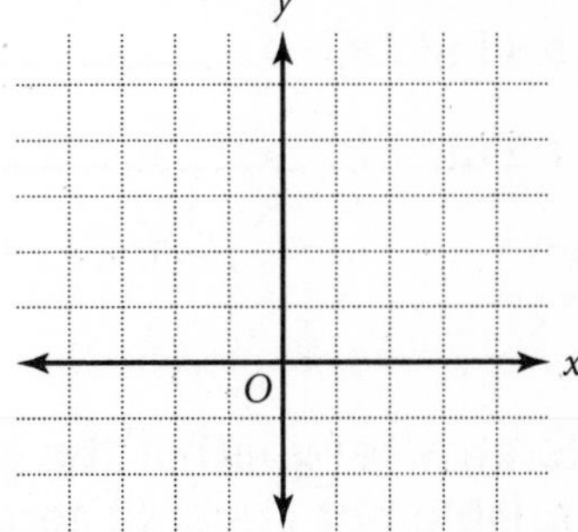

For Exercises 5–6, point *P* and scale factor *n* are given. Find the equation of the line though the preimage and image points.

5. $P(3, 0)$; $n = -3$ ____________
6. $P(-4, 2)$; $n = \frac{1}{2}$ ____________

NAME ______________________ CLASS ______________ DATE ______________

Quick Warm-Up: Assessing Prior Knowledge
8.2 Similar Polygons

1. Given $\triangle ABC \cong \triangle DEF$, name the congruent angles and sides. ______________________

____

2. The dilation image of a segment with a length of 4 has a length of 6. What is the scale factor? ______________

3. Solve the equation $6x = 45$. ______________

Lesson Quiz
8.2 Similar Polygons

Determine whether each pair of figures is similar.

1. ______________________ 2. ______________________

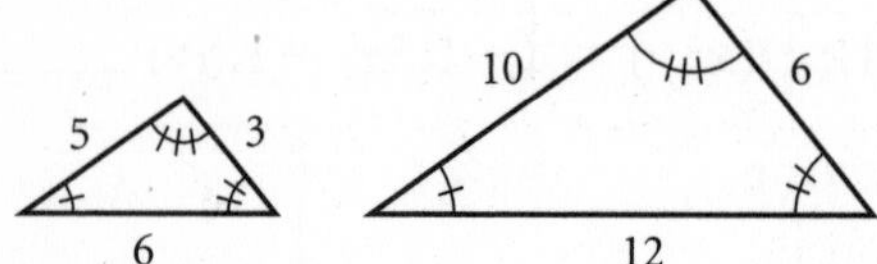

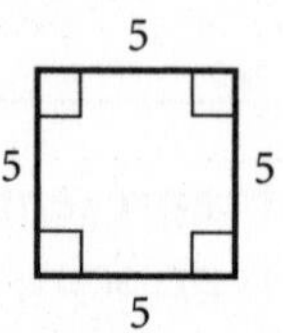

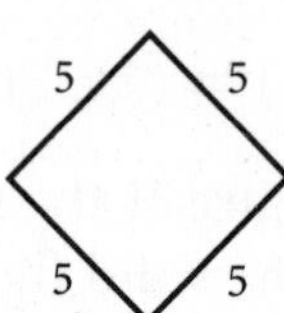

Solve for *x*.

3. $\frac{2}{3} = \frac{x}{9}$ ______________ 4. $\frac{5}{x} = \frac{x}{20}$ ______________ 5. $\frac{4}{x+1} = \frac{7}{x-2}$ ______________

Given: parallelogram *ABCD* ~ parallelogram *PQRS*

6. Find *DC*. ______________________

7. Find *SR*. ______________________

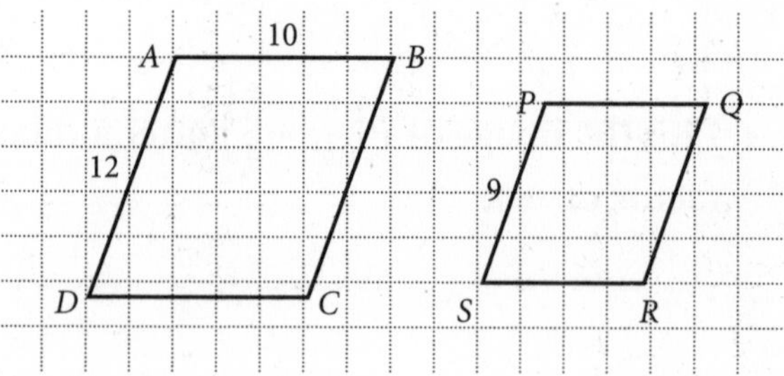

Determine whether the given proportion is true for all values of the variable not equal to zero. If not, give a numerical counterexample.

8. If $\frac{a}{b} = \frac{x}{y}$, then $\frac{b}{a} = \frac{y}{x}$ ______________ 9. If $\frac{a}{b} = \frac{x}{y}$, then $\frac{a+y}{b} = \frac{x+y}{y}$ ______________

NAME ______________________ CLASS ____________ DATE ____________

Quick Warm-Up: Assessing Prior Knowledge
8.3 Triangle Similarity Postulates

1. List the triangle congruence postulates. ______________________

2. What is the ratio of corresponding sides for two congruent triangles? ____________

3. What is the ratio of corresponding angles for two congruent triangles? ____________

Lesson Quiz
8.3 Triangle Similarity Postulates

1. State the Angle-Angle Similarity Postulate. ______________________

____

2. State the Side-Angle-Side Similarity Postulate. ______________________

____

Determine whether each pair of triangles can be proven similar. If so, write the postulate used.

3.

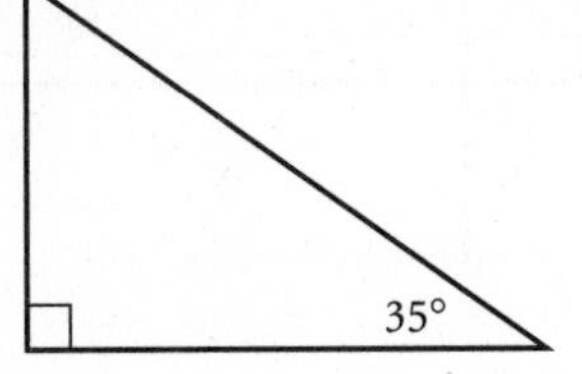

35°

____

____

4.

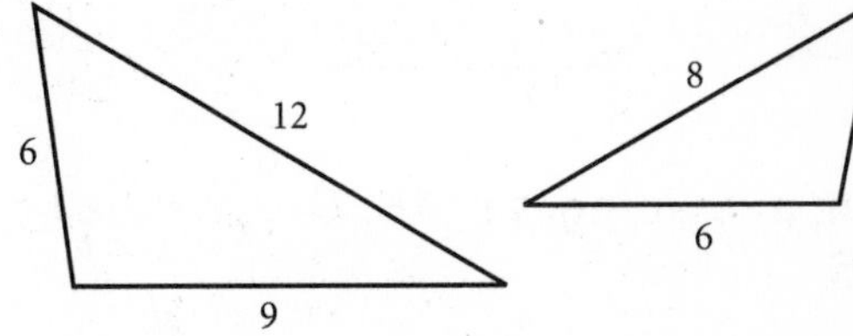

____

____

△ABC ~ △MNO.

5. Find *MN*.

____

6. What is the ratio of the perimeter of △*MNO* to the perimeter of △*ABC*?

____

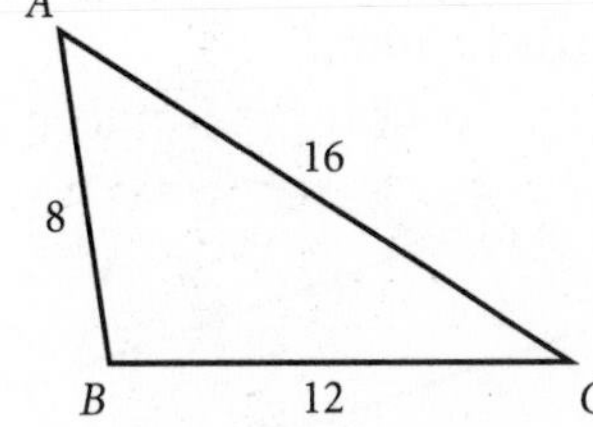

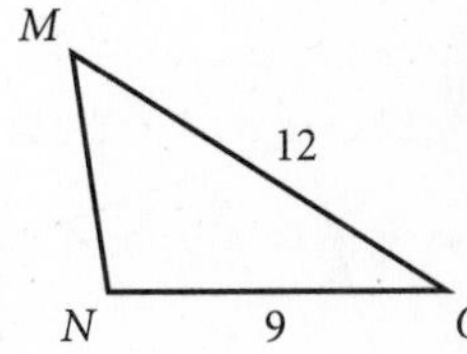

NAME ______________________ CLASS ______________ DATE ______________

Mid-Chapter Assessment

Chapter 8 (Lessons 8.1–8.3)

Write the letter that best answers the question or completes the statement.

______ 1. The point $(-3, 2)$ is transformed by a dilation with a scale factor of 3. What are the coordinates of the image?

a. $(3, -2)$ b. $(2, -3)$ c. $(9, -6)$ d. $(-9, 6)$

______ 2. The point $(8, -12)$ is transformed to the point $(-2, 3)$. What is the scale factor of the dilation?

a. $-\frac{1}{4}$ b. $\frac{1}{2}$ c. -2 d. 4

______ 3. If $\frac{p}{q} = \frac{r}{s}$, which of the following is true? (Assume that p, q, r, and $s \neq 0$.)

a. $\frac{r}{q} = \frac{p}{s}$ b. $\frac{p}{r} = \frac{q}{s}$ c. $\frac{p}{s} = \frac{r}{q}$ d. $\frac{s}{q} = \frac{p}{r}$

______ 4. If $\triangle ABC \sim \triangle XYZ$, which of the following is true?

a. $\frac{AB}{XY} = \frac{BC}{YZ}$ b. $\frac{AC}{XY} = \frac{BC}{YZ}$ c. $\frac{BC}{XY} = \frac{AB}{YZ}$ d. $\frac{AC}{XZ} = \frac{BC}{XY}$

5. Plot the image of $P(2, 1)$ after a dilation with a scale factor of 3.

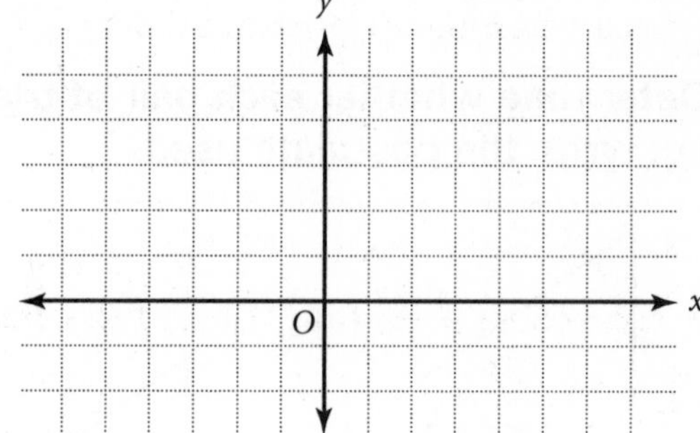

Solve each proportion for *x*.

6. $\frac{6}{15} = \frac{8}{x}$ ______________________

7. $\frac{4}{x + 2} = \frac{7}{21}$ ______________________

Determine whether each pair of triangles can be proven similar. If so, write the postulate used.

8.

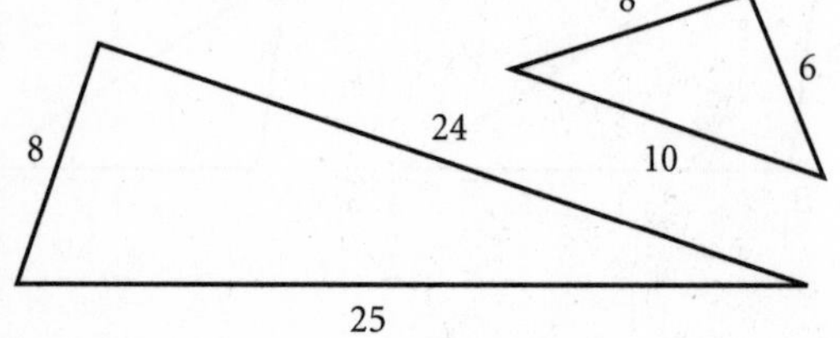

____

9.

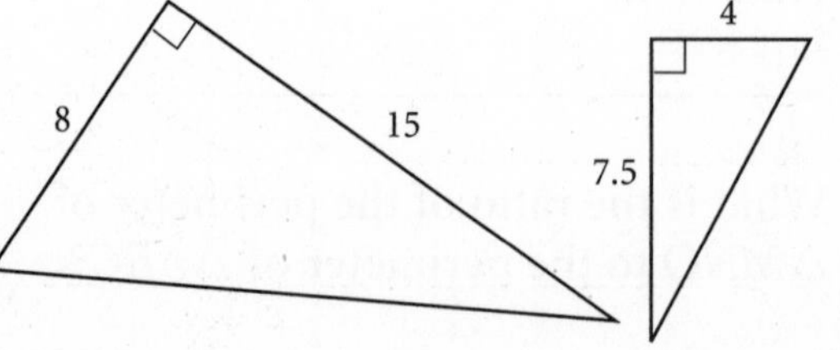

____

NAME ______________________ CLASS ______________ DATE ______________

Quick Warm-Up: Assessing Prior Knowledge
8.4 The Side-Splitting Theorem

State the three ways to determine triangle similarity.

____

____

____

Lesson Quiz
8.4 The Side-Splitting Theorem

In Exercises 1–2, complete each statement.

1. A line ______________ to one side of a triangle divides the other two sides ______________.

2. Three or more ______________ divide two intersecting transversals proportionally.

3. State the converse of the Side-Splitting Theorem. ____

____

In Exercises 4–9, find the value of x in each figure.

4.
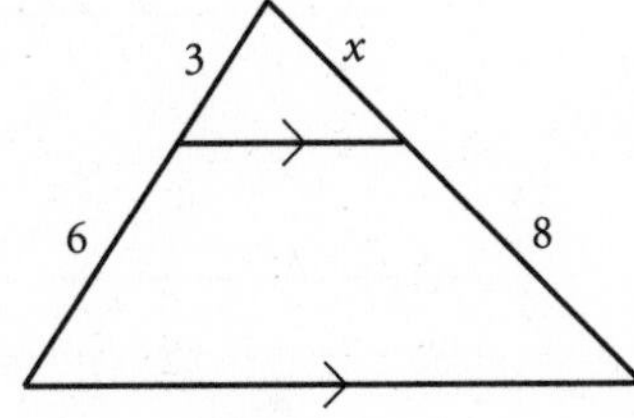

____

5.
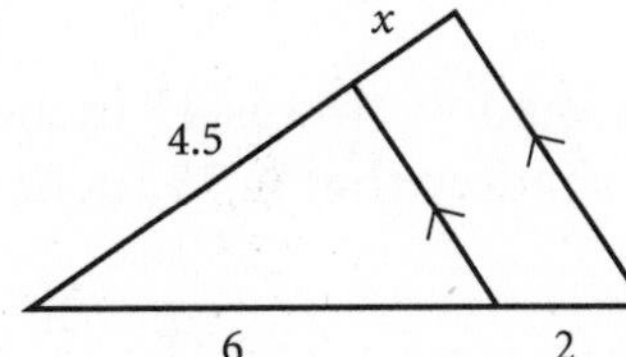

____

6.
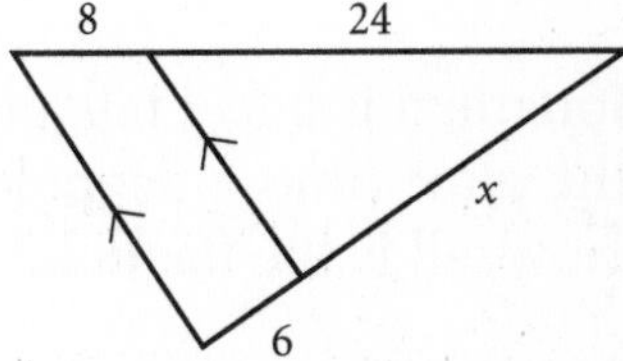

____

7.
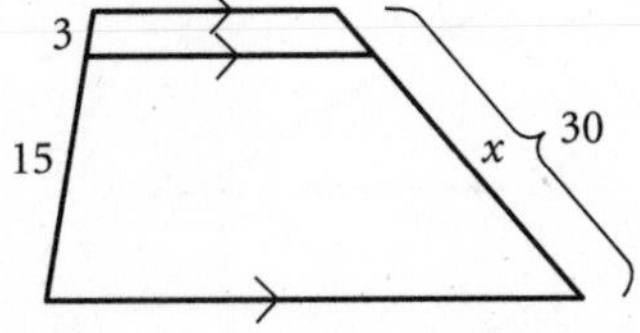

____

8.
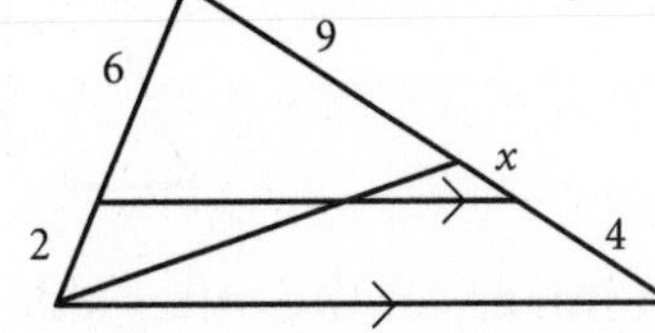

____

9.
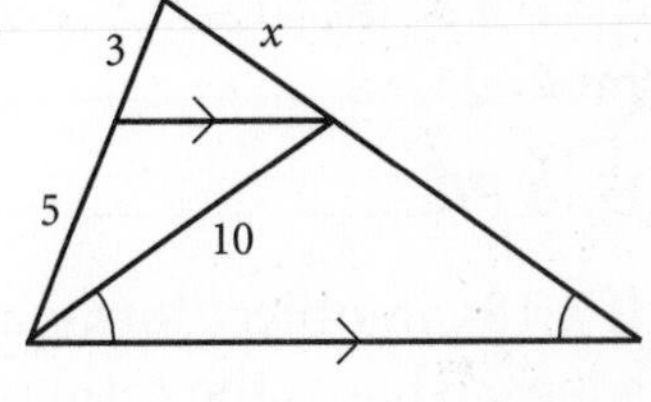

____

Quick Warm-Up: Assessing Prior Knowledge

8.5 Indirect Measurement and Additional Similarity Theorems

1. Solve the proportion $\frac{3}{x} = \frac{15}{20}$. ______________
2. Given $\triangle ABC \sim \triangle DEF$, find the ratio of their sides. ______________
3. $\triangle ABC$ has side lengths of 5, 5, and 6. Find the length of the altitude from the vertex angle. ______________

Lesson Quiz

8.5 Indirect Measurement and Additional Similarity Theorems

In Exercises 1–2, complete each statement.

1. If two triangles are similar, then their corresponding altitudes have ______________________________.
2. An ______________ of a triangle divides the opposite side into two segments that have the same ratio as the other two sides.
3. The triangles at right are similar. What is the value of x?

   ______________________________.

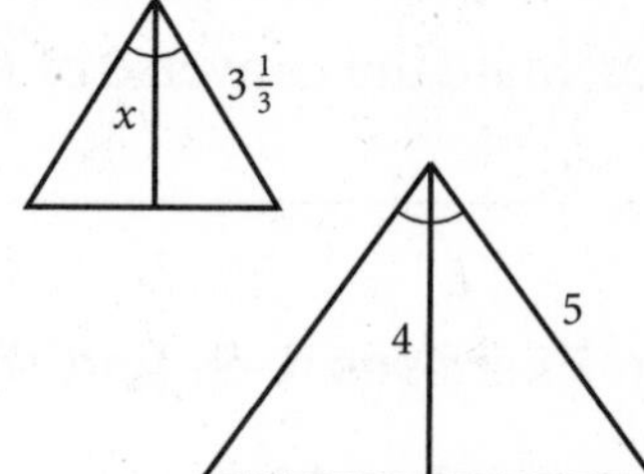

4. Jonathan is 6 feet tall and casts a shadow that is 45 inches long. At the same time, a flagpole casts a shadow that is 75 inches long. How tall is the flagpole? ______________

Given: $\triangle ABC \sim \triangle PQR$, $AD = DC$, and $PS = SR$

5. Find BE. ______________
6. Find QS. ______________
7. Find PR. ______________

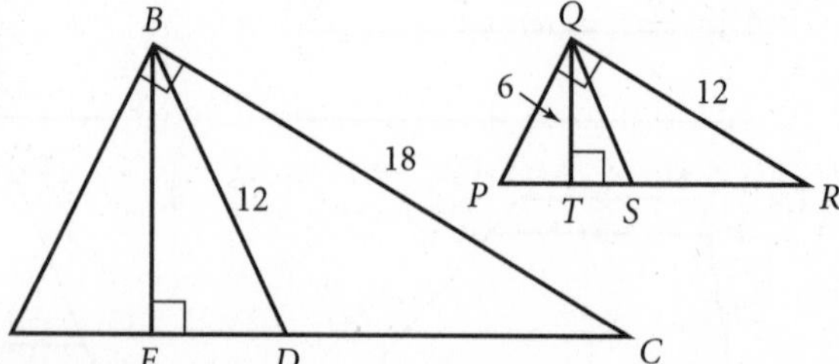

8. If a 10-cm object forms a 15-cm image on the other side of a lens, what is the ratio of the object's size to the image's size? ______________
9. Jan placed a lens 24 cm away from an object that is 18 cm tall. An image formed 6 cm away from the lens. How tall was the image? ______________

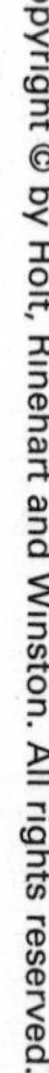

Quick Warm-Up: Assessing Prior Knowledge

8.6 Area and Volume Ratios

1. Find the area of a circle with a radius of 4 cm. ____________
2. Find the volume of a rectangular prism with dimensions of 4 cm by 6 cm by 7 cm. ____________
3. Find the volume of a right cylinder with a radius of 5 cm and a height of 6 cm. ____________

Lesson Quiz

8.6 Area and Volume Ratios

The two cylinders shown at right are similar. Find the ratio between the following:

1. the circumference of their bases ____________
2. their lateral areas ____________
3. their volumes ____________

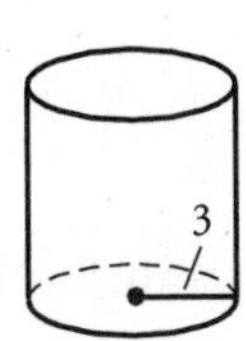

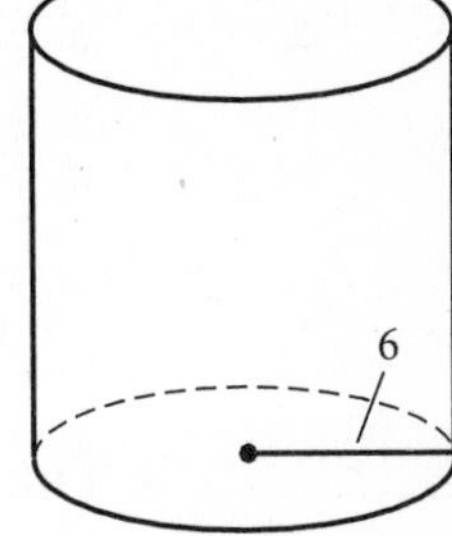

The side lengths of two squares are 4 and 12. Find the ratio of the following:

4. their perimeters ____________
5. their areas ____________

The ratio of the areas of two similar triangles is 9:16. Find the ratio of the following:

6. their altitudes ____________
7. their perimeters ____________
8. Each side of a regular pentagon measures 4 in. Each side of a similar regular pentagon measures 10 in. The area of the smaller pentagon is about 27.5 in^2.

   What is the area of the larger pentagon? ____________
9. A rectangular photo is 4 in. wide. The photo is enlarged by a scale factor of 1.5. The area of the new photo is 54 in^2.

   What is the area of the original photo? ____________

Chapter Assessment

Chapter 8, Form A, page 1

Write the letter that best answers the question or completes the statement.

______ 1. Find the image of $P(6, 12)$ after it is transformed by the dilation $D(x, y) = \left(-\frac{2}{3}x, -\frac{2}{3}y\right)$

a. $(9, 18)$ b. $(-9, -18)$ c. $(4, 8)$ d. $(-4, -8)$

______ 2. The scale factor of a dilation is $\frac{3}{2}$. What is the equation of the line through $P(2, -4)$ and its image?

a. $y = 2x$ b. $y = -2x$ c. $x = 2y$ d. $x = -2y$

______ 3. The dashed-line preimage has been transformed to form the solid-line image. What is the scale factor of the dilation?

a. $\frac{2}{5}$ b. 2

c. $\frac{5}{2}$ d. 5

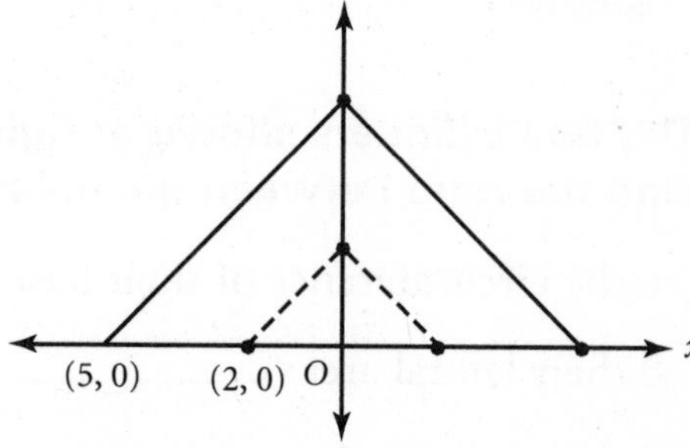

______ 4. The two triangles below are similar. If $YZ = 6$, then $AB =$ ______.

a. 2.5 b. 6 c. 10 d. 18

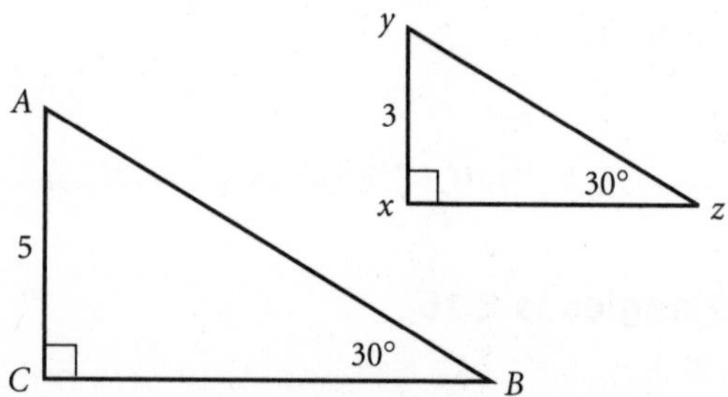

______ 5. The perimeter of $\triangle GHI$ is 18, the perimeter of $\triangle PQR$ is 30, and $\triangle GHI \sim \triangle PQR$. If GH is 10, what is PQ?

a. 3.60 b. 6.00 c. 16.67 d. 27.78

______ 6. Which proportion illustrates the Side-Splitting Theorem?

a. $\frac{AD}{AB} = \frac{AL}{AE}$ b. $\frac{AD}{EL} = \frac{AB}{AL}$

c. $\frac{AD}{AE} = \frac{AL}{AB}$ d. $\frac{AD}{DB} = \frac{AE}{EL}$

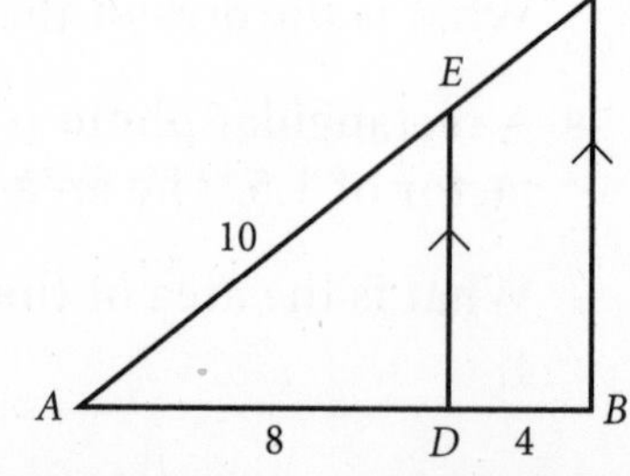

NAME ______________________ CLASS __________ DATE __________

Chapter Assessment

Chapter 8, Form A, page 2

_______ 7. In $\triangle ADE$ on the previous page, if $ED = 6$, what is BL?

a. 3 b. 5 c. 9 d. 12

_______ 8. What is the value of x in the figure at the right?

a. 8 b. 12
c. 14 d. 16

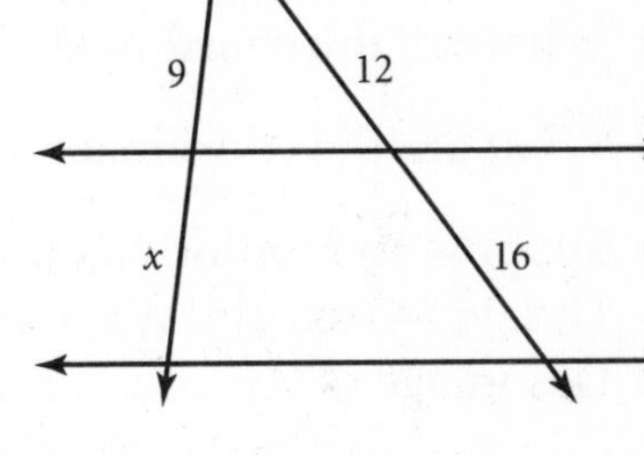

_______ 9. The shadow of a man 6 feet tall is 30 inches long. At the same time of day, a building casts a shadow 125 inches long. How tall is the building?

a. 15 feet b. 25 feet c. 30 feet d. 50 feet

_______ 10. Dan placed a lens 16 cm away from an 8-cm object. An image formed 10 cm away from the lens. How tall was the image?

a. 5 cm b. 10 cm c. 15 cm d. 20 cm

_______ 11. In the figure at right, $\triangle MNO \sim \triangle HIJ$. What is the length of $\overline{NQ}$?

a. 6 b. 10
c. 12 d. 15

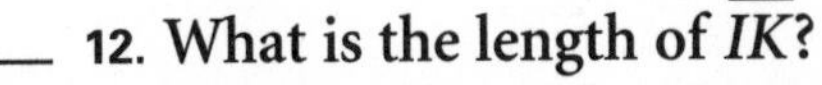

_______ 12. What is the length of $\overline{IK}$?

a. 9 b. 10
c. 12 d. 15

_______ 13. $MN =$

a. 10 b. 11
c. 12 d. 15

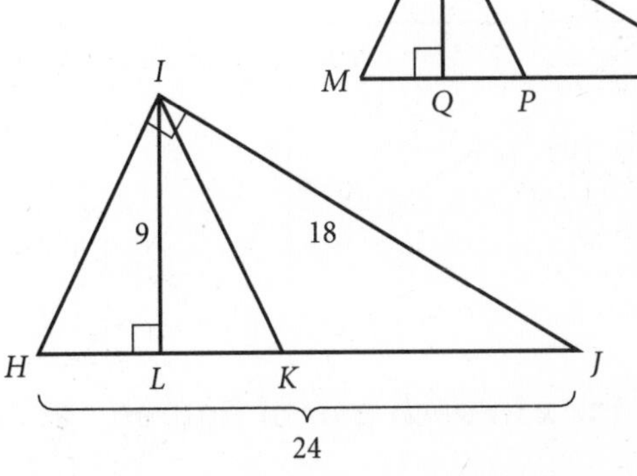

_______ 14. Two spheres have radii of 3 cm and 5 cm. What is the ratio between the areas of their great circles?

a. 3:5 b. 6:10 c. 9:25 d. 27:125

_______ 15. It cost \$144 to refinish a floor that is 9 feet by 12 feet. At the same rate, how much will it cost to refinish a floor that is 12 feet by 16 feet?

a. \$81 b. \$108 c. \$256 d. \$576

_______ 16. The area of one side of a cube is 36 ft². If the edges of the cube are tripled, what is the volume of the new cube?

a. 36 ft³ b. 196 ft³ c. 324 ft³ d. 5832 ft³

Chapter Assessment

Chapter 8, Form B, page 1

The dashed figure at right represents the preimage of a dilation. The solid figure represents its image.

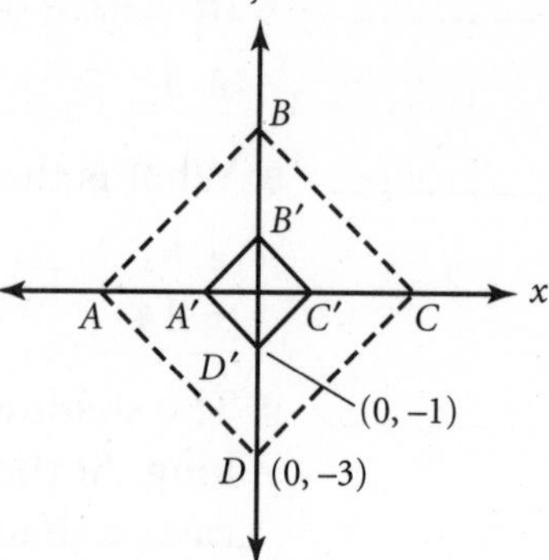

1. What is the scale factor of the dilation? ______________
2. What are the coordinates of the image of B? ______________
3. What are the coordinates of the preimage of C'? ______________
4. Suppose that point A is transformed by the dilation $D(x, y) = (2x, 2y)$. What are the coordinates of this image of A? ______________
5. If $\frac{x}{18} = \frac{12.5}{45}$, then $x =$ ______________.
6. If $\frac{x}{45} = \frac{12.5}{18}$, then $x =$ ______________.

Can the pair of triangles below be proven similar? Explain why or why not.

7.

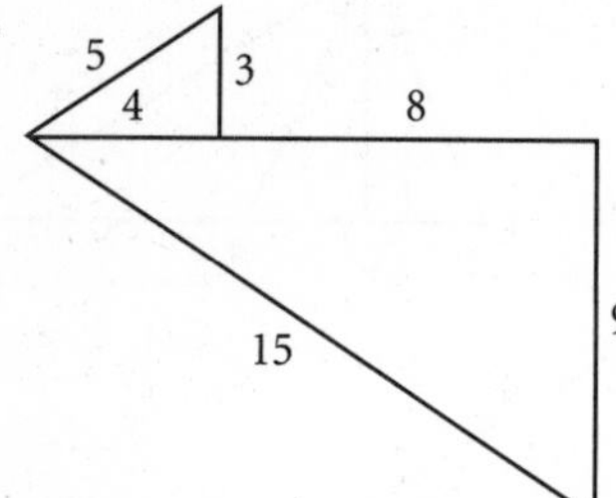

____

____

____

Solve for x in each pair of figures.

8.

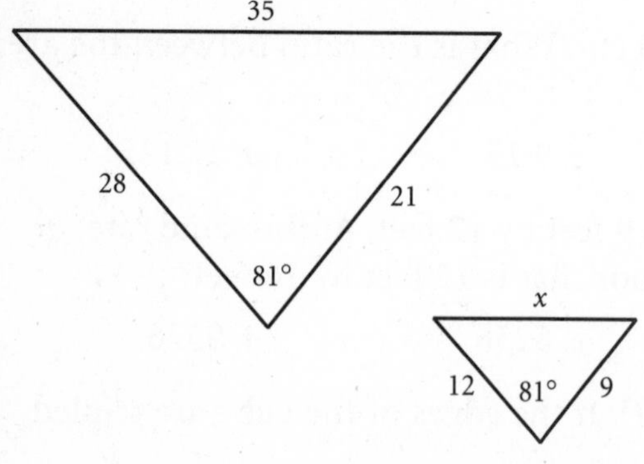

____

9.

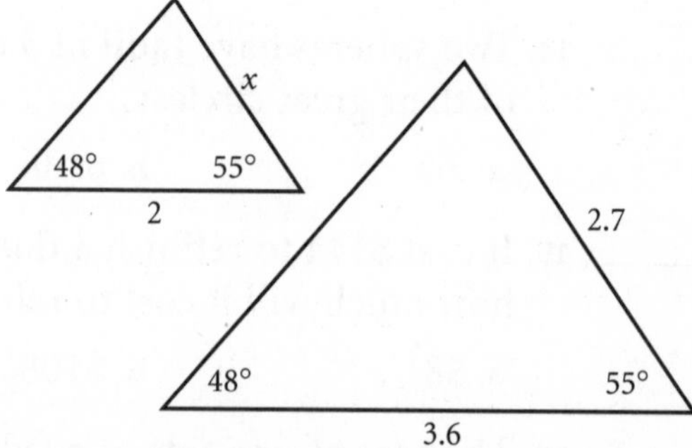

____

NAME ______________________ CLASS ____________ DATE ____________

Chapter Assessment

Chapter 8, Form B, page 2

10. In $\triangle ABC$ and $\triangle DEF$, the length of $\overline{AB}$ is 12 and the length of $\overline{DE}$ is 9. If $\triangle ABC \sim \triangle DEF$ and the perimeter of $\triangle ABC$ is 36, what is the perimeter of $\triangle DEF$? ____________

Solve for *x* and *y* in each figure.

11.

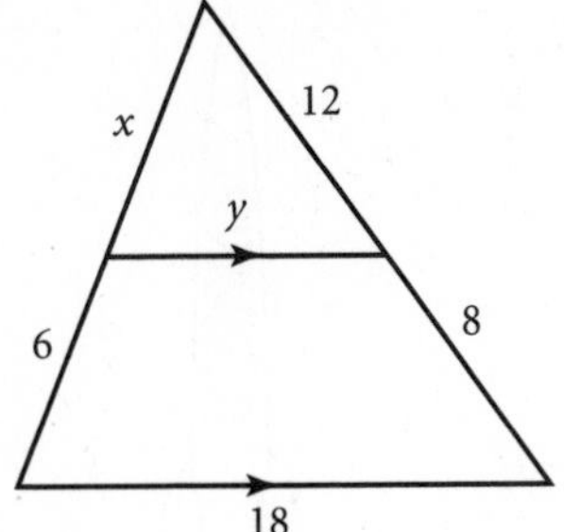

12.

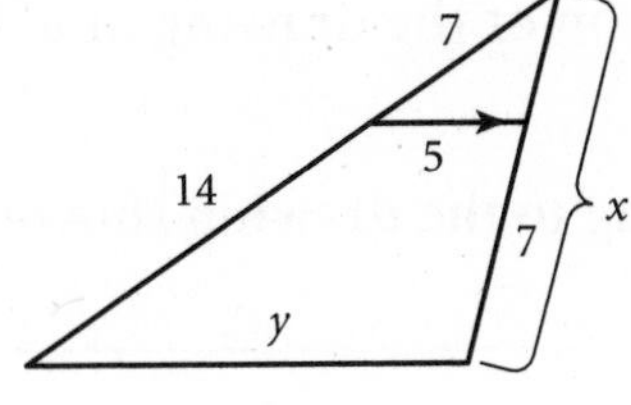

13.

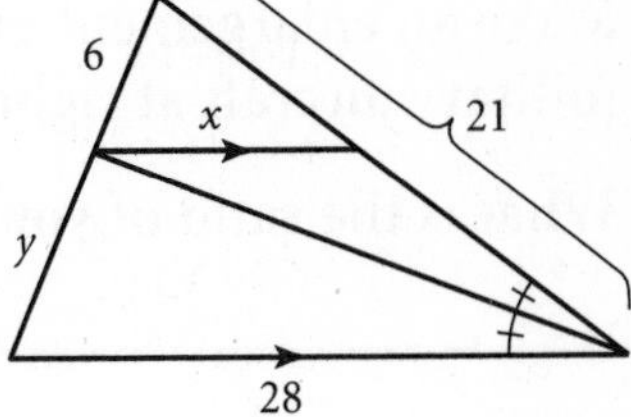

____________ ____________ ____________

Given: $\triangle PQR \sim \triangle JKL$, $PS = SR$, and $JM = ML$

14. Find QS. ____________

15. Find KN. ____________

16. Find PR. ____________

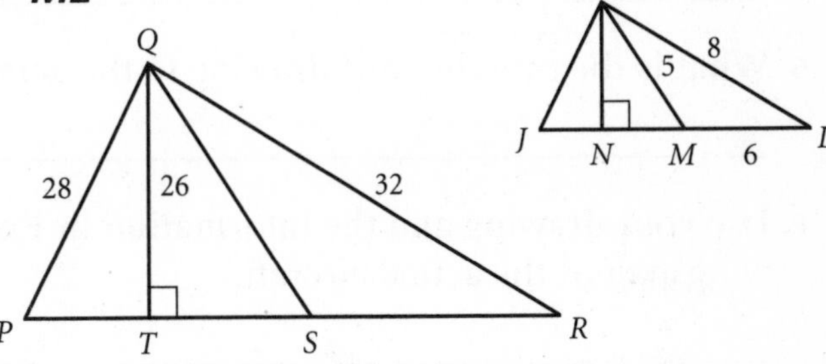

17. Aaron placed an object 18 cm away from a lens. A 12-cm image formed 8 cm away from the lens. How tall was the object? ____________

18. A yardstick casts a 2-foot shadow. How long is the shadow cast by a $13\frac{1}{2}$-foot tree? ____________

19. Roberto wants to enlarge a 10-cm by 15-cm photo by 30%. Find the area of the enlarged photo. ____________

20. Two pipes are similar cylinders whose lengths have a ratio of $\frac{3}{2}$. The smaller pipe can hold $2\frac{2}{3}$ gallons of water. How much water can the larger pipe hold? ____________

NAME ______________________ CLASS __________ DATE __________

Alternative Assessment

Scale Drawings, Chapter 8, Form A

TASK: Draw and interpret a scale drawing.

HOW YOU WILL BE SCORED: As you work through the task, your teacher will be looking for the following:

- whether you can apply a ratio to create a scale drawing
- how well you describe the use of ratios in scale drawings

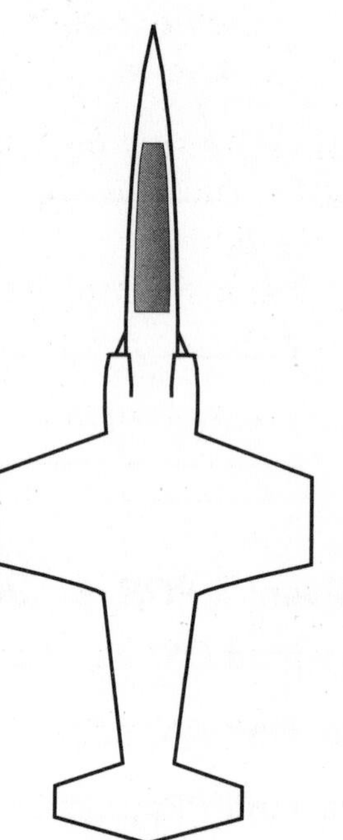

1. Make an enlargement or reduction of the drawing of a military aircraft at right.
2. What is the ratio of your drawing to the drawing shown?

   ____

3. Explain your technique for enlarging or reducing the drawing.

   ____

4. The actual length of the aircraft is 54 ft 9 in. What ratio was used for this drawing?

   ____

5. What is the ratio of your drawing to the actual aircraft?

   ____

6. Use your drawing and the information in Exercise 4 to find the wingspan of the actual aircraft.

   ____

7. If the width of the tail is 12 feet, what width should the tail in your scale drawing have? Measure to see whether or not the tail is that width. Explain any differences.

   ____

   ____

SELF-ASSESSMENT: What does it mean when the ratio of the actual object to its model is less than 1? What does it mean when the ratio is greater than 1?

NAME ______________________ CLASS __________ DATE __________

Alternative Assessment

Indirect Measurements, Chapter 8, Form B

TASK: Explain how indirect measurements can be made by using ratios.

HOW YOU WILL BE SCORED: As you work through the task, your teacher will be looking for the following:

- how well you can describe the process of finding a measurement indirectly
- whether or not you can write a proportion for finding unknown lengths

For each situation below:

a. Explain how you could use indirect measurement to estimate the length.
b. Name the properties or theorems you used in your reasoning.
c. Write a proportion for finding the length.

1. You want to know the height of a tree.

____

____

____

____

____

____

2. You want to reinforce the sides of a tent with braces. The tent is shaped like a square pyramid. You know the length of the four poles that form its base and the height of the four poles that form the frame. How long should you make the braces?

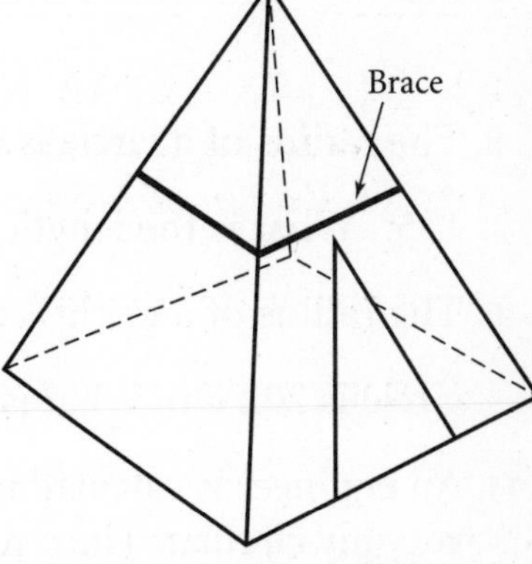

____

____

____

____

____

____

SELF ASSESSMENT: Write an actual problem with numbers for each situation above. Solve it by using your techniques and ratios.

NAME ______________________ CLASS ______________ DATE ______________

Quick Warm-Up: Assessing Prior Knowledge

9.1 Chords and Arcs

Define each term.

1. circle ______________________
2. radius ______________________
3. diameter ______________________

Lesson Quiz

9.1 Chords and Arcs

Complete each statement.

1. If two central angles are congruent, then their arcs are ______________________.
2. The measure of a central angle of a circle is $A°$. The length of the intercepted arc is ______________________.

In Exercises 3–8, find the degree measure of each arc by using the central angle measures of ⊙*P*.

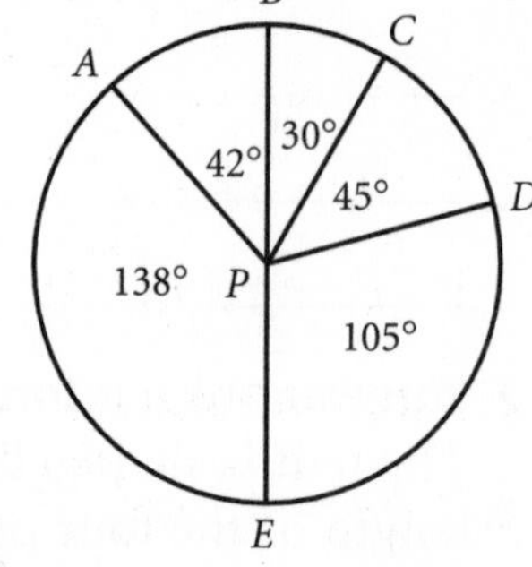

3. $m\overparen{AB}$ = ______________
4. $m\overparen{BD}$ = ______________
5. $m\overparen{CE}$ = ______________
6. $m\overparen{ACD}$ = ______________
7. $m\overparen{BAE}$ = ______________
8. $m\overparen{AED}$ = ______________

9. The radius of a circle is 8 m. The measure of $\angle APB$ is 135°. What is the length of $\overparen{APB}$ to the nearest hundredth? ______________
10. The radius of a circle is 60 cm. The length of an arc of the circle is 20π cm. What is the degree measure of the arc? ______________
11. An engineer is calculating the length of a section of a driveway that is roughly circular. The central angle of the section measures 120° and the radius of the circle is 45 feet. What is the length of the section of driveway? ______________
12. The radius of a circle is 12 m. The length of an arc of the circle is 20π m. What is the degree measure of the arc? ______________

Quick Warm-Up: Assessing Prior Knowledge

9.2 Tangents to Circles

How many points are described in each situation below?

1. the intersection of a circle and one of its chords ____________
2. the intersection of a circle and one of its radii ____________
3. the intersection of a circle and one of its central angles ____________

Lesson Quiz

9.2 Tangents to Circles

Complete each statement.

1. A line that intersects a circle at exactly one point is called a ____________.
2. A radius that is perpendicular to a chord of a circle ____________ the chord.

In $\odot Q$, $\overline{QS} \perp \overline{PR}$ at T.

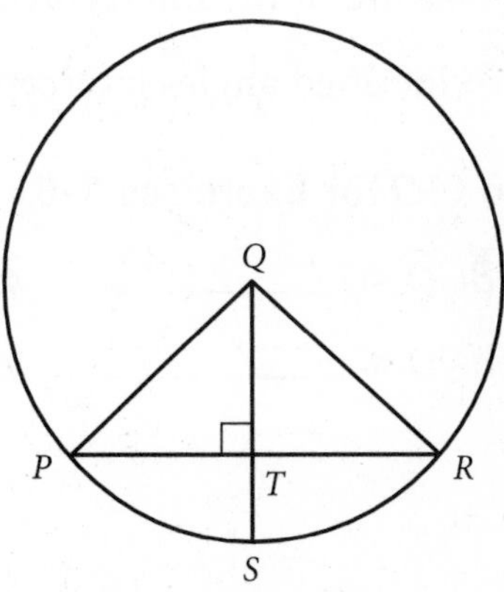

3. $\overline{PT} \cong$ ____________
4. $\overline{QP} \cong$ ____________
5. If $QS = 13$ and $QT = 5$, $PT =$ ____________.
6. If $QS = 17$ and $PR = 16$, $QT =$ ____________.
7. If $PR = 8$ and $QT = 3$, $QS =$ ____________.
8. If $PR = 12$ and $QS = 10$, $TS =$ ____________.

In $\odot O$, $\overline{MN}$ is tangent at M.

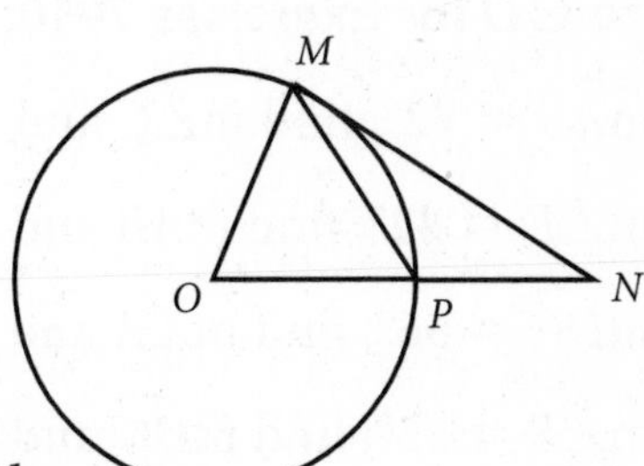

9. If $MN = 15$ and $ON = 17$, then $MO =$ ____________.
10. If $MN = 21$ and $MO = 20$, then $PN =$ ____________.
11. If $m\angle MNO = 32°$, then $m\widehat{MP} =$ ____________.
12. A chord of a circle is 10 inches long and 12 inches from the center of the circle. Find the length of the radius. ____________
13. A chord of a circle is 3 feet long and 18 inches from the center of the circle. Find the length of the radius to the nearest hundredth of a foot. ____________

NAME ______________________ CLASS ____________ DATE ____________

Quick Warm-Up: Assessing Prior Knowledge

9.3 *Inscribed Angles and Arcs*

A circle and an angle are drawn in the same plane. The vertex of the angle is on the circle.

1. Find and sketch all the possible ways the two figures can be arranged.

____

2. For each arrangement, give the number of intersection points.

____

Lesson Quiz

9.3 *Inscribed Angles and Arcs*

Complete each statement.

1. The measure of an angle inscribed in a circle is equal to ____________ the measure of the intercepted arc.
2. If two incribed angles intercept the same arc, then they have the ____________.

Refer to ⊙*Q* for Exercises 3–6.

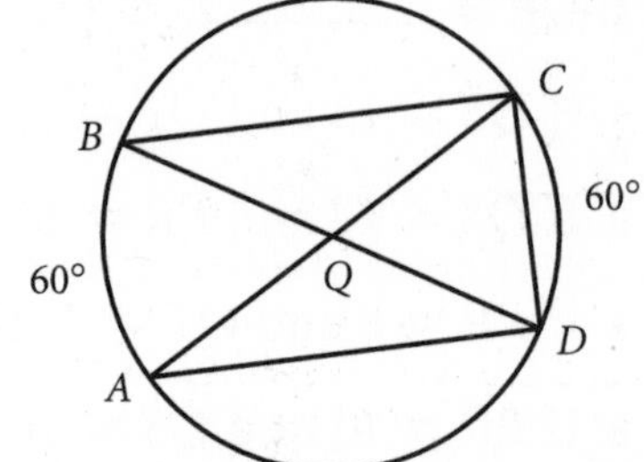

3. $m\angle BCD =$ __________
4. $m\angle ADB =$ __________
5. $m\angle ACD =$ __________
6. $m\overset{\frown}{BAC} =$ __________

Refer to ⊙*O* for Exercises 7–10.

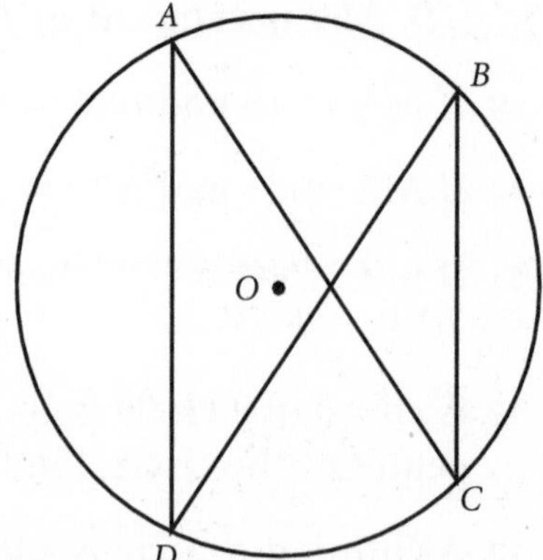

7. If $m\overset{\frown}{AB} = 72°$, find $m\angle C$ and $m\angle D$. __________
8. If $m\angle D = 42°$, find $m\overset{\frown}{AB}$ and $m\angle C$. __________
9. If $m\overset{\frown}{DC} = 56°$, find $m\angle A$ and $m\angle B$. __________
10. If $m\angle B = 55°$, find $m\overset{\frown}{DC}$ and $m\angle A$. __________

11. Quadrilateral *ABCD* is inscribed in a circle. Why are the opposite angles of the quadrilateral supplementary? __________

NAME ______________________ CLASS ______________ DATE ______________

Mid-Chapter Assessment

Chapter 9 (Lessons 9.1–9.3)

Match the line segment with its name.

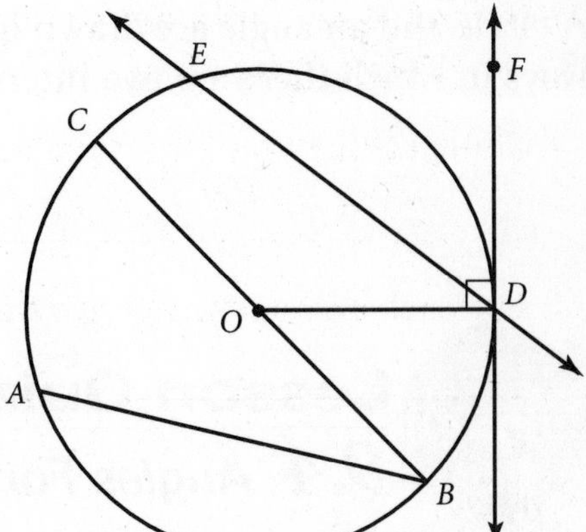

1. $\overline{CB}$ ______________ a. radius
2. $\overline{AB}$ ______________ b. secant
3. $\overline{OD}$ ______________ c. tangent
4. $\overleftrightarrow{ED}$ ______________ d. chord
5. $\overleftrightarrow{FD}$ ______________ e. diameter

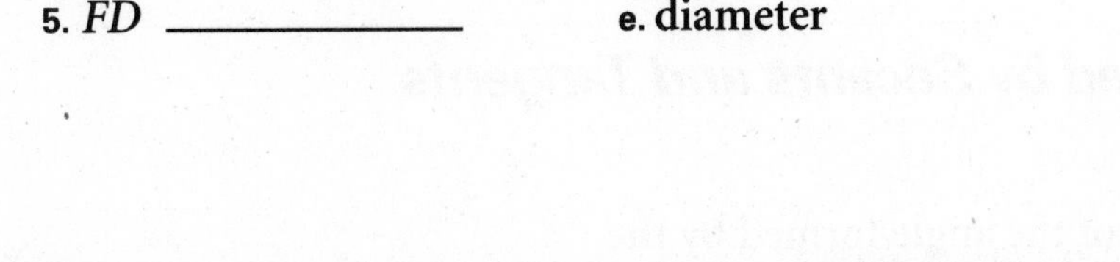

Write the letter that best answers the question or completes the statement. In ⊙*Q*, m$\overparen{PS}$ = 60° and m∠*PRT* = 55°.

______ 6. m∠*PQS* =

a. 30° b. 60°
c. 120° d. 180°

______ 7. m∠*PRS* =

a. 30° b. 60°
c. 90° d. 120°

______ 8. m$\overparen{PT}$ =

a. 55° b. 80° c. 110° d. 135°

______ 9. m∠*QSR* =

a. 30° b. 55° c. 60° d. 110°

______ 10. If *QP* = 18, then *PS* = ______.

a. 9 b. 8 c. 18 d. 72

In ⊙*G*, $\overline{HI}$ is tangent to ⊙*G* at *H*, and the radius is 10.

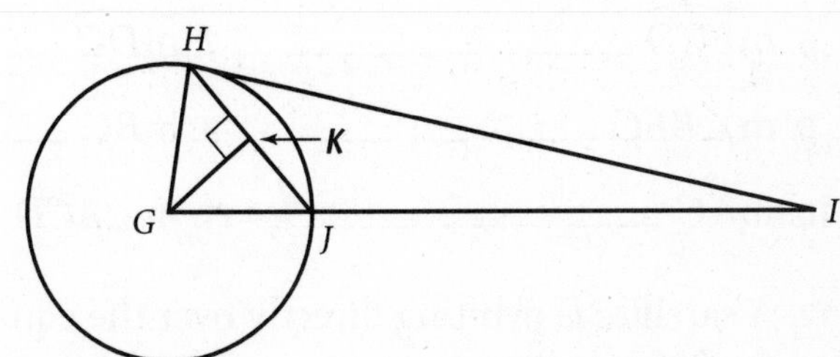

11. If *GK* = 8, then *HJ* = ______________.

12. If *HI* = 24, then *JI* = ______________.

13. If *JI* = 4.5, then *HJ* = ______________.

NAME ______________________ CLASS ______________ DATE ______________

Quick Warm-Up: Assessing Prior Knowledge

9.4 Angles Formed by Secants and Tangents

A circle and an angle are drawn in the same plane. Find all the possible ways in which there are two intersection points.

Lesson Quiz

9.4 Angles Formed by Secants and Tangents

1. Explain how to find the measure of the angle formed by the intersection of a tangent and a secant at the point of tangency. ______________
2. Explain how to find the measure of the angle formed by the intersection of two chords. ______________

In Exercises 3–5, find the value of *x*.

3.

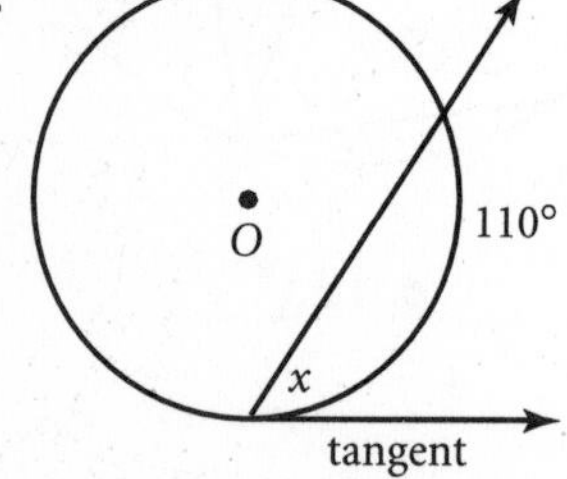

____

4.

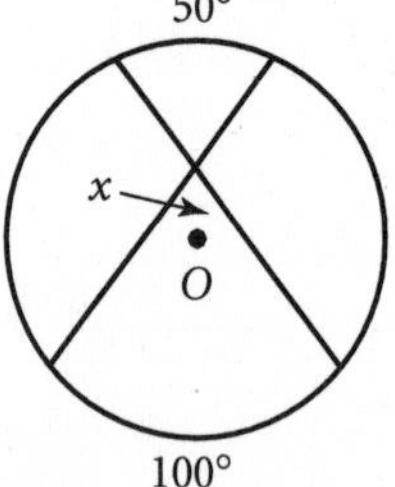

____

5.

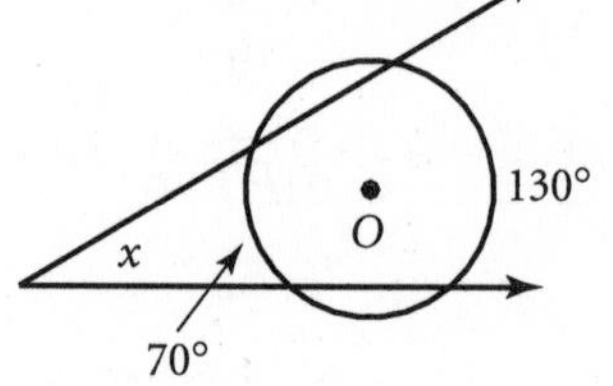

____

$\overline{FD}$ is tangent to the given circle at point *D* and m∠*BDF* = 85°. Find each of the following:

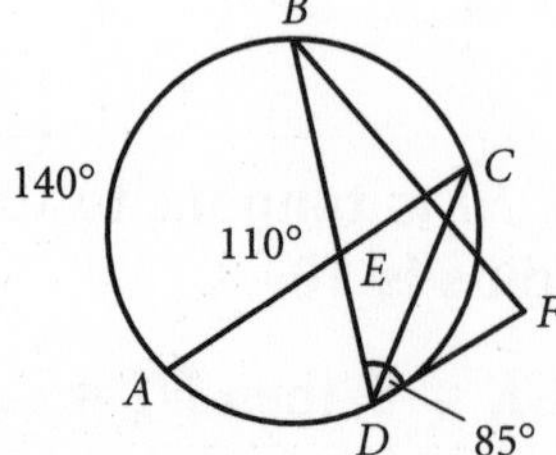

6. m$\overset{\frown}{BCD}$ ______________
7. m$\overset{\frown}{DC}$ ______________
8. m∠*BEC* ______________
9. m$\overset{\frown}{BC}$ ______________
10. m$\overset{\frown}{AC}$ ______________
11. m∠*ACD* ______________

12. A satellite is orbiting directly over the equator. The measure of the angle formed by the tangents from the satellite to Earth is 20°. Find the measure of the arc at the equator that is visible from the satellite. ______________

Quick Warm-Up: Assessing Prior Knowledge

9.5 Segments of Tangents, Secants, and Chords

Use a compass and straightedge or geometry graphics software to perform the following constructions:

1. Construct $\odot P$ and two secants that pass through the circle. The lines intersect at point X outside the circle.
2. Start from point X and move along one secant line toward the circle. Label the first intersection point C and the second intersection point A.

Lesson Quiz

9.5 Segments of Tangents, Secants, and Chords

$\overline{AB}$ is tangent to $\odot O$ at A. Complete each statement.

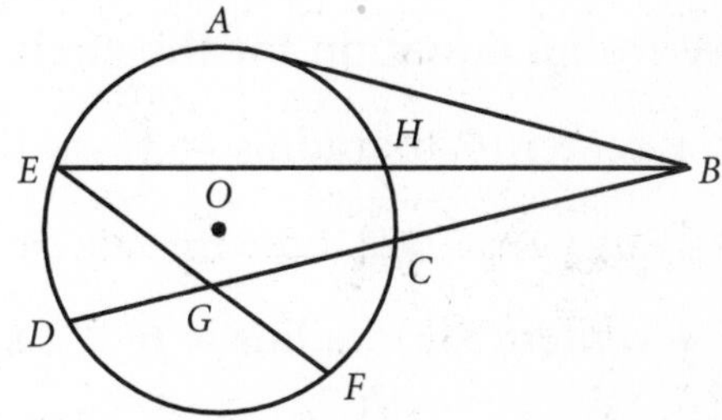

1. $(AB)^2 = DB \times$ ______________________
2. $EG \times GF = DG \times$ ______________________
3. $EB \times$ ______________ $= DB \times$ ______________

In Exercises 4–9, find the value of *x*.

4. 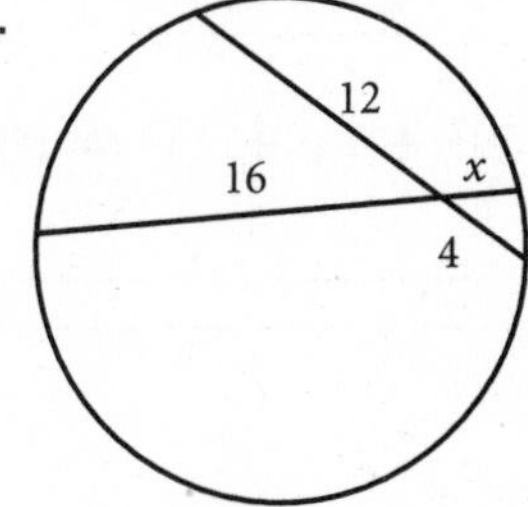

____

5. 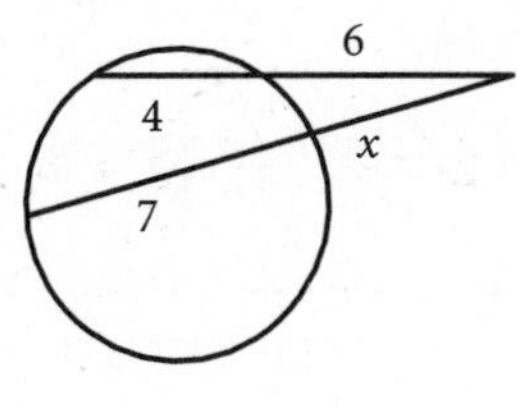

____

6. 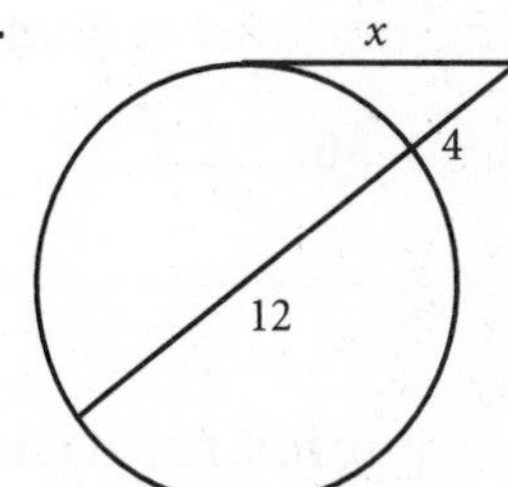

____

7. 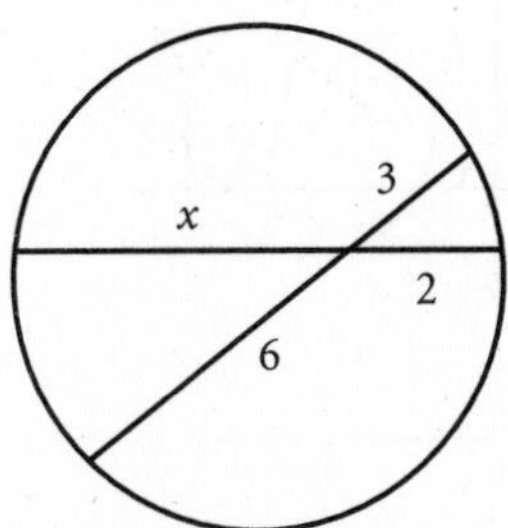

____

8. 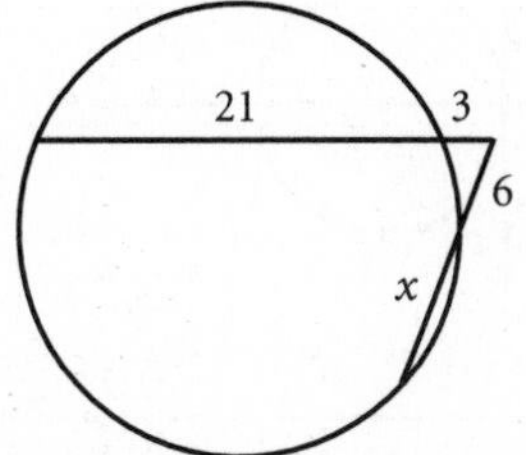

____

9. 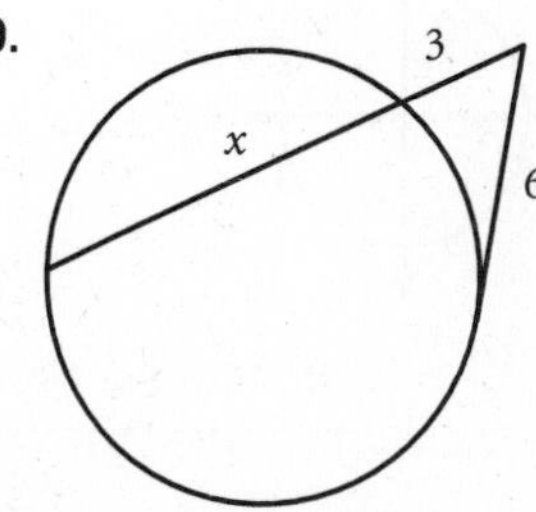

____

NAME ______________________ CLASS ______________ DATE ______________

Quick Warm-Up: Assessing Prior Knowledge

9.6 Circles in the Coordinate Plane

Solve for *x*.

1. $x^2 + 4^2 = 5^2$ ______________
2. $x^2 + 12^2 = 13^2$ ______________
3. $3^2 + x^2 = 6^2$ ______________
4. $y^2 + x^2 = 4^2$ ______________

Lesson Quiz

9.6 Circles in the Coordinate Plane

Find the *x*- and *y*-intercepts for each equation.

1. $x^2 + y^2 = 49$ ______________
2. $x^2 + y^2 = 81$ ______________
3. $x^2 + y^2 = 144$ ______________

Write an equation for the circle with the given center and radius.

4. center $(0, 0)$; radius $= 3$ ______________
5. center $(4, 0)$; radius $= 8$ ______________
6. center $(-3, 2)$; radius $= 15$ ______________
7. center $(3, 5)$; radius $= 6$ ______________
8. center $(2, -4)$; radius $= 2$ ______________
9. center $(-5, -1)$; radius $= 1$ ______________

Find the center and radius of each circle.

10. $x^2 + y^2 = 36$ ______________
11. $x^2 + (y - 3)^2 = 25$ ______________
12. $(x - 2)^2 + (y + 1)^2 = 100$ ______________

Write an equation for each circle.

13. 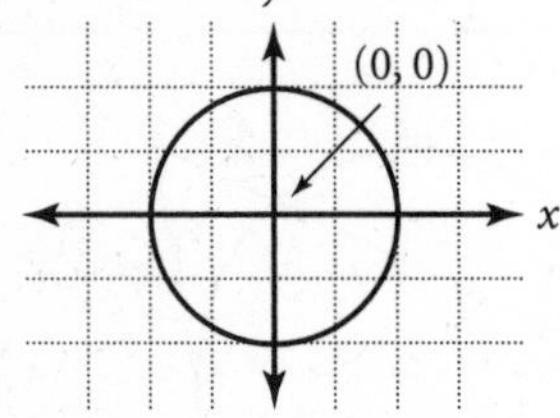

____

14. 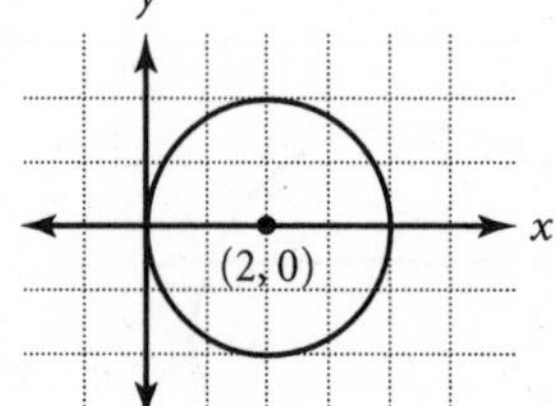

____

15. 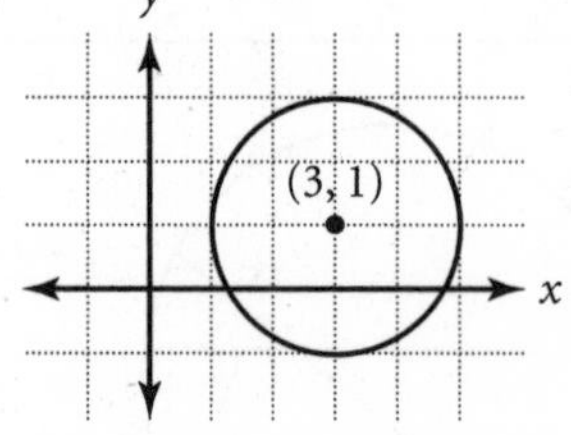

____

16. Find an equation for a circle centered at $(-2, 1)$ and containing the point $(1, 1)$. ______________

NAME ______________________ CLASS ____________ DATE ____________

Chapter Assessment

Chapter 9, Form A, page 1

Write the letter that best answers the question or completes the statement.

In ⊙*B*, $\overline{CD}$ is tangent to *B* at *C* and m$\widehat{AC}$ = 70°.

______ **1.** m∠*ABC* =

a. 35° **b.** 55° **c.** 70° **d.** 140°

______ **2.** m∠*ACD* =

a. 35° **b.** 55° **c.** 70° **d.** 140°

______ **3.** m∠*BCA* =

a. 35° **b.** 55° **c.** 70° **d.** 140°

4. In a circle, the length of an arc intercepted by a central angle is 3π and the radius of the circle is 9 inches. What is the measure of the central angle?

a. 30° **b.** 60° **c.** 90° **d.** 120°

5. A regular octagon is inscribed in a circle with a radius of 12 inches. What is the length of one arc of the circle intercepted by one side of the octagon?

a. 45° **b.** 8 inches **c.** 3π inches **d.** 12π inches

In ⊙*Q*, the radius is 15, *DG* = 6, and *JK* = 24.

______ **6.** *FH* =

a. 9 **b.** 12 **c.** 15 **d.** 24

______ **7.** *FK* =

a. 24 **b.** 36 **c.** 39 **d.** 54

8. If m$\widehat{HJ}$ = 70°, then m∠*K* = ________.

a. 20° **b.** 30° **c.** 40° **d.** 90°

______ **9.** *QD* =

a. 3 **b.** 6 **c.** 9 **d.** 12

Chapter Assessment

Chapter 9, Form A, page 2

For Exercises 10–11, refer to ⊙*P*.

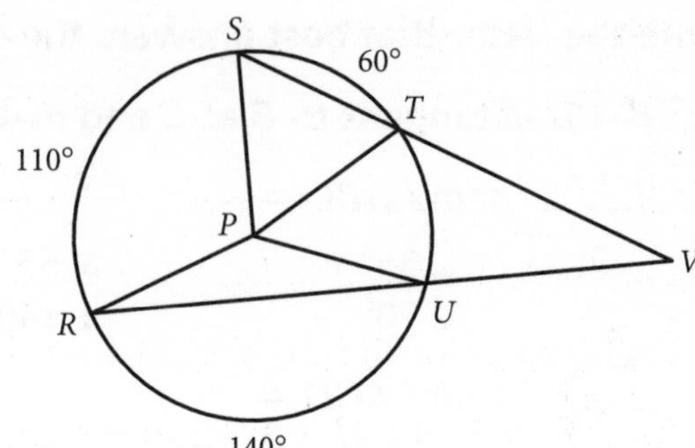

______ 10. m$\angle RPU$ =

a. 30° b. 50°
c. 80° d. 140°

______ 11. m$\angle TVU$ =

a. 30° b. 50°
c. 80° d. 100°

For Exercises 12–14, use the given circle.

______ 12. AF =

a. 6 b. 8
c. 12 d. 15

______ 13. AC =

a. 6 b. 8
c. 10 d. 14

______ 14. DC =

a. 3 b. 7 c. 12 d. 16

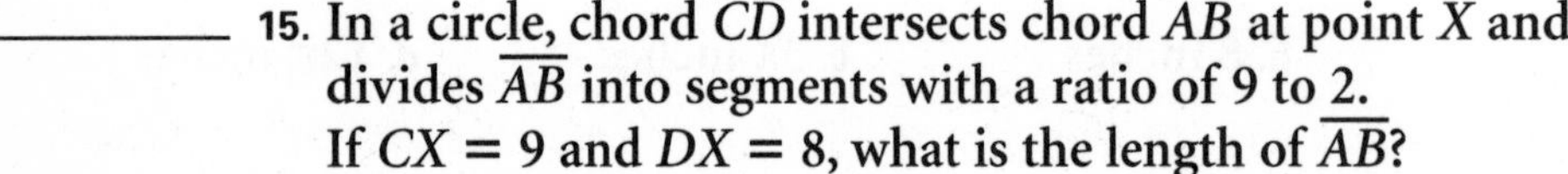

______ 15. In a circle, chord $\overline{CD}$ intersects chord $\overline{AB}$ at point X and divides $\overline{AB}$ into segments with a ratio of 9 to 2. If $CX = 9$ and $DX = 8$, what is the length of $\overline{AB}$?

a. 2 b. 4 c. 22 d. 72

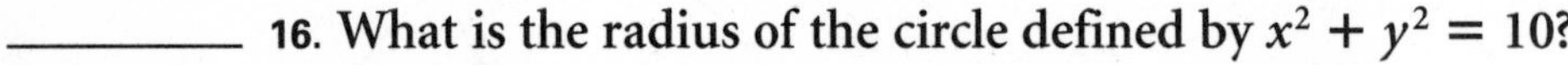

______ 16. What is the radius of the circle defined by $x^2 + y^2 = 10$?

a. $\sqrt{10}$ b. 10 c. $\sqrt{20}$ d. 100

______ 17. The center of the circle defined by $(x - 5)^2 + y^2 = 25$ is ______.

a. $(-5, 0)$ b. $(5, 0)$ c. $(0, -5)$ d. $(0, 5)$

______ 18. What is the equation of a circle centered at $(3, -2)$ with a radius of 4?

a. $(x + 3)^2 + (y - 2)^2 = 4$ b. $(x + 3)^2 + (y - 2)^2 = 16$
c. $(x - 3)^2 + (y + 2)^2 = 4$ d. $(x - 3)^2 + (y + 2)^2 = 16$

______ 19. What are the x-intercepts of the circle defined by $(x - 5)^2 + y^2 = 25$?

a. $x = 20$ and $x = 3$ b. $x = 25$ c. $x = 5$ d. $x = 0$ and $x = 10$

Chapter Assessment

Chapter 9, Form B, page 1

In ⊙*O*, $\overline{EH}$ is a diameter, m∠*HOI* = 42°, m$\widehat{FG}$ = 42°, and *HI* = 9.

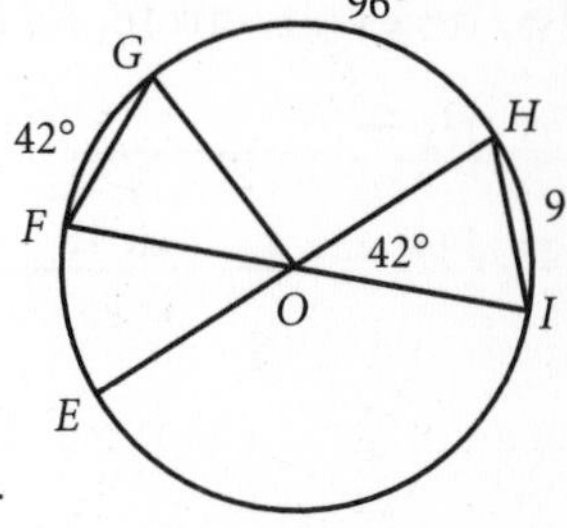

1. Find m∠*GOH*. ________ 2. Find m$\widehat{HI}$. ________
3. Find *FG*. ________ 4. Find m$\widehat{EI}$. ________
5. Find m∠*EHI*. ________ 6. Find m∠*FIH*. ________
7. If the radius of the circle is 30, find the length of $\widehat{HI}$. ________
8. If the length of $\widehat{GH}$ is 8π, find the circumference of the circle. ________

In ⊙*O*, *KM*⊥*JL*.

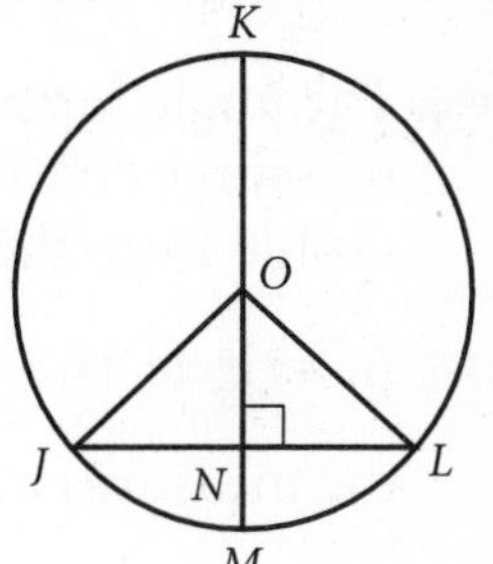

9. If the radius is 13 and *ON* = 5, find *JN*. ________
10. If *ON* = 8 and *JL* = 30, find the radius. ________
11. If the radius is 15 and *JL* = 24, find *NM*. ________
12. If the radius is 8 and *ON* = 6, find *JL* to the nearest tenth. ________
13. If *ON* = 6 and *JN* = 1, find the radius to the nearest tenth. ________

In the given circle, m$\widehat{UR}$ = 140°, m$\widehat{RS}$ = 100°, and m$\widehat{ST}$ = 30°.

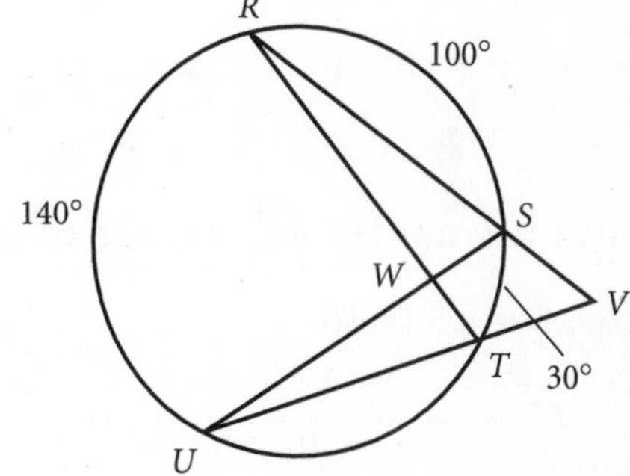

14. Find m∠*RSU*. ________ 15. Find m∠*RVU*. ________
16. Find m∠*USV*. ________ 17. Find m∠*RWS*. ________
18. If m$\widehat{RU}$ = 120° and m∠*RVU* = 35°, find m$\widehat{ST}$. ________

Find m$\widehat{AB}$ in each circle.

19.

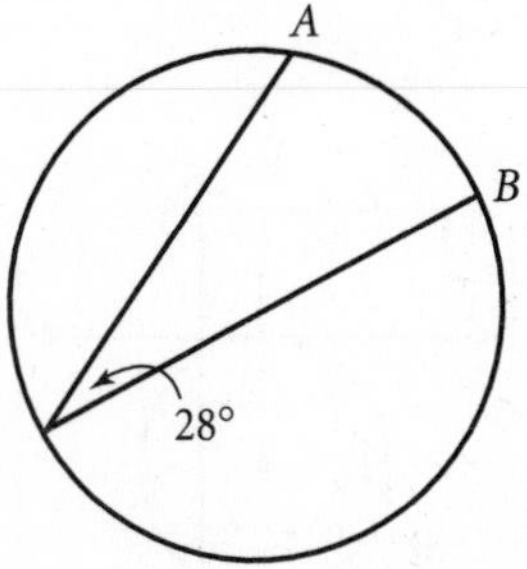

____

20.

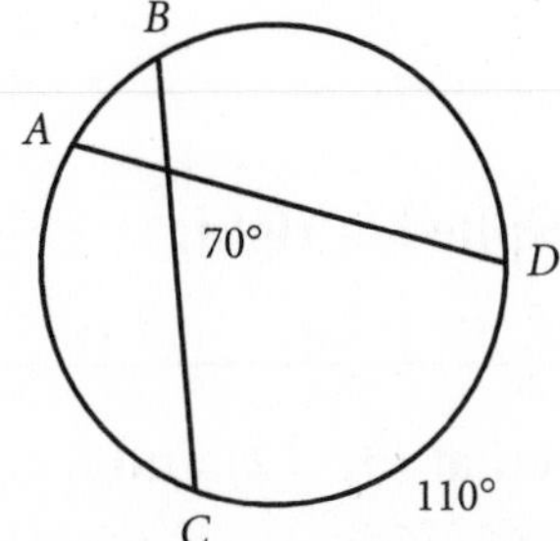

____

21.

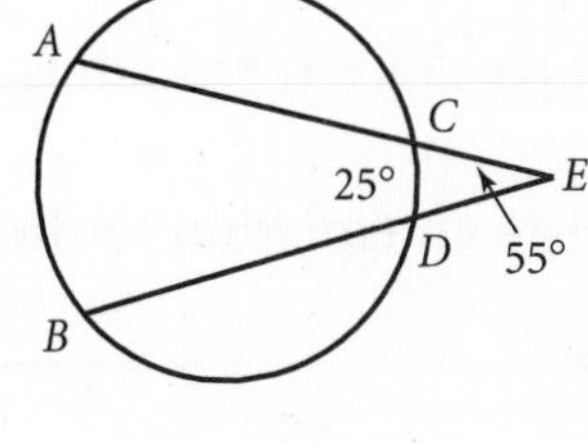

____

Chapter Assessment

Chapter 9, Form B, page 2

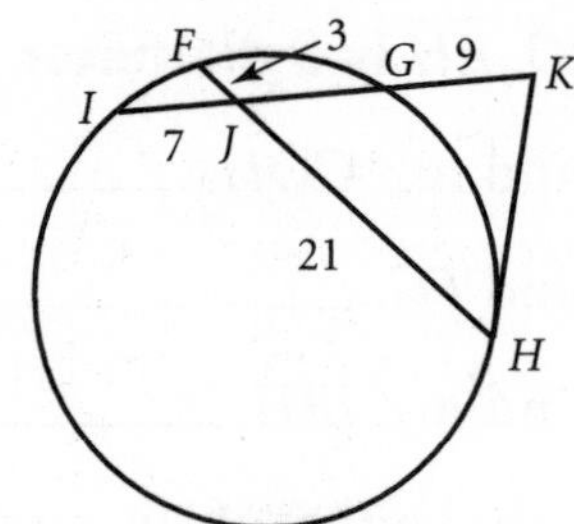

In the given circle, $\overline{KH}$ is tangent at point H.

22. $JG =$ ______________________

23. $HK =$ ______________________

24. A regular hexagon is inscribed in a circle. If one side of the hexagon is 10 cm, what is the circumference of the circle? ______________

25. The angle formed by two tangents from a communications satellite measures 20°. What is the measure of the arc on the Earth that is visible from the satellite? ______________

26. In a circle, two chords intersect at a 90° angle. What is the sum of the measures of two of the intercepted arcs? Explain why. ______________

____

Find the *x*- and *y*-intercepts for each equation.

27. $x^2 + y^2 = 16$ ______________ 28. $x^2 + y^2 = 20$ ______________

Find the center and radius of each circle.

29. $x^2 + y^2 = 36$ 30. $x^2 + (y - 2)^2 = 49$ 31. $(x - 1)^2 + (y + 3)^2 = 10$

______________ ______________ ______________

Write an equation of a circle with the given center and radius.

32. center at $(4, 0)$; radius $= 5$ ______________________

33. center at $(-1, 5)$; radius $= 2$ ______________________

34. Write an equation for the circle graphed at right.

____

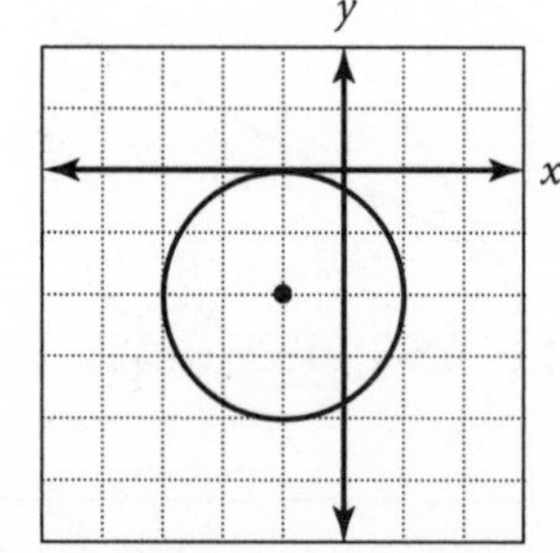

35. Find an equation of a circle centered at $(3, -2)$ and containing the point $(0, -2)$. ______________________

Alternative Assessment

Logos, Chapter 9, Form A

TASK: Analyze a logo.

HOW YOU WILL BE SCORED: As you work through the task, your teacher will be looking for the following:

- how well you can describe the figures and formation of the logo

Analyze the logo shown at right. You may want to label points so that you can express relationships geometrically.

1. Describe three different minor arcs in the logo. Describe one major arc. What are the degree measures of the arcs? What are the lengths of the arcs in terms of the circumference of the circle?

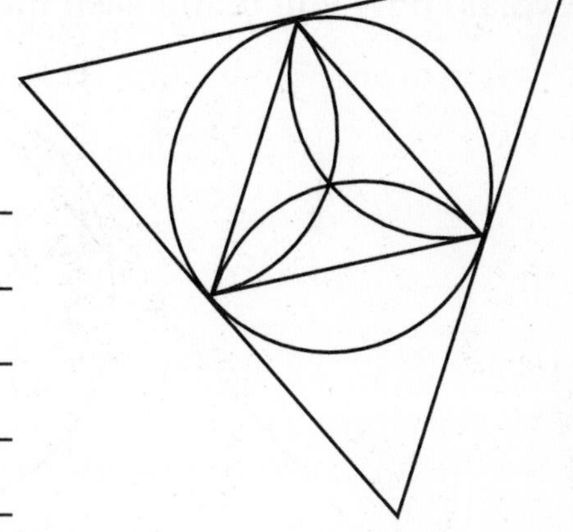

2. Describe the relationships between the segments and the circle by answering these questions. Which of the segments are chords? Which are secants? Which are tangents?

3. Assume that you are talking to a friend on the telephone. Explain how to draw the logo.

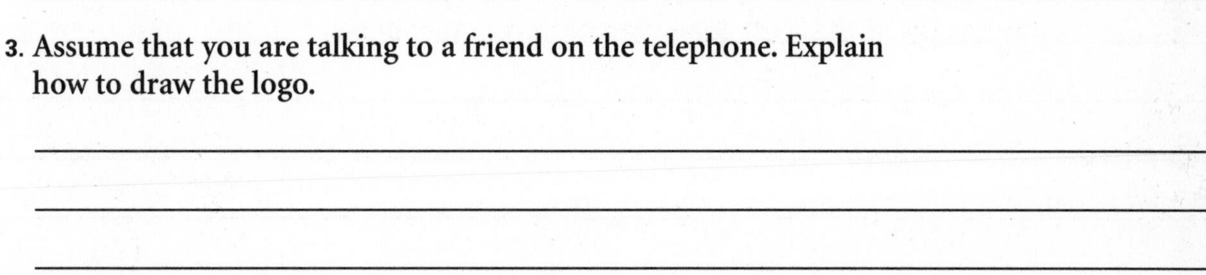

SELF-ASSESSMENT: Create your own logo. Describe how to draw the logo.

NAME ______________________ CLASS ______________ DATE ______________

Alternative Assessment

Angles and Arcs, Chapter 9, Form B

TASK: Find the measures of angles and arcs.

HOW YOU WILL BE SCORED: As you work through the task, your teacher will be looking for the following:

- whether you can find the measures of angles and arcs
- how well you can explain your reasoning

In the figure below, m∠*EDF* = 30° and m$\widehat{BD}$ = 80°. Find the measure of the remaining arcs and angles in the figure. Explain how you found each measure.

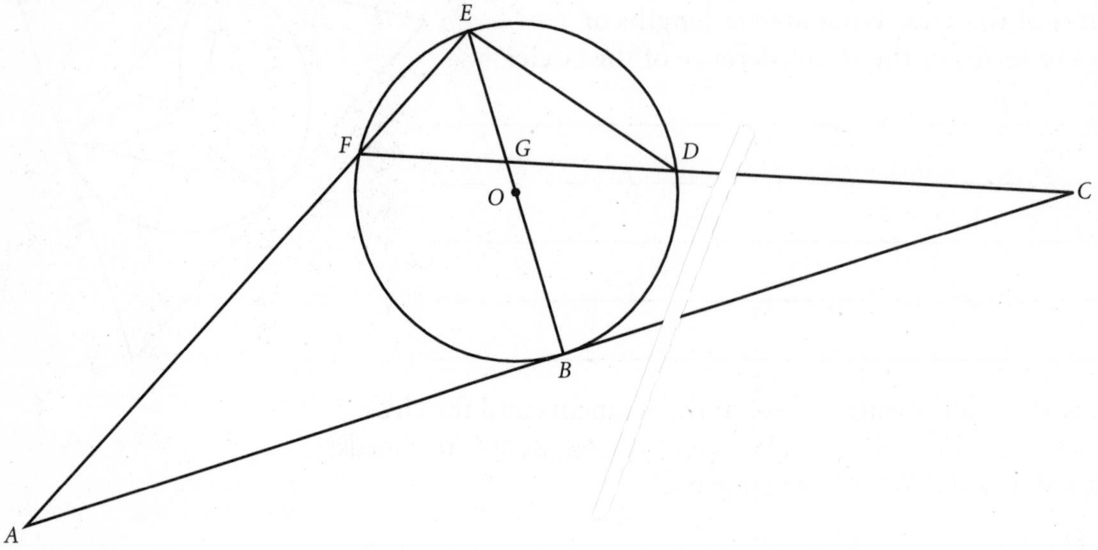

E F G D O C B A

____

____

____

____

____

____

____

____

SELF-ASSESSMENT: Draw another tangent to ⊙*O* that intersects tangent $\overline{AC}$. Explain how you can find the measure of the angle formed by the two tangents.

Quick Warm-Up: Assessing Prior Knowledge

10.1 Tangent Ratios

Construct any right triangle. Label the right angle *C*. Use *A* and *B* for the other two vertices. Name each part of the triangle indicated below.

1. the hypotenuse ________
2. the leg opposite $\angle A$ ________
3. the leg adjacent to $\angle A$ ________
4. the leg opposite $\angle B$ ________
5. the leg adjacent to $\angle B$ ________

Lesson Quiz

10.1 Tangent Ratios

Complete each statement below.

1. In right triangle *ABC*, the tangent of acute ∠A is the ratio of ____________________.
2. As the measure of an angle increases, the tangent of the angle ____________________.

Find the tangent for each angle. Round your answers to the nearest hundredth.

3. 45° ____________
4. 30° ____________
5. 60° ____________

Find the inverse tangent for each ratio. Round your answers to the nearest degree.

6. $\frac{9}{10}$ ____________
7. 1.4 ____________
8. 11.4301 ____________

Solve for *x*. Round your answers to the nearest hundredth.

9.

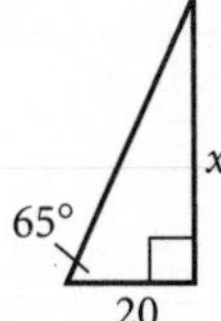

____

10.

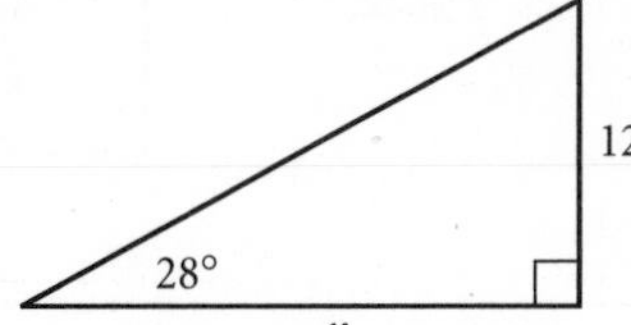

____

11.

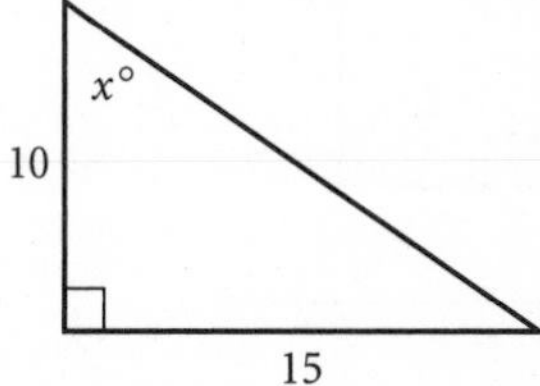

____

12. A wire from the top of a 10-foot pole makes an angle of 35° with the ground. Find the distance between the base of the pole and the place where the wire meets the ground. ____________

NAME ______________________ CLASS ____________ DATE ____________

Quick Warm-Up: Assessing Prior Knowledge
10.2 Sines and Cosines

If one acute angle in a right triangle increases in 5° increments, what happens to the other acute angle?

____

____

Lesson Quiz
10.2 Sines and Cosines

Find each of the following for the given triangle:

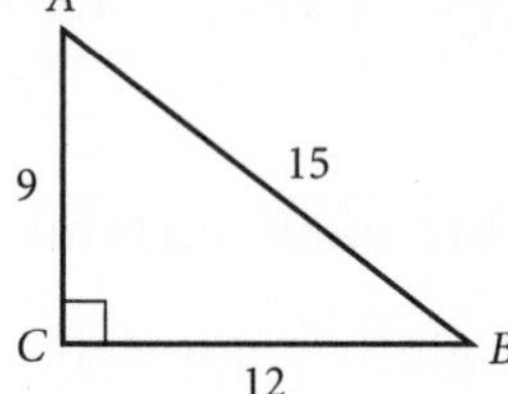

1. $\sin B$ ________ 2. $\cos A$ ________

3. $\cos B$ ________ 4. $m\angle A$ ________

Find each of the following. Round your answers to the nearest hundredth.

5. $\sin 75°$ ________ 6. $\cos 10°$ ________ 7. $\cos 87°$ ________

Find each of the following angles. Round your answers to the nearest degree.

8. $\sin^{-1} 0.44$ ________ 9. $\cos^{-1} 0.139$ ________ 10. $\sin^{-1} 0.848$ ________

Solve for *x*. Round your answers to the nearest hundredth.

11.

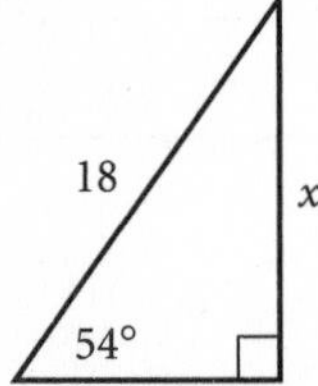

12.

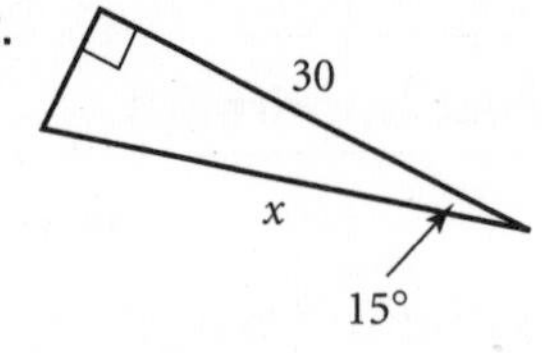

13.

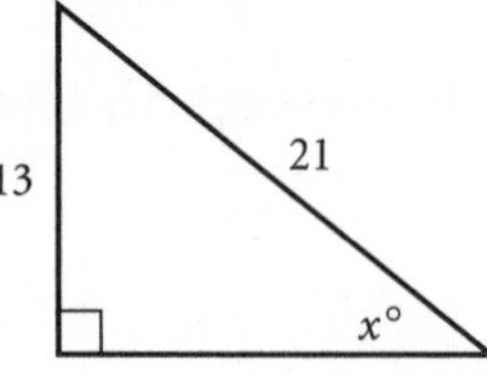

14.

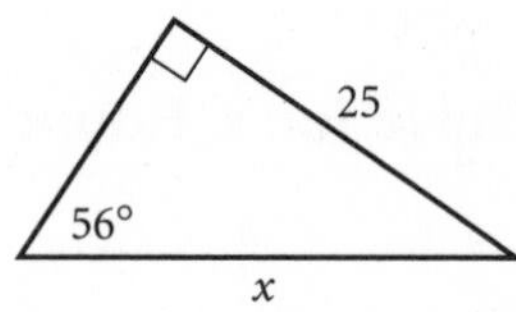

________ ________ ________ ________

15. When a ladder is placed against the side of a house, it should be placed at an angle of no more than 75° with the ground for safety. If you place a 9-foot ladder against the side of a house, how far away from the house can you safely place the bottom of the ladder? ________

NAME ______________________ CLASS ______________ DATE ______________

Quick Warm-Up: Assessing Prior Knowledge
10.3 Extending the Trigonometric Ratios

Use a protractor to draw an angle with each measure indicated below.

1. 90° 2. 180° 3. 270° 4. 360° 5. 0°

Lesson Quiz
10.3 Extending the Trigonometric Ratios

Find each of the following. Round your answers to the nearest thousandth.

1. sin 150° ______ 2. cos 150° ______ 3. sin 215° ______
4. cos 215° ______ 5. sin 235° ______ 6. cos 235° ______
7. sin 315° ______ 8. cos 315° ______ 9. sin 330° ______

Find the *x*- and *y*-coordinates of a point on the unit circle at the given angle. Round your answers to the nearest thousandth.

10. 45° ______ 11. 135° ______ 12. 240° ______
13. 270° ______ 14. 315° ______ 15. 330° ______

Give two values between 0° and 180° for an angle θ with the given value of sin θ. Round your answers to the nearest degree.

16. 0.259 ______ 17. 0.423 ______ 18. 0.5 ______
19. 0.866 ______ 20. 0.966 ______ 21. 0.985 ______
22. 0.819 ______ 23. 0.766 ______ 24. 0.707 ______

25. Let θ be an angle of rotation. If $P(x, y)$ is a point on the unit circle that is rotated $\theta°$ about the origin, find sin θ. ______

NAME ______________________ CLASS ______________ DATE ______________

Mid-Chapter Assessment

Chapter 10 (Lessons 10.1–10.3)

Write the letter that best answers the question or completes the statement. For Exercises 1–4, use the given right triangle.

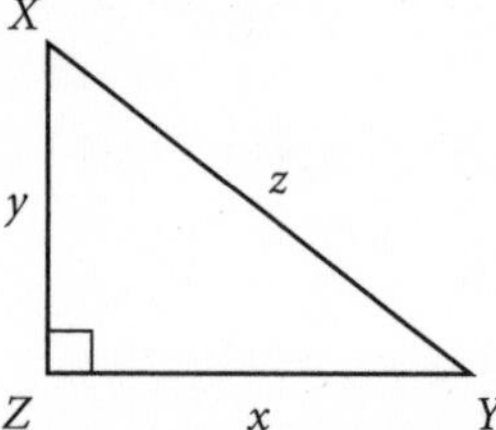

______ 1. What is tan Y?

a. $\frac{y}{x}$ b. $\frac{x}{y}$

c. $\frac{y}{z}$ d. $\frac{x}{z}$

______ 2. If $m\angle Y = 40°$ and $x = 30$, what is the value of y to the nearest unit?

a. 25 b. 36 c. 39 d. 47

______ 3. If $m\angle X = 70°$ and $x = 30$, what is the value of z to the nearest unit?

a. 11 b. 28 c. 32 d. 88

______ 4. If $x = 30$ and $z = 71$, then $m\angle Y =$ ______.

a. 23° b. 25° c. 65° d. 67°

Solve for *x*. Round your answers to the nearest tenth.

5. 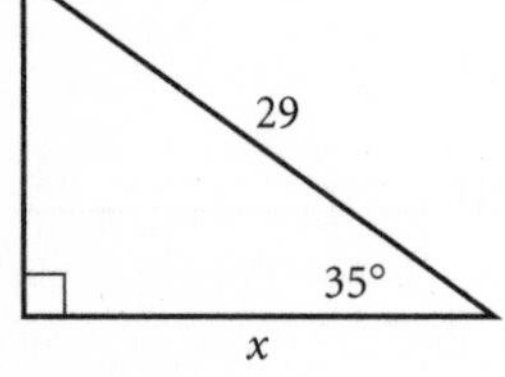

6. 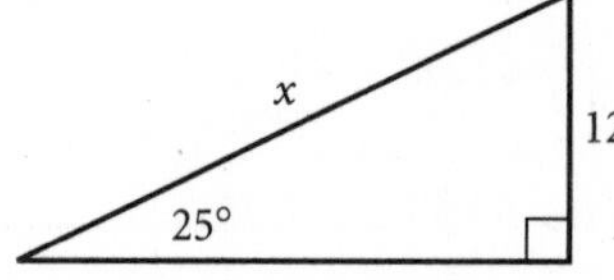

7. 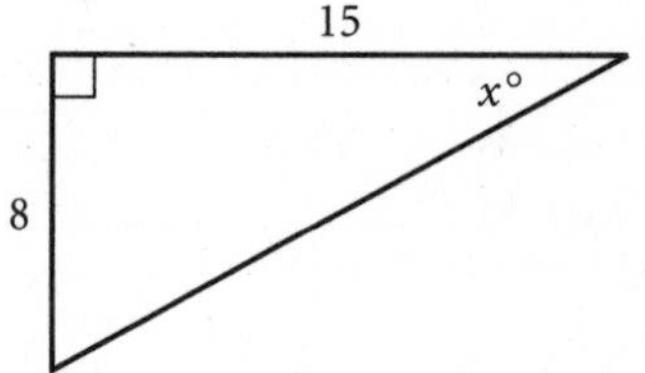

______ ______ ______

8. A skateboarding club wants to build a ramp. How much should the ramp rise over each 10 meters of run for an angle of 12°?

______

9. Find $\cos^{-1} 0.643$ to the nearest degree.

______

10. Find sin 240° to the nearest thousandth.

______

11. $P(0.5, 0.866)$ is a point on the unit circle formed by the angle of rotation θ. $\cos \theta =$

______

12. Find two values of θ between 0° and 180° for which $\cos \theta = 0.92$.

______

NAME ______________________ CLASS ____________ DATE ____________

Quick Warm-Up: Assessing Prior Knowledge

10.4 The Law of Sines

1. Draw any non-right triangle. Label the vertices *A*, *B*, and *C*. Use *a*, *b*, and *c* to label the sides, with side *a* opposite $\angle A$, side *b* opposite $\angle B$, and side *c* opposite $\angle C$.

2. List all of the different ways to name two angles and one side of the triangle drawn.

Lesson Quiz

10.4 The Law of Sines

Complete the following statement:

To use the law of sines in a triangle, you need to know the measure of either

1. ______________________ and ______________________ or

2. ______________________ and ______________________.

Find the measure of each indicated part. Round your answers to the nearest tenth. In Exercise 4, assume that $\angle C$ is acute.

3. Find *a*.

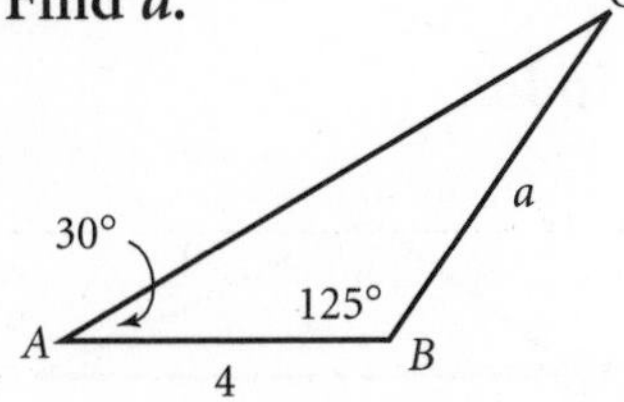

____

4. Find $m\angle C$.

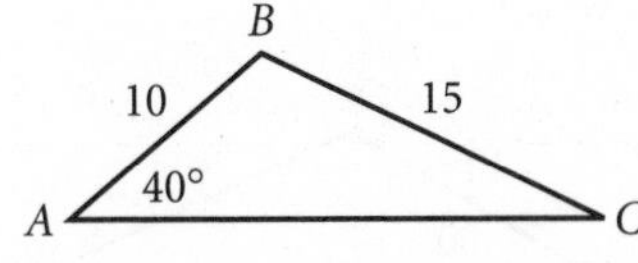

____

5. Find *b*.

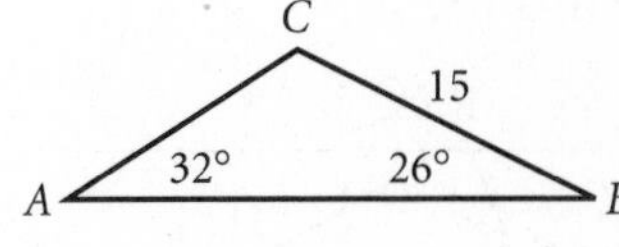

____

6. Find the third side length of all possible triangles with $a = 8$, $c = 10$, and $m\angle A = 48°$. Round your answers to the nearest tenth. ______________

7. Two ranger stations located 10 km apart receive a distress call from a camper. Electronic equipment indicates that the camper's position is 71° from the first station and 100° from the second station. Both angles have one side along the line segment connecting the stations. Which one is closer to the camper? How close is it? ______________

____

NAME ______________________ CLASS ______________ DATE ______________

Quick Warm-Up: Assessing Prior Knowledge

10.5 The Law of Cosines

1. Draw any non-right triangle. Label the vertices A, B, and C. Use a, b, and c to label the sides, with side a opposite $\angle A$, side b opposite $\angle B$, and side c opposite $\angle C$.

2. List all the possible ways to name two sides and the included angle of this triangle. ______________

Lesson Quiz

10.5 The Law of Cosines

Complete the following statement:

To use the law of cosines in a triangle, you need to know the measure of either

1. ______________________________ and ______________________________

or

2. ____.

Find the measure of each indicated part. Assume that all angles are acute. Round your answers to the nearest tenth.

3. Find $m\angle B$.

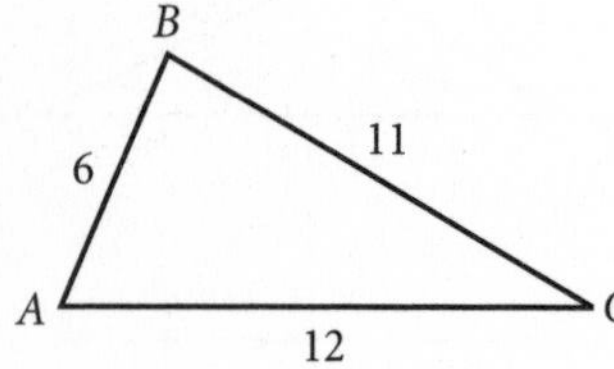

____

4. Find a.

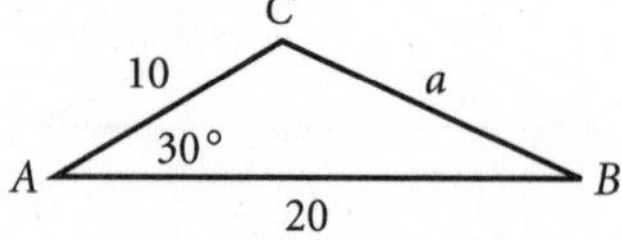

____

5. Find b.

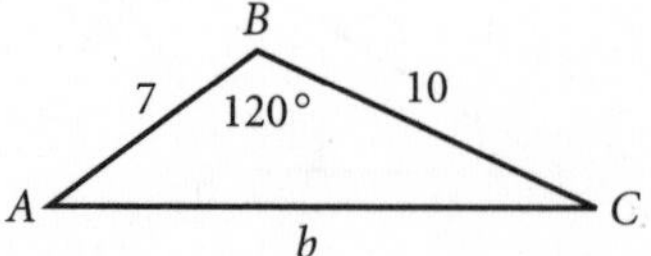

____

6. Find $m\angle A$ if $a = 6$, $b = 9$, and $c = 10$. ______________

7. Two markers, A and B, lie at opposite ends of a pond. From a stake on the shore at point P, the distance to marker A is 257 feet and the distance to marker B is 321 feet. A surveyor uses her transit to find that $m\angle APB$ is 97.3°. To the nearest foot, what is the distance between the markers? ______________

NAME ______________________ CLASS ____________ DATE ____________

Quick Warm-Up: Assessing Prior Knowledge

10.6 Vectors in Geometry

1. Draw an *xy* coordinate system on graph paper. Choose any number *a* and draw a segment from (0, 0) to (*a*, *a*).
2. What is the length of the segment you just drew? ______________________

Lesson Quiz

10.6 Vectors in Geometry

Complete each statement.

1. When two vectors are combined, the resultant vector is called the ______________________.
2. The vector sum is the ______________________ of a parallelogram formed by the two vectors.
3. Vectors with the same direction are ______________________.
4. Vectors at right angles are ______________________.

Find the sum of each pair of vectors.

5.

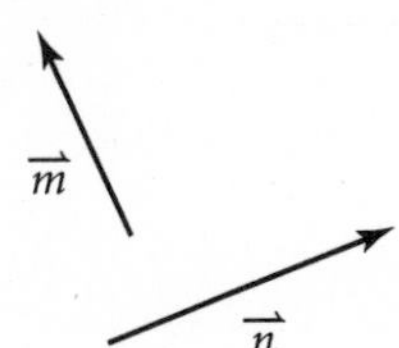

6.

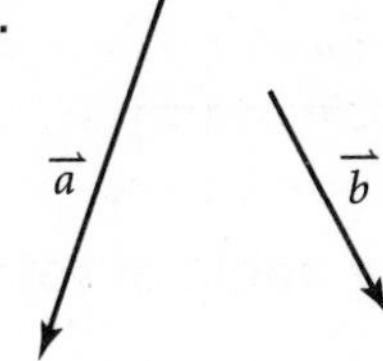

7. Use the parallelogram method to draw vector $\vec{c}$, the resultant of $\vec{a}$ and $\vec{b}$.
8. Find the angles of the parallelogram. ______________________
9. Find the magnitude of $\vec{c}$, the resultant vector, to the nearest unit. ______________________
10. To the nearest degree, find the angle that the resultant vector $\vec{c}$ makes with $\vec{b}$.

____

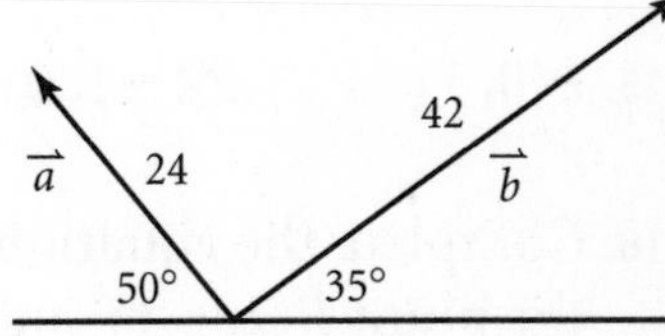

NAME ______________________ CLASS ____________ DATE ____________

Quick Warm-Up: Assessing Prior Knowledge
10.7 Rotations in the Coordinate Plane

1. Draw an xy coordinate system. Draw $\triangle DEF$ in the first quadrant with vertices at intersection points of the grid lines.

2. Rotate $\triangle DEF$ 180° counterclockwise about the origin.

Lesson Quiz
10.7 Rotations in the Coordinate Plane

For point P and the given angle of rotation, find the image point P' of the rotation. Round your answers to the nearest tenth.

1. $P(2, -1)$; 90° ____________
2. $P(3, 2)$; 270° ____________
3. $P(0, -3)$; 180° ____________
4. $P(-1, -1)$; 360° ____________
5. $P(2, 0)$; 450° ____________
6. $P(-2, 3)$; 630° ____________
7. $P(10, 1)$; 45° ____________
8. $P(-1, 10)$; 150° ____________
9. $P(10, -10)$; 240° ____________
10. $P(5, -5)$; 15° ____________

Given P and P', determine the angle of rotation (90°, 180°, 270°, or 360°).

	Point	Image	Angle of rotation
11.	$P(3, 1)$	$P'(-1, 3)$	____________
12.	$P(2, -5)$	$P'(-5, -2)$	____________
13.	$P(-4, 3)$	$P'(4, -3)$	____________
14.	$P(-1, -2)$	$P'(-1, -2)$	____________
15.	$P(0, 1)$	$P'(-1, 0)$	____________

16. Complete the equation below to show how to use matrices to rotate the point $P(2, -1)$ counterclockwise 45° about the origin.

$$\begin{bmatrix} \cos 45° & \_\_\_\_ \\ \_\_\_\_ & \_\_\_\_ \end{bmatrix} \times \begin{bmatrix} \_\_\_\_ \\ \_\_\_\_ \end{bmatrix} = \begin{bmatrix} \_\_\_\_ & \_\_\_\_ \\ \_\_\_\_ & \_\_\_\_ \end{bmatrix}$$

NAME ______________________________ CLASS ______________ DATE ______________

Chapter Assessment

Chapter 10, Form A, page 1

Write the letter that best answers the question or completes the statement. For Exercises 1–3, refer to $\triangle ACB$.

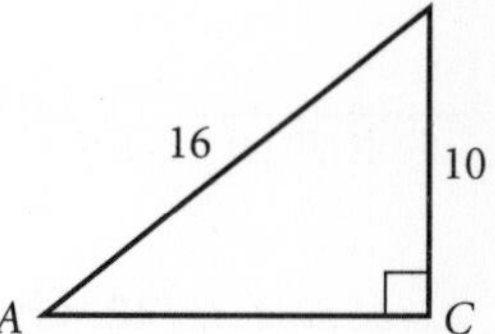

_______ 1. Which equation can be used to find $m\angle A$?

a. $\sin A = \frac{10}{16}$ b. $\cos A = \frac{10}{16}$

c. $\tan A = \frac{10}{16}$ d. $\cos A = \frac{10}{16}$

_______ 2. What is $m\angle B$ to the nearest degree?

a. 32° b. 39° c. 51° d. 90°

_______ 3. What is AC to the nearest unit?

a. 10 b. 12 c. 16 d. 20

_______ 4. To the nearest degree, the inverse tangent of 0.667 is _______.

a. 30° b. 33° c. 34° d. 60°

_______ 5. To the nearest thousandth, sin 240° is _______.

a. −0.500 b. 0.500 c. 0.866 d. −0.866

_______ 6. A kite string is 60 feet long and makes an angle of 42° with the ground. To the nearest foot, how high above the ground is the kite?

a. 40 feet b. 44 feet

c. 54 feet d. 90 feet

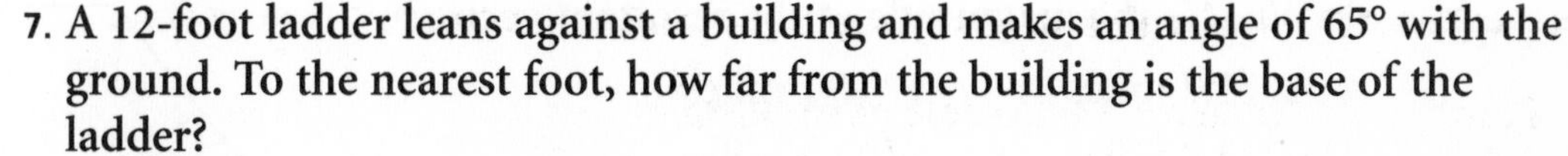

_______ 7. A 12-foot ladder leans against a building and makes an angle of 65° with the ground. To the nearest foot, how far from the building is the base of the ladder?

a. 3 feet b. 4 feet

c. 5 feet d. 6 feet

_______ 8. A 6-foot person walks 75 feet from a tree. The angle formed by the person's line of sight and the horizontal is 25°. About how tall is the tree?

a. 34 feet b. 35 feet

c. 41 feet d. 50 feet

For Exercises 9–13, find the indicated measure.

_______ 9. Given $m\angle A = 110°$, $m\angle B = 45°$, and $a = 10$, what is the value of b to the nearest unit?

a. 7 b. 8 c. 12 d. 14

Chapter Assessment

Chapter 10, Form A, page 2

______ **10.** Given $m\angle A = 42°$, $a = 22$, and $b = 12$, what is $m\angle B$, to the nearest degree?

a. 0.4° **b.** 12° **c.** 21° **d.** 66°

______ **11.** Given $m\angle A = 110°$, $a = 125$, and $b = 80$, what is $m\angle C$ to the nearest degree?

a. 33° **b.** 37° **c.** 70° **d.** 102°

______ **12.** Given $m\angle A = 120°$, $b = 3$, and $c = 10$, what is the value of a to the nearest unit?

a. 8 **b.** 9 **c.** 10 **d.** 12

______ **13.** Given $a = 5$, $b = 7$, and $c = 10$, what is $m\angle A$ to the nearest degree?

a. 28° **b.** 62° **c.** 81° **d.** 85°

______ **14.** A ranger in an observation tower sights a bear 15 miles due north and campers 19 miles to the southeast. If the angle between the two lines of sight is 104°, how far is the bear from the campers, to the nearest mile?

a. 6 miles **b.** 21 miles

c. 27 miles **d.** 33 miles

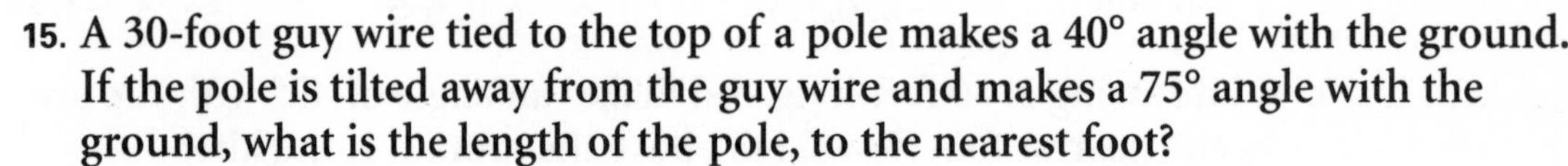

______ **15.** A 30-foot guy wire tied to the top of a pole makes a 40° angle with the ground. If the pole is tilted away from the guy wire and makes a 75° angle with the ground, what is the length of the pole, to the nearest foot?

a. 19 feet **b.** 20 feet

c. 45 feet **d.** 89 feet

______ **16.** What is the magnitude of the resultant vector?

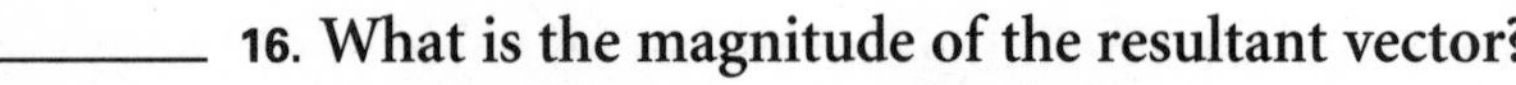

a. 11.7 miles **b.** 16.1 miles

c. 24.5 miles **d.** 93.2 miles

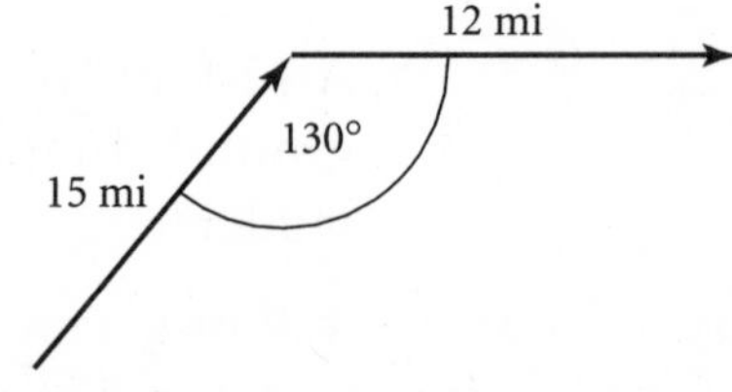

______ **17.** To the nearest degree, the angle that the resultant vector makes with the horizontal is ______.

a. 2° **b.** 28° **c.** 72° **d.** 108°

______ **18.** If point $P(2, 0)$ is rotated by 45° (counterclockwise) about the origin, its image is ______.

a. (1.414, 1.414) **b.** (−1.414, −1.414)

c. (1.414, −1.414) **d.** (−1.414, 1.414)

______ **19.** The coordinates of $P(-2, 1)$ are rotated to $P'(-1, -2)$. What is the angle of rotation?

a. 360° **b.** 270° **c.** 180° **d.** 90°

Chapter Assessment

Chapter 10, Form B, page 1

In right △*ACB*, *AC* = 15, *BC* = 8, and *AB* = 17.

1. Find sin *A*. ______________ 2. Find cos *B*. ______________

3. Find tan *B*. ______________ 4. Find m∠*A* ______________

5. To the nearest degree, $\sin^{-1} 0.906$ is ______________.

6. To the nearest thousandth, sin 190° is ______________.

Solve for *x*. Round your answers to the nearest unit or degree.

7.

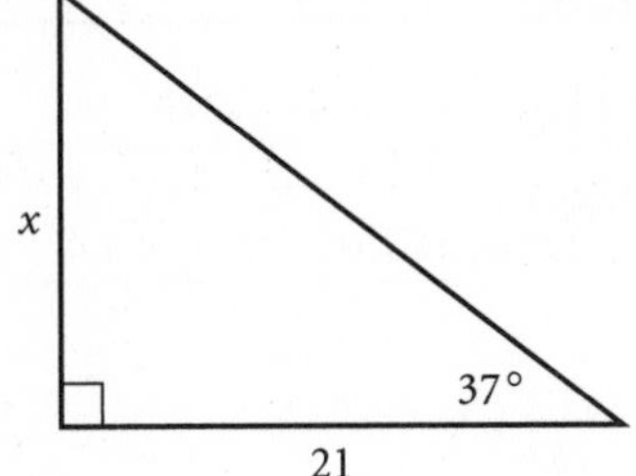

____

8.

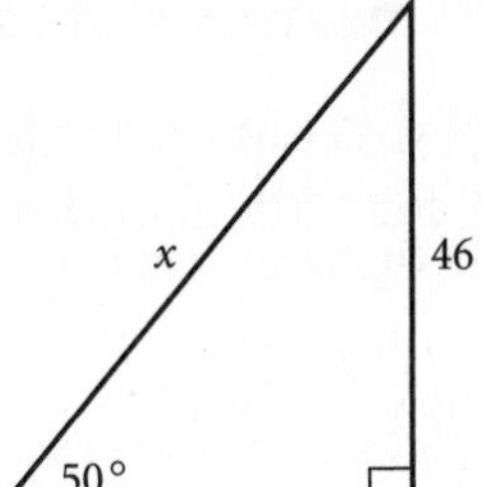

____

9.

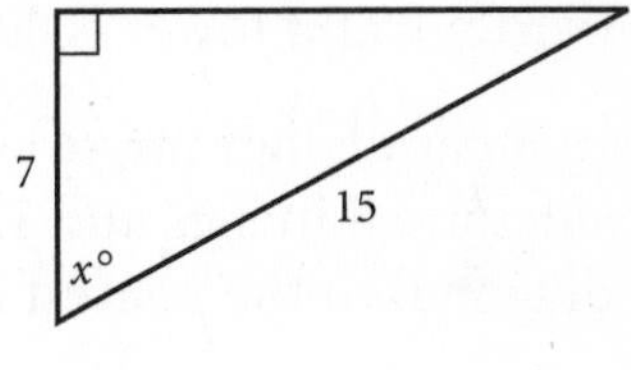

____

10. Find the area of the triangle to the nearest unit.

____

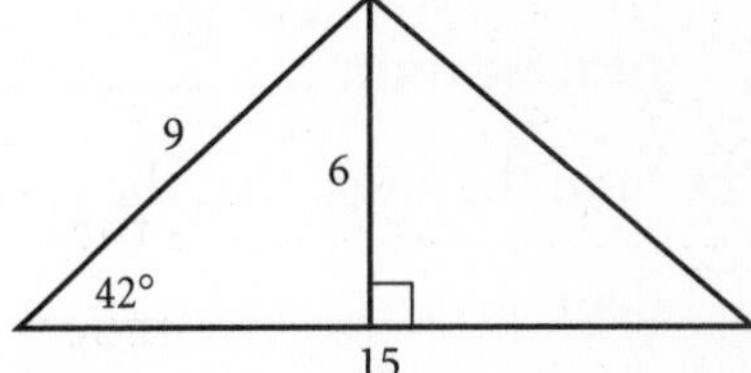

11. A tree casts a 20-foot shadow when the sun makes an angle of 42° with the ground. Find the height of the tree to the nearest foot. ______________

12. An airplane takes off at an angle of 15° with the ground. To the nearest 10 feet, how far has the airplane flown when it

has covered a horizontal distance of 1800 feet? ______________

13. A pilot is in a plane 6 miles from a point directly over the airport. The pilot observes the airport at a 10° angle to the horizontal.

Find the altitude of the plane to the nearest 100 feet? ______________

NAME ______________________ CLASS __________ DATE __________

Chapter Assessment

Chapter 10, Form B, page 2

For Exercises 14–17, find the indicated measure.

14. Given m$\angle A = 35°$, m$\angle B = 120°$, and $c = 12$, $b =$ ______________________.

15. Given $a = 8$, $b = 12$, and $c = 7$, m$\angle A =$ ______________________.

16. Given m$\angle A = 115°$, $b = 21$, and $c = 12$, $a =$ ______________________.

17. Given $a = 12$, m$\angle B = 63°$, and $b = 14$, m$\angle C =$ ______________________.

18. A tree and a house are 50 feet apart on one side of a river. Across the river, a tower is located so that the angle formed by the tower, tree (vertex), and house is 70° and the angle formed by the tree, house (vertex), and tower is 80°. How wide is the river, to the nearest foot? ______________

19. In a parallelogram, two adjacent sides meet at an angle of 35°. The sides are 8 inches and 12 inches long. Find the length of the longer diagonal to the nearest inch. ______________

20. In the figure, draw the resultant vector $\vec{c}$ at right.

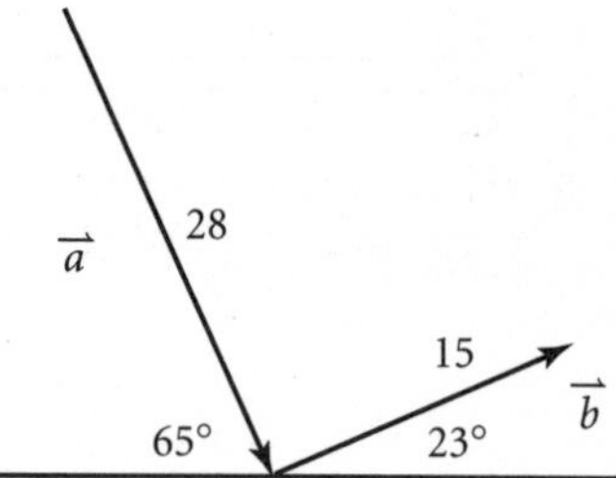

21. Find the measure of the angle between the vectors.

____

22. Find the magnitude of the resultant vector to the nearest unit. ______________________

23. Find the angle that the resultant vector $\vec{c}$ makes with $\vec{a}$ to the nearest degree. ______________________

24. Point $P(-1, 5)$ is rotated 180°. Find its image point.

____

25. Point $P(4, -2)$ is rotated 270°. Find its image point.

____

26. Point $P(2, 3)$ is rotated 90°. Find its image point.

____

27. The image of point $P(3, -1)$ is $P'(-1, -3)$. Find the angle of rotation.

____

28. The image of point $P(4, 5)$ is $P'(-5, 4)$. Find the angle of rotation.

____

29. The image of point $P(-6, 2)$ is $P'(6, -2)$. Find the angle of rotation.

____

NAME ______________________ CLASS ____________ DATE ____________

Alternative Assessment

Estimating Distances, Chapter 10, Form A

TASK: Estimate distance by using trigonometry.

HOW YOU WILL BE SCORED: As you work through the task, your teacher will be looking for the following:

- whether you can estimate distance by using trigonometry
- how well you describe the process you use to estimate distance

Suppose that you are sailing past a lighthouse in a boat on the ocean. You have a book that gives the height of the lighthouse and the height above sea level of the cliff on which it is built.

1. How could you estimate your distance from shore?

2. How could you determine the angle between the top of the lighthouse and sea level from your viewpoint?

3. Suppose, from your point of view, that you estimate the angle from the top of the lighthouse to sea level to be about 30°. If the lighthouse is 120 feet tall and the cliff rises 30 feet above sea level, how far are you from shore?

4. If you were 800 feet from shore, what would be the angle between the top of the lighthouse and the base of the cliff from your point of view?

SELF-ASSESSMENT: How might a person in the lighthouse estimate the distance from the shore to a boat?

NAME ______________________ CLASS ____________ DATE ____________

Alternative Assessment

Row Your Boat, Chapter 10, Form B

TASK: Determine a route across a river.

HOW YOU WILL BE SCORED: As you work through the task, your teacher will be looking for the following:

- how well you assess the problem situation
- whether you can establish a direction for rowing across the river

Suppose that you want to row across a river and land directly across from your starting point.

1. Draw vectors to illustrate this situation.

2. What information wo'.d help you determine the angle at which you will row so that you ' .nd directly across from your starting point?

____

____

3. Assume that you have made the necessary measurements or estimates or that you know the essential facts. Assign values and determine the direction you should row. Show that your choice will result in your landing directly across from your starting point.

SELF-ASSESSMENT: Draw a vector diagram to illustrate the effect of an upstream wind on the situation described above.

NAME ______________________ CLASS __________ DATE __________

Quick Warm-Up: Assessing Prior Knowledge

11.1 Golden Connections

1. Solve the proportion $\frac{6}{x} = \frac{12}{8}$. ______________________

2. Use the quadratic formula to solve the equation $x^2 - 2x - 4 = 0$. ______________________

3. Use a calculator to evaluate $\frac{1 + \sqrt{5}}{2}$. ______________________

Lesson Quiz

11.1 Golden Connections

1. What is a golden rectangle? ______________________

____

____

2. What is the golden ratio in the golden rectangle shown at right?

____

a

b *b*

b *a* – *b*

3. Let $a = 1$. Find the value of the golden ratio.

____

Use the golden ratio to find the value of *x* for each of the golden rectangles shown below. Round your answers to the nearest tenth.

4.

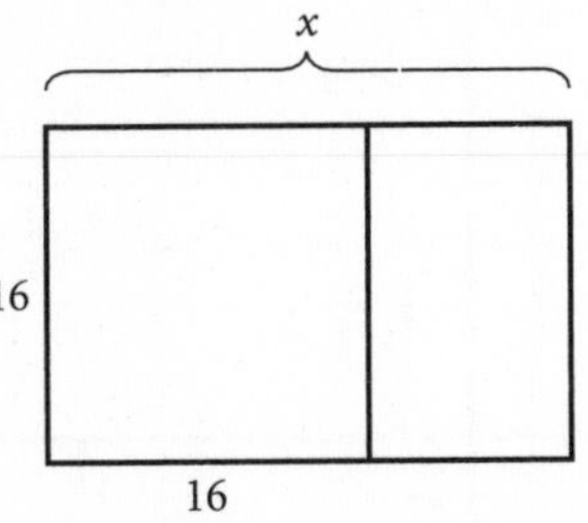

____

5.

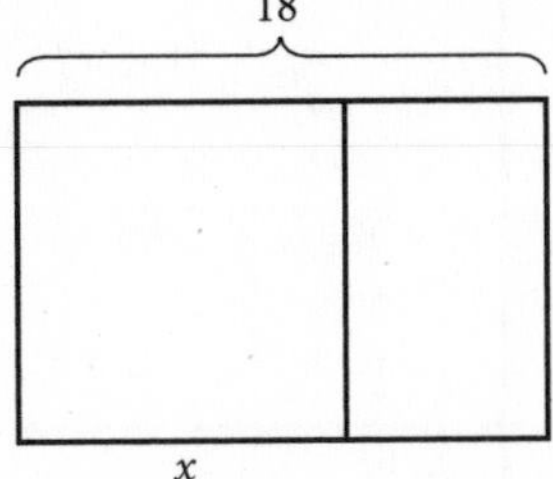

____

6.

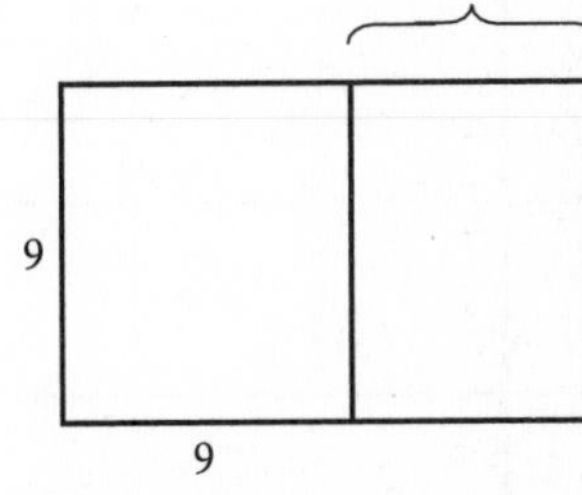

____

NAME ______________________ CLASS __________ DATE __________

Quick Warm-Up: Assessing Prior Knowledge

11.2 Taxicab Geometry

1. Points $(-4, 4)$, $(1, 4)$, $(1, -1)$, and $(-4, -1)$ are vertices of a quadrilateral. Name the type of quadrilateral. ______________

2. Find the shortest distance between points $(1, 1)$ and $(5, 7)$. ______________

3. Find the distance between points $(1, 1)$ and $(5, 7)$, traveling along only the grid lines of graph paper. ______________

Lesson Quiz

11.2 Taxicab Geometry

1. Explain what is meant by the taxicab distance between two points. ______________

2. What is a circle with a radius of r in taxicab geometry? ______________

3. What is the minimum number of pathways between two points that are 4 taxicab units apart? ______________

4. Find the circumference of a taxicab circle with a radius of 2. ______________

Determine the taxidistance between the two points on each grid below.

5. 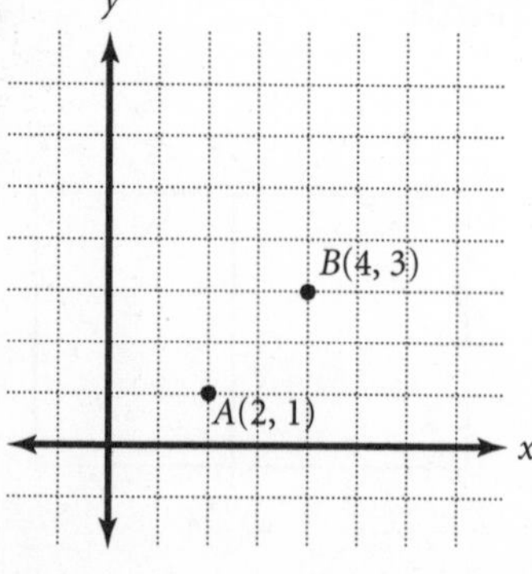

____

6. 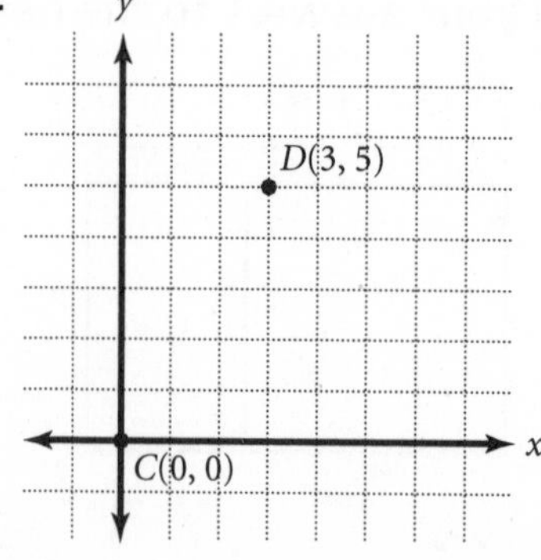

____

7. 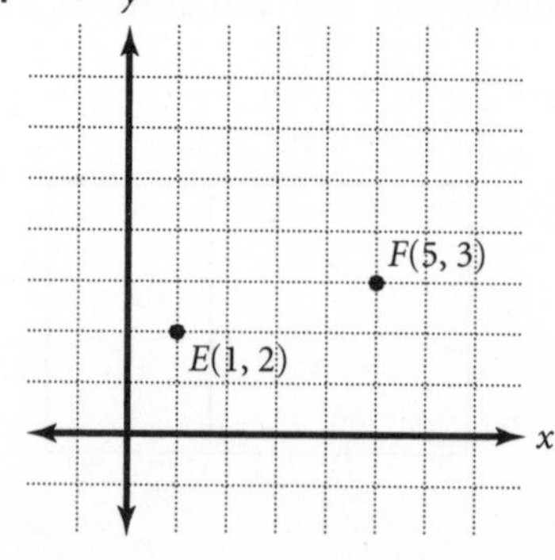

____

NAME ______________________ CLASS __________ DATE __________

Quick Warm-Up: Assessing Prior Knowledge

11.3 Graph Theory

1. How is the term *network* commonly used? ______________________

____

2. How is the number of vertices of a polygon related the number of sides? ______________________

____

3. Define the term *traverse*. ______________________

Lesson Quiz

11.3 Graph Theory

1. What is a graph? ______________________

____

2. What is an Euler path? ______________________

____

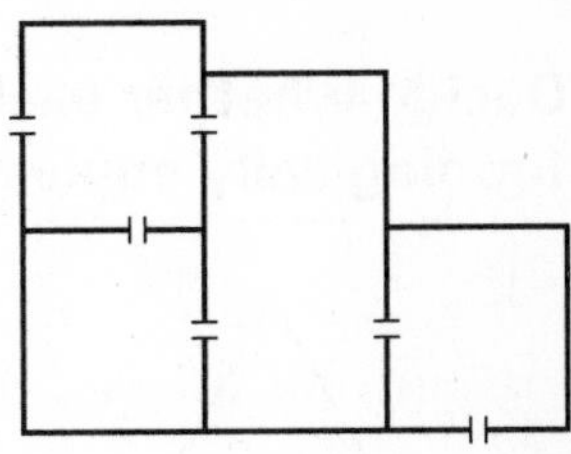

3. Complete the following: A graph contains an Euler path if and only if there are at most __________ odd vertices.

4. Draw a graph that represents all possible routes through the rooms shown at right.

5. Does the graph contain an Euler circuit? __________

Determine whether each graph contains an Euler path.

6. 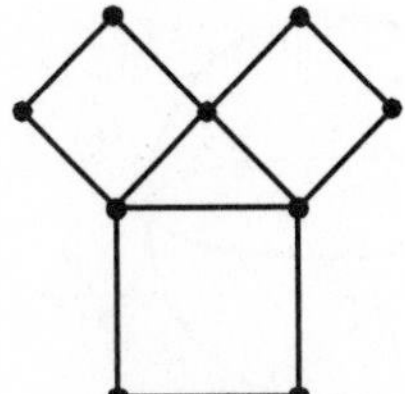

____

7. 

____

8. 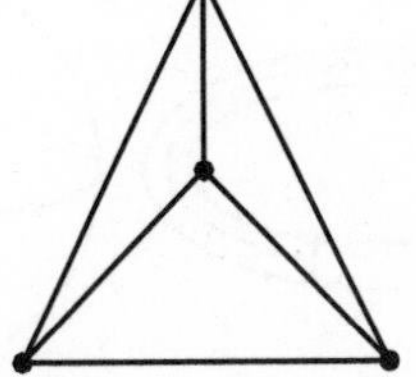

____

NAME ______________________ CLASS ______________ DATE ______________

Quick Warm-Up: Assessing Prior Knowledge

11.4 Topology: Twisted Geometry

1. Imagine a line drawn from the North Pole to a point on the equator. Is the line straight? Why or why not? Is it the shortest distance? Explain.

____

2. Imagine a balloon with a message printed on it. Draw a picture to show how the message would look if the balloon were deflated.

Lesson Quiz

11.4 Topology: Twisted Geometry

1. In topology, when are two shapes considered equivalent? ______________

____

2. How can you tell if a curve in a plane is a simple closed curve? ______________

____

3. What is Euler's formula for polyhedrons with V vertices, E edges, and F faces?

____

Decide whether each of the following pairs of figures is topologically equivalent. Justify your reasoning.

4.

____

5.

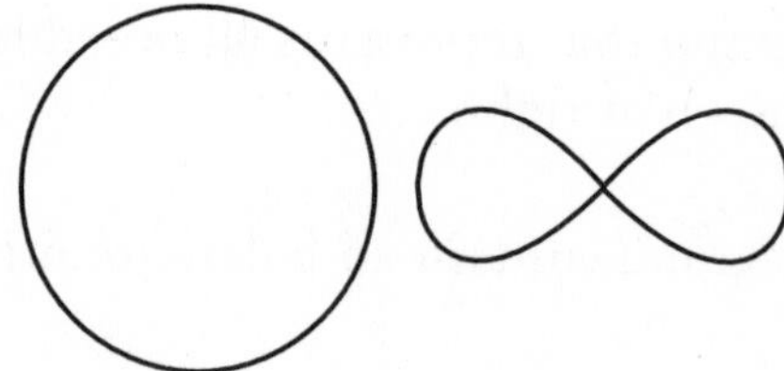

____

6.

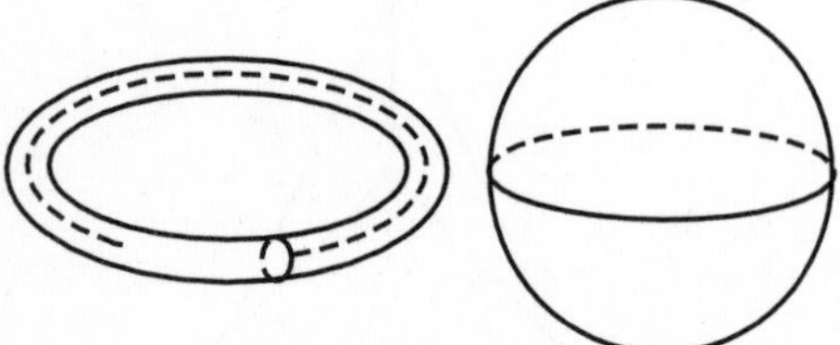

____

7.

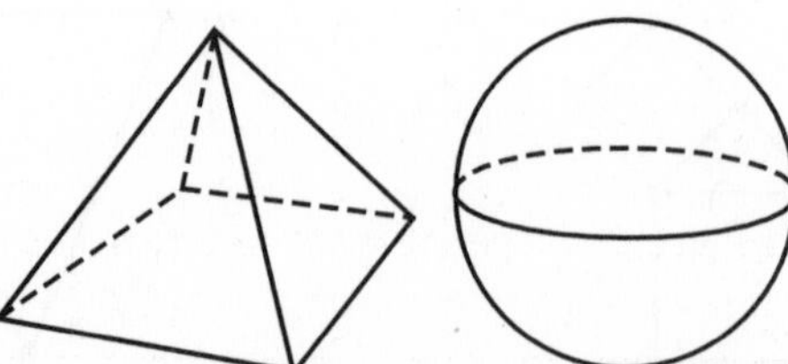

____

NAME ____________________ CLASS ____________ DATE ____________

Mid-Chapter Assessment

Chapter 11 (Lessons 11.1–11.4)

Write the letter that best answers the question or completes the statement.

______ 1. Approximately what value of x would make the given figure a golden rectangle?

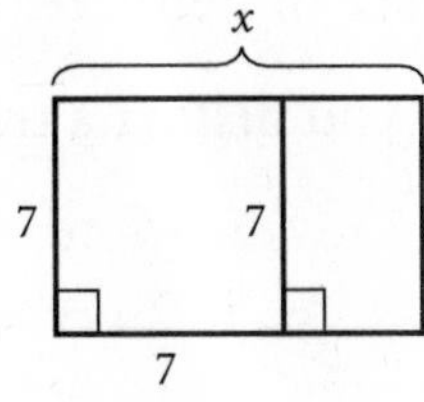

a. 4 b. 7
c. 11 d. 14

______ 2. What is the minimum number of pathways between two points if the taxidistance between the points is 4 units?

a. 1 b. 3 c. 4 d. 11

______ 3. How many points are located on a taxicab circle with a radius of 2?

a. 2 b. 4 c. 6 d. 8

______ 4. A polyhedron has 4 vertices and 4 faces. How many edges does it have?

a. 2 b. 4 c. 6 d. 8

______ 5. Which of these shapes is *not* topologically equivalent to the others?

a.  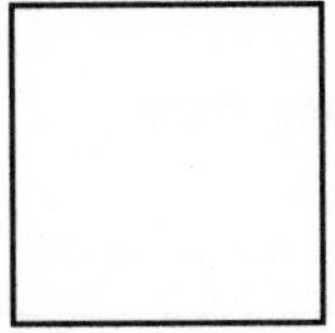 b. 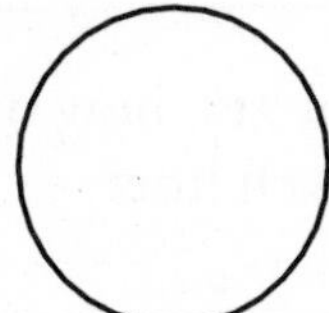 c. 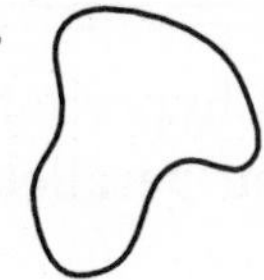 d. 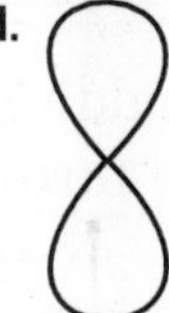

6. Find the taxidistance between points A and B.

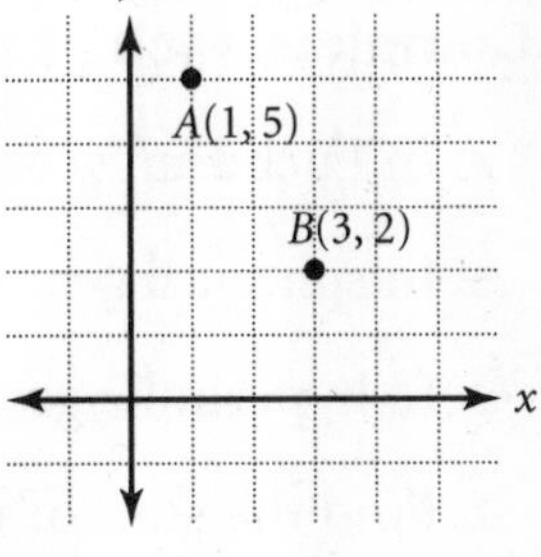

____

7. Does the figure at right contain an Euler path? Explain.

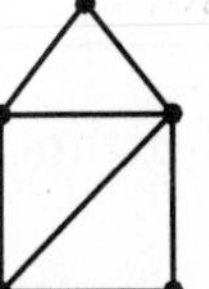

____

8. Are the figures at right topologically equivalent? Explain.

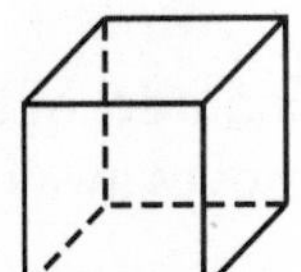

____

Quick Warm-Up: Assessing Prior Knowledge
11.5 *Euclid Unparalleled*

1. Consider the measure of the angles of a triangle drawn on the surface of a sphere. Is the measure greater than or less than 180°? ______________.

2. Can you distort a triangle so that the sum of its angles is less than 180°?

____

Lesson Quiz
11.5 *Euclid Unparalleled*

1. What are logically equivalent statements? ______________________________

____

2. What is a line in spherical geometry? ______________________________

____

3. In the "saddle" model of hyperbolic geometry, how many lines may be drawn through a point parallel to a given line?

____

Complete each of the following.

4. In Euclidean geometry, the sum of the measures of the angles of a triangle is ______________.

5. In spherical geometry, the sum of the measures of the angles of a triangle is ______________.

6. In hyperbolic geometry, the sum of the measures of the angles of a triangle is ______________.

7. Find the sum of the angles of the spherical triangle shown at right.

____

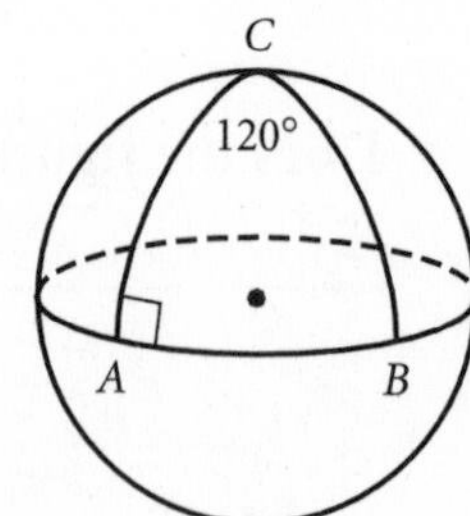

8. Name a pair of orthogonal lines on the sphere. ______________

9. In Poincaré's model of hyperbolic geometry, what happens to an object as it moves away from the center of the universe?

____

NAME ______________________ CLASS ____________ DATE ____________

Quick Warm-Up: Assessing Prior Knowledge
11.6 Fractal Geometry

Describe the mathematical qualities of plants such as ferns and broccoli. ______________________

____

Lesson Quiz
11.6 Fractal Geometry

1. What is a fractal? ______________________

____

2. Explain the meaning of *iteration*. ______________________

____

3. Draw three iterations of Cantor dust.

4. Write the first five rows of Pascal's triangle.

5. Draw the next iteration of the following fractal:

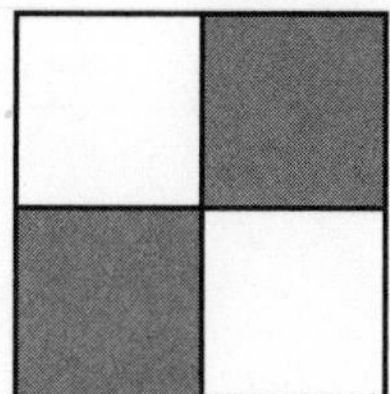

6. How many line segments are formed in the first iteration of the Koch snowflake? __________

NAME ______________________ CLASS ____________ DATE ____________

Quick Warm-Up: Assessing Prior Knowledge
11.7 Other Transformations: Projective Geometry

1. List the transformations that preserve the size and shape of an object. ______________________

____

2. Which is preserved by dilations: size or shape? ______________________

3. List the transformations that preserve the property that parallel lines are transformed into parallel lines. ______________________

Lesson Quiz
11.7 Other Transformations: Projective Geometry

Write *true* or *false* for each statement.

1. In an affine transformation, parallel lines transform to parallel lines. ____________
2. In an affine transformation, distance properties are preserved. ____________
3. In projective geometry, only an unmarked straightedge can be use to draw a figure. ____________
4. If one triangle is a projection of another triangle, then the intersections of the lines containing the corresponding sides of the two triangles are collinear. ____________

Sketch the image for each affine transformation on the grid provided.

5. $T(x, y) = (2x, -y)$

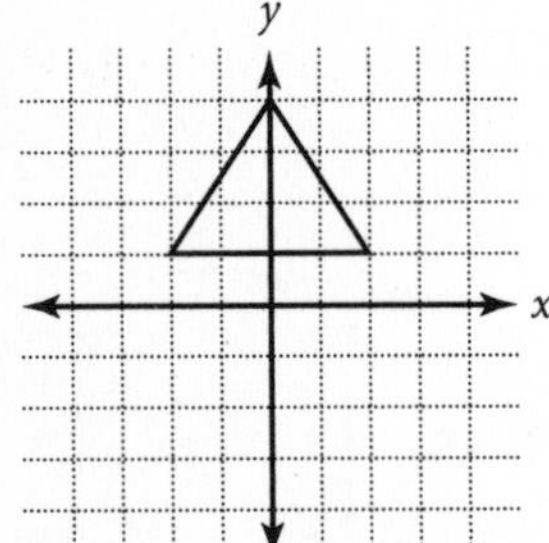

6. $T(x, y) = (0.5x, 2y)$

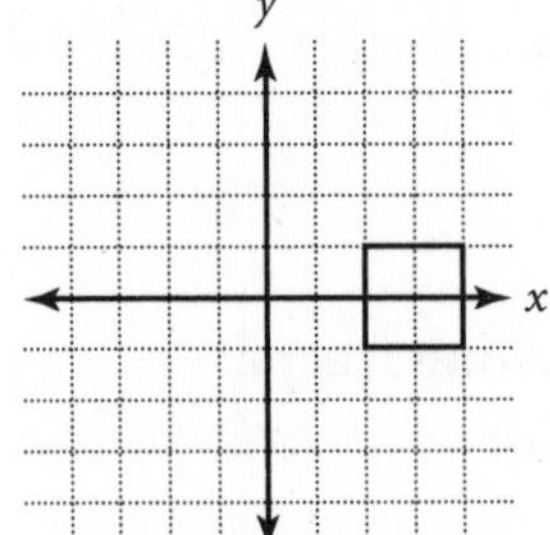
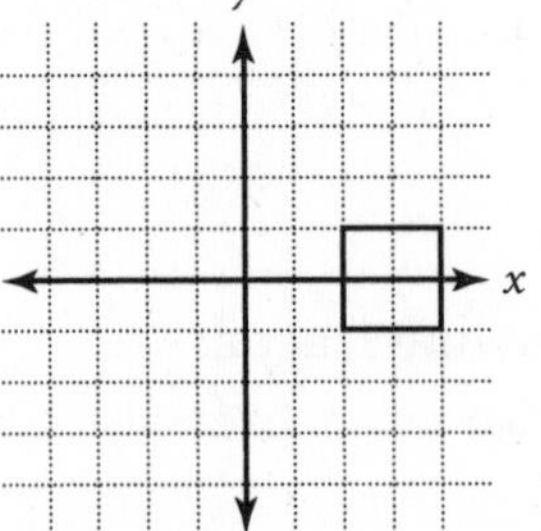

7. $T(x, y) = (-x, -2y)$

8. If M is the center of projection of ℓ_1 onto ℓ_2, what is the projection of C onto ℓ_2?

____

Chapter Assessment

Chapter 11, Form A, page 1

Write the letter that best answers the question or completes the statement.

______ 1. Approximately what value of x would make the figure at right a golden rectangle?

a. 10 b. 8

c. 15 d. 16

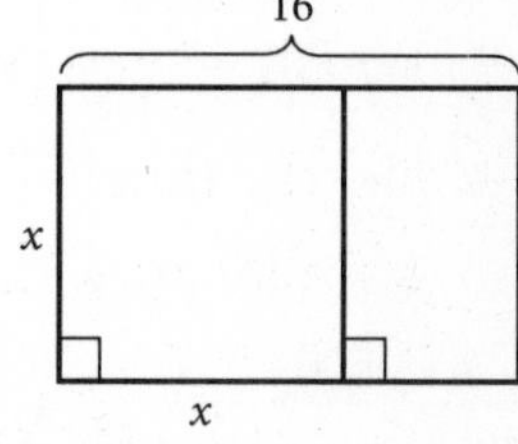

______ 2. Approximately what value of x would make the figure at right a golden rectangle?

a. 3 b. 7

c. 12 d. 19

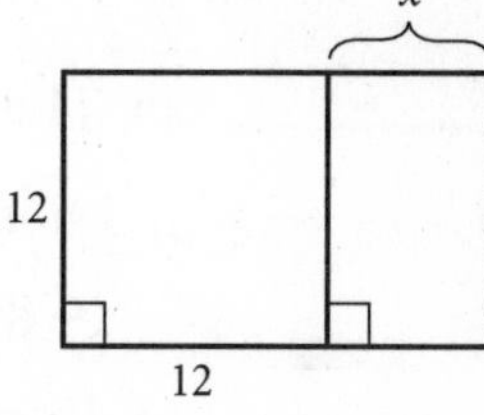

______ 3. What is the taxidistance between points A and B?

a. 3 b. 4

c. 5 d. 7

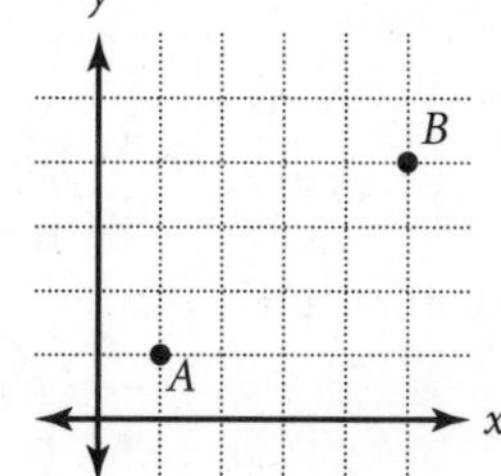

______ 4. If the taxidistance between two points is 3 units, what is the minimum number of pathways between the two points?

a. 3 b. 4 c. 6 d. 9

______ 5. What is the circumference of a taxicab circle with a radius of 4?

a. 4 b. 8 c. 16 d. 32

______ 6. A graph contains an Euler path if and only if there are at most how many odd vertices?

a. 0 b. 1

c. 2 d. 3

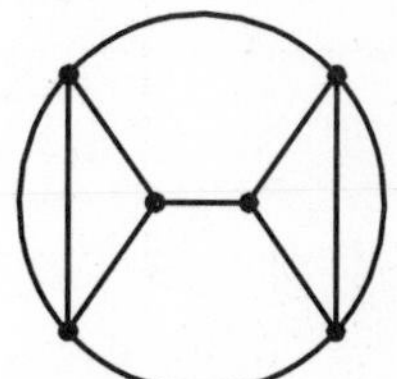

______ 7. How many odd vertices are shown in the graph at right?

a. 0 b. 2

c. 4 d. 6

______ 8. Properties that stay the same no matter how a figure is transformed are called ______.

a. vertices b. invariants

c. fractals d. iterations

Chapter Assessment

Chapter 11, Form A, page 2

______ 9. Which pair of figures is *not* topologically equivalent?

a.  

b. 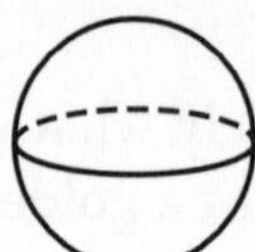 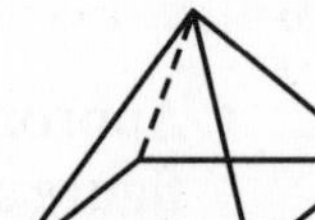

c.  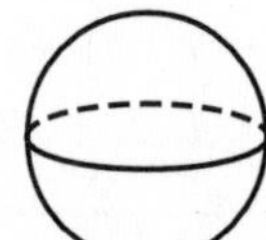

d.  

______ 10. The sum of the angles of a spherical triangle is ______.

a. less than 180° b. equal to 180°

c. greater than 180° d. cannot be determined

______ 11. In spherical geometry, a line is defined as a ______.

a. radius b. chord c. semicircle d. great circle

______ 12. Which of the following shows a triangle in Poincaré's model of hyperbolic geometry?

a. 

b. 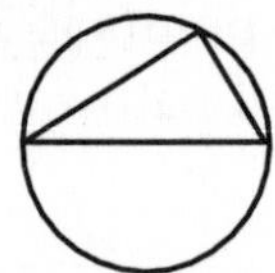

c. 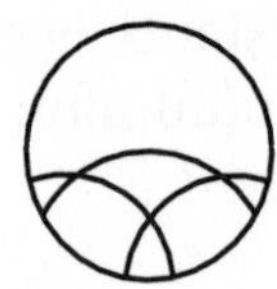

d. 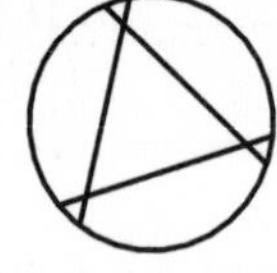

______ 13. Which of the following is *not* an example of a fractal?

a. Cantor dust b. Sierpenski gasket

c. Koch snowflake d. Klein bottle

______ 14. What is the projection of point A onto ℓ_4?

a. A b. B

c. C d. D

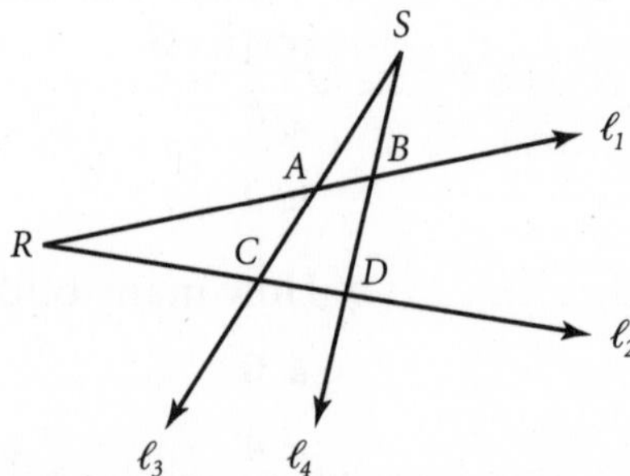

______ 15. Which of the following is a nonrigid, affine transformation?

a. $T(x, y) = (2x, 2y)$ b. $T(x, y) = (x + 1, y - 3)$

c. $T(x, y) = (-x, -y)$ d. $T(x, y) = (3x, 2y)$

Chapter Assessment

Chapter 11, Form B, page 1

Find *x* in each of the following golden rectangles. Round your answers to the nearest unit.

1. 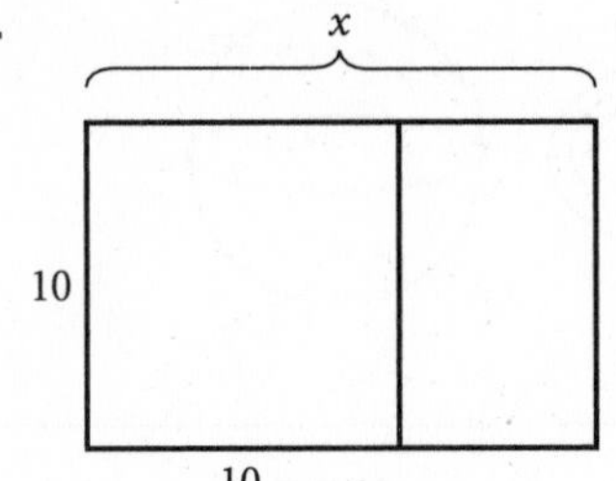

2. 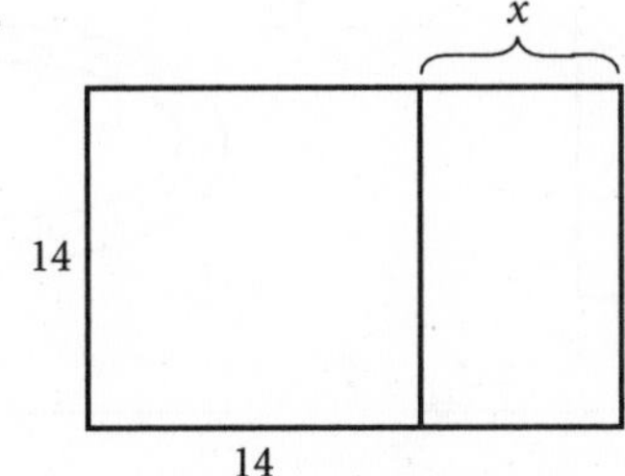

3. 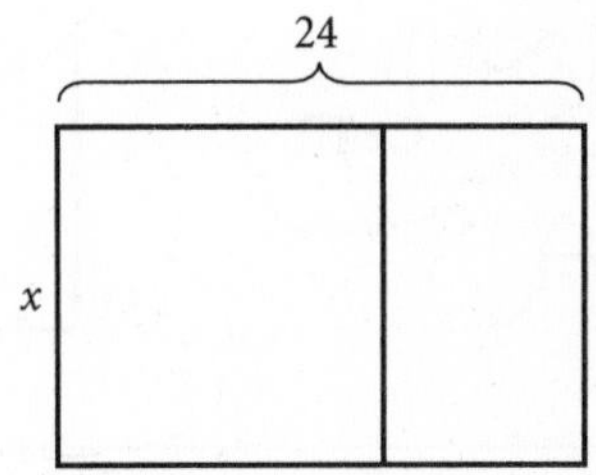

______ ______ ______

Find the taxidistance between the two points on each grid.

4. 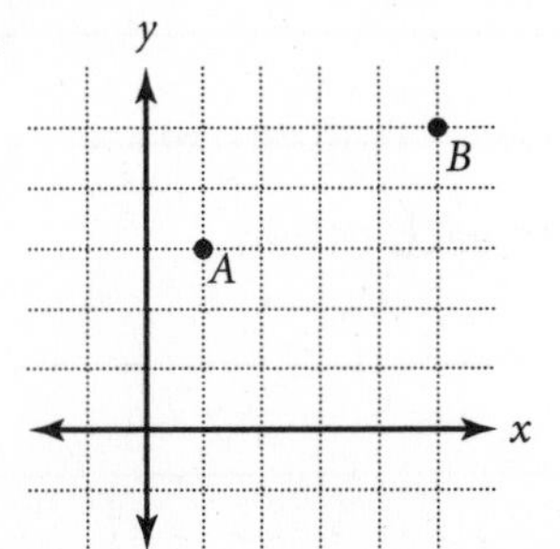

______

5. 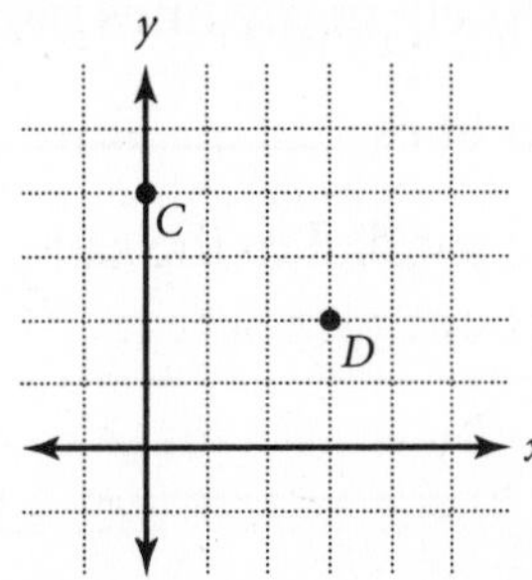

______

6. Find the taxidistance between the points $(-3, 5)$ and $(2, -1)$. ______

7. What is the minimum number of pathways between two points with a taxidistance of 4? ______

8. A taxicab circle has a radius of 5 units. What is the circumference? ______

9. The figure at right shows a floor plan of a house. Draw an Euler path that represents the routes through the house.

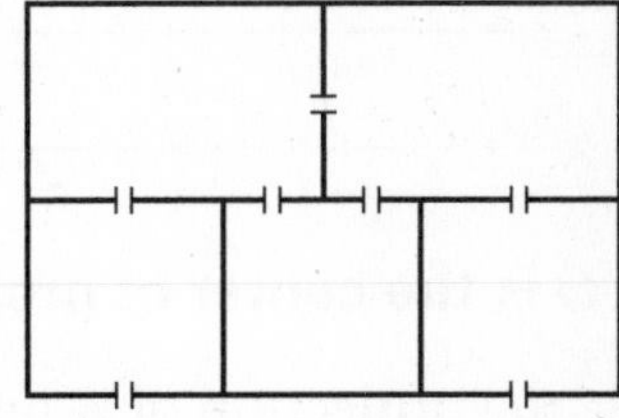

10. Sketch the image for the affine transformation $T(x, y) = (3x, 2y)$ on the grid provided.

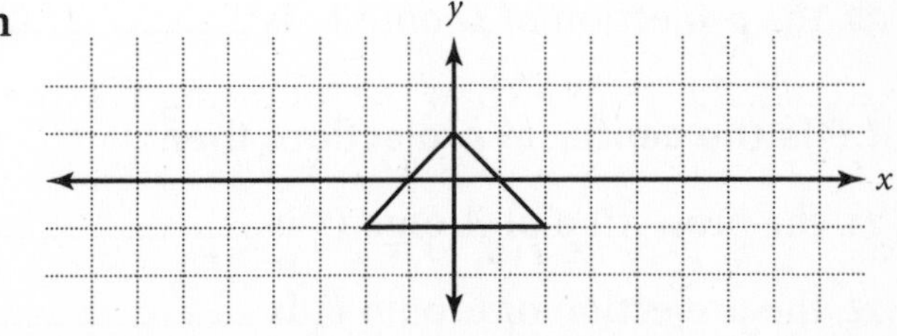

NAME ______________________ CLASS ____________ DATE ____________

Chapter Assessment

Chapter 11, Form B, page 2

Decide whether each of the following pairs is topologically equivalent. Explain.

11.  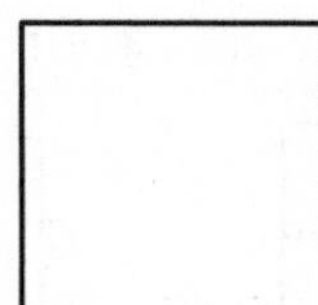

____

____

12. 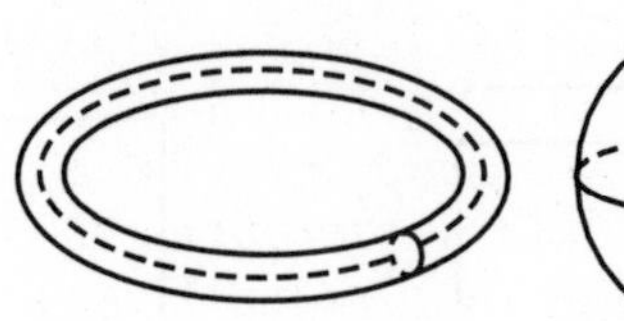 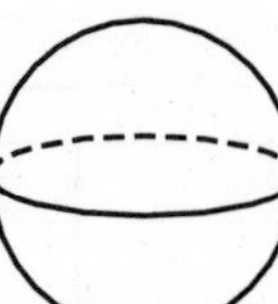

____

____

Identify the type of geometry for which each condition exists.

13. If line ℓ and point P not on line ℓ lie in the same plane, then there exist infinitely many lines parallel to ℓ through P. ______________________

14. There are no parallel lines. ______________________

15. Straight lines that are parallel to the same straight lines are parallel to each other.

____

____

Determine whether each graph contains an Euler path. Explain.

16. 

____

____

17. 

____

____

18. 

____

____

If *O* is the center of projection, then

19. the projection of A onto ℓ_3 is ____________.

20. the projection of D onto ℓ_2 is ____________.

If *P* is the center of projection, then

21. the projection of A onto ℓ_4 is ____________.

22. the projection of C onto ℓ_4 is ____________.

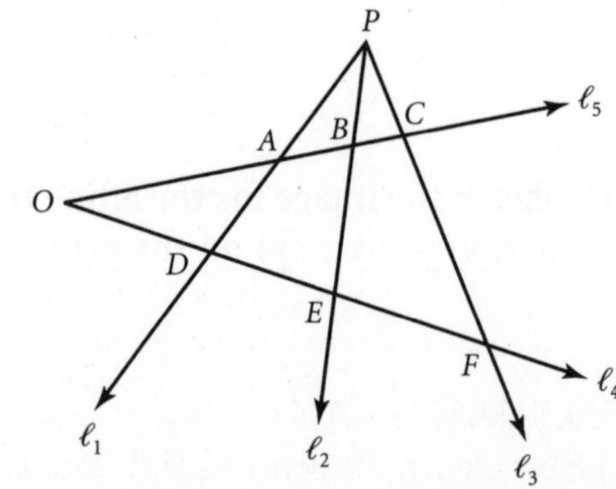

NAME ______________________ CLASS __________ DATE __________

Alternative Assessment

Neighborhood Watch, Chapter 11, Form A

TASK: Illustrate and explain topological equivalence and traversability of networks.

HOW YOU WILL BE SCORED: As you work through the task, your teacher will be looking for the following:

- whether you can draw topologically equivalent shapes
- how well you explain topological equivalence and network traversability

1. Draw two shapes that are topologically equivalent.

2. Explain how you know that the shapes are topologically equivalent.

____

____

3. Draw a shape that is not topologically equivalent to the shapes you drew in Exercise 1.

4. Once a week, you patrol the neighborhood represented at right as part of the neighborhood watch. Can you patrol the neighborhood in such a way that you walk each street only once? If so, show how and describe any restrictions on where you must begin and end your patrol. If not, explain why not.

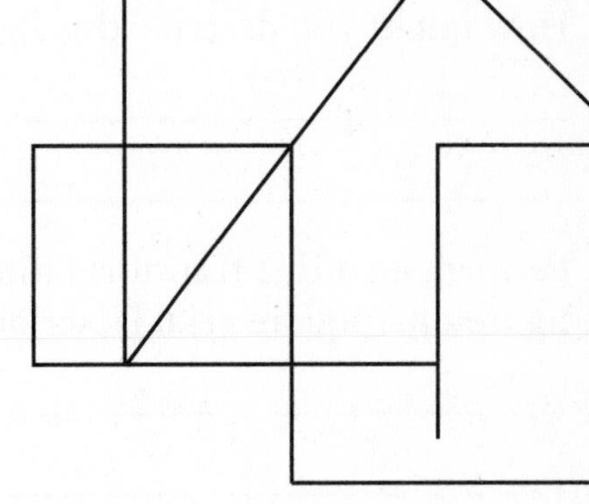

____

____

____

____

____

SELF-ASSESSMENT: Draw a map of a local neighborhood. Can you walk the whole neighborhood and walk each street just once? Why or why not?

Alternative Assessment

Where's the Money? Chapter 11, Form B

TASK: Figure out where Jessie hid his money.

HOW YOU WILL BE SCORED: As you work through the task, your teacher will be looking for the following:

- whether you can determine where Jessie hid his money
- how well you can explain your reasoning and describe the technique Jessie used to draw the figure

If you can figure this figure out, you will know where I hid my money.

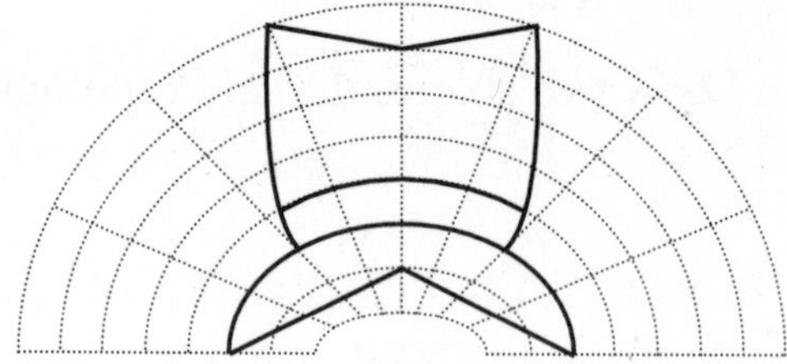

1. Where did Jessie hide the money?

____

2. How did you figure out where the money is?

____

____

3. What technique did Jessie use to draw the figure?

____

____

4. How might you describe this figure by using coordinates?

____

____

5. Perform an affine transformation on the figure by redrawing the figure on a square grid. Describe your transformation.

____

____

SELF-ASSESSMENT: Draw a figure that Jessie can use to determine where you will meet him.

Quick Warm-Up: Assessing Prior Knowledge
12.1 Truth and Validity in Logical Arguments

1. Convert the following statement to if-then form: A frog is an amphibian.

____

2. Write the converse of the following statement: If Jamie scores 95% on the test, then she will get an A in the course.

____

Lesson Quiz
12.1 Truth and Validity in Logical Arguments

In Exercises 1–4 use the following conditional to answer each question: If it is cold, Juanita will wear a coat. State whether the argument is valid or invalid.

1. Write a *modus ponens* argument with the given conditional as one of its premises. _______________
2. Write a *modus tollens* argument with the given conditional as one of its premises. _______________
3. Write an *invalid* argument in the *affirming the consequent* form, with the given conditional as one of its premises.

   ____
4. Write an *invalid* argument in the *denying the antecedent* form, with the given conditional as one of its premises.

   ____

Determine whether each conclusion is valid based on the premises at right:

If the sun is shining, Ben will go to the lake.
If Ben goes to the lake, he will go swimming.

5. The sun is shining. Ben will go swimming. _______________
6. Ben did not go to the lake. Ben will not go swimming. _______________
7. Ben did not go swimming. The sun is not shining. _______________
8. Ben did not go to the lake. The sun is not shining. _______________

Quick Warm-Up: Assessing Prior Knowledge

12.2 And, Or, and Not in Logical Arguments

1. Rewrite the following two sentences as one sentence connected by the word *and;* It is raining today. School is in session.

____

2. Rewrite the two sentences in Exercise 1 as one sentence connected by the word *or.*

____

3. Add the word *not* in order to negate the following sentence: Albert got permission to see a movie.

____

Lesson Quiz

12.2 And, Or, and Not in Logical Arguments

1. When will a conjunction be true? ______________
2. When will a disjunction be false? ______________

Write the negation of each statement.

3. Omar failed math. ______________
4. The light is on. ______________

For each expression below, indicate whether the statement represented by the symbols is true or false.

p: $3 > 2$ q: $3 = 2 + 1$ r: $2 > 3$

5. p AND q ______________
6. q OR r ______________
7. q AND r ______________
8. p OR q ______________
9. $\sim q$ ______________
10. $\sim q$ OR r ______________
11. $\sim r$ ______________
12. p AND $\sim r$ ______________

NAME ______ CLASS ______ DATE ______

Quick Warm-Up: Assessing Prior Knowledge

12.3 A Closer Look at If-Then Statements

1. Write the following statement as a conditional: Warm-blooded animals are mammals.

______

2. Write the converse of the conditional that you wrote in Exercise 1.

______

Lesson Quiz

12.3 A Closer Look at If-Then Statements

Write the converse, inverse, and contrapositive of the following conditional:

If p and q are even numbers, then pq is an even number.

1. Converse: ______
2. Inverse: ______
3. Contrapositive: ______

Determine whether each statement in Exercises 1–3 is true or false.

4. Converse ______
5. Inverse ______
6. Contrapositive ______

Write each statement in if-then form.

7. Practicing the piano every day will improve your skill.

______

8. The lake freezes at 30°F.

______

9. Kate is a member of the student organization.

______

NAME ______________________ CLASS ____________ DATE ____________

Mid-Chapter Assessment

Chapter 12 (Lessons 12.1–12.3)

Write the letter that best answers the question or completes the statement.

For Exercises 1–3, use the following form: If p then q
not q
Therefore, not p

_______ 1. This form of argument is called _____.

a. modus ponens   b. modus tollens

c. affirming the antecedent   d. denying the consequent

_______ 2. Statement p is called _____.

a. an inverse   b. a conclusion

c. a premise   d. a contrapositive

_______ 3. Which is the converse of the conditional?

a. If q then p.   b. if not p, then not q.

c. If not q, then not p.   d. if q, then not p.

4. Write the truth table for a disjunction.

For Exercises 5–9, write the statement expressed by each set of symbols by using the statements:

p: The sun is shining.
q: Joe goes swimming.
r: Joe is wet.

5. $p \Rightarrow q$ ________________________________

6. p and q ________________________________

7. p or $\sim r$ ________________________________

8. $\sim p \Rightarrow r$ ________________________________

9. $\sim q \Rightarrow \sim p$ ________________________________

NAME ______________________ CLASS ____________ DATE ____________

Quick Warm-Up: Assessing Prior Knowledge
12.4 Indirect Proof

1. Write the contrapositive of the following statement: If the defendant was in his car at 10:00 P.M., then he was in an automobile accident.

____

2. If a triangle is a right triangle, then is it possible for it also to be an obtuse triangle? Explain.

____

Lesson Quiz
12.4 Indirect Proof

Fill in the missing statements and reasons in the following proof:

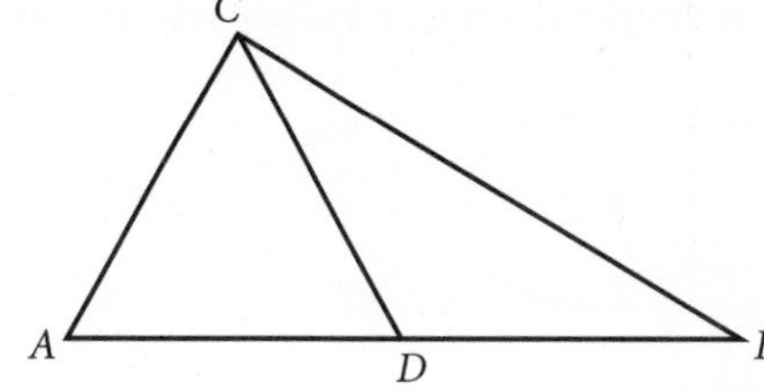

Given: $\triangle ABC$ is scalene.
$\overline{CD}$ is the median of $\overline{AB}$.

Prove: $\overline{CD}$ is not perpendicular to $\overline{AB}$.

Statements	Reasons
1. Assume that ______________.	By assumption
∠CDA and ∠CDB are right angles.	2. ______________
3. ______________	All right angles are congruent.
$\overline{CD}$ is the median of $\overline{AB}$.	4. ______________
$\overline{AD} \cong \overline{DB}$	5. ______________
$\overline{CD} \cong \overline{CD}$	6. ______________
$\triangle ACD \cong \triangle BCD$	7. ______________
$\overline{AC} \cong \overline{CB}$	8. ______________
This contradicts the fact that $\overline{AC} \ncong \overline{CB}$.	9. ______________ ______________
10. Therefore, ______________	The assumption is false in a proof by contradiction.

Quick Warm-Up: Assessing Prior Knowledge

12.5 Computer Logic

1. Give examples of electronic devices with on-off switches. ______________________
____

2. Write a truth table for p AND q.

3. Write a truth table for $\sim p$.

Lesson Quiz

12.5 Computer Logic

Write a logical expression for each network below.

1.

2.

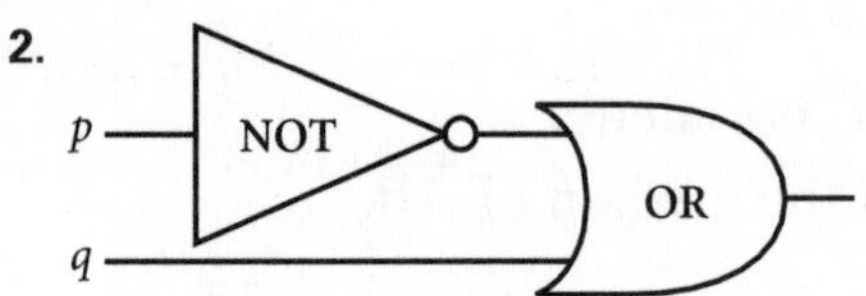

Construct a network of logic gates for each of the following expressions.

3. $\sim(p \text{ OR } q)$

4. p AND $(\sim q)$

Complete the input-output table for this network.

	p	q	$\sim q$	p OR $\sim q$
5.	1	1	________	________
6.	1	0	________	________
7.	0	1	________	________
8.	0	0	________	________

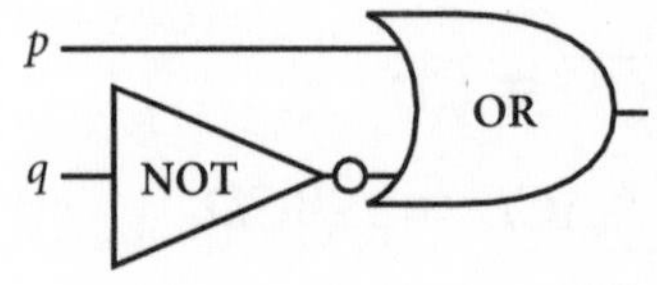

9. Construct an input-output table for this network on your own paper.

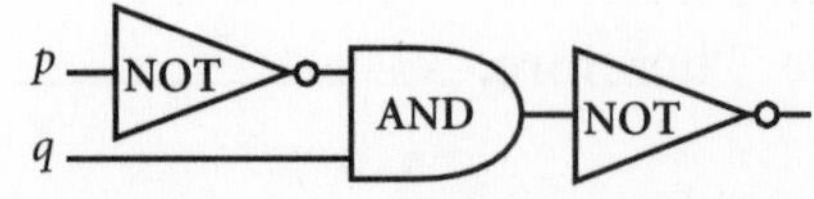

Chapter Assessment

Chapter 12, Form A, page 1

Write the letter that best answers the question or completes the statement.

For Exercises 1–2, use the following: If p then q
p
Therefore, q

______ 1. This form of argument is called ______.

a. *modus ponens*
b. *modus tollens*
c. affirming the antecedent
d. denying the consequent

______ 2. Statement q is called ______.

a. an inverse
b. a conclusion
c. a premise
d. a contrapositive

______ 3. What conclusion can be drawn from the premises below?
If an animal is a mammal, then it lives in trees.
A cow does not live in trees.

a. A cow is a mammal.
b. A cow is not a mammal.
c. A cow is not an animal.
d. cannot be determined

______ 4. Which is the converse of the following statement?
If Sue plants an apple seed, a tree will grow.

a. If Sue does not plant an apple seed, a tree will not grow.
b. If a tree grows, Sue planted an apple seed.
c. If a tree does not grow, Sue did not plant an apple seed.
d. If Sue does not plant an apple seed, a tree will grow.

______ 5. Which of the following results is *not* a true statement for $p \Rightarrow q$?

a. p is true and q is true.
b. p is false and q is true.
c. p is true and q is false.
d. p is false and q is false.

______ 6. Given the statements p, "the temperature is 90°F," and q, "the apartment is hot," which is the symbolic representation of, "the temperature is 90°F and the apartment is not hot"?

a. p AND q
b. p AND $\sim q$
c. $\sim p$ AND q
d. $\sim p$ AND $\sim q$

______ 7. In an indirect proof, which logical form of the statement to be proven is assumed?

a. inverse
b. converse
c. disjunction
d. negation

Chapter Assessment

Chapter 12, Form A, page 2

______ 8. In an indirect proof for the statement "An obtuse triangle cannot contain a right triangle," the contradiction used to prove the statement is ______.

a. the triangle is obtuse b. the obtuse angle measures 90°
c. the triangle is acute d. all the angles are acute

______ 9. If p is true and q is false, which of the following statements is true?

a. p AND q b. $\sim p$ OR q c. p AND $\sim q$ d. $p \Rightarrow q$

______ 10. If p is true and q is true, which of the following statements is false?

a. $\sim q \Rightarrow p$ b. $\sim p$ OR q c. p AND q d. $p \Rightarrow \sim q$

______ 11. Which is the logical expression for this network?

a. $\sim(p$ AND $q)$
b. $(\sim p)$ AND q
c. $\sim[(\sim p)$ AND $q]$
d. $(\sim p)$ AND $(\sim q)$

______ 12. Which is the logical expression for this network?

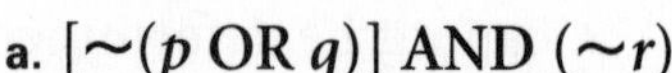

a. $[\sim(p$ OR $q)]$ AND $(\sim r)$
b. $[p$ OR $(\sim q)]$ AND $(\sim r)$
c. $[p$ AND $(\sim r)]$ OR $(\sim q)$
d. $[\sim(p$ AND $r)]$ OR $(\sim q)$

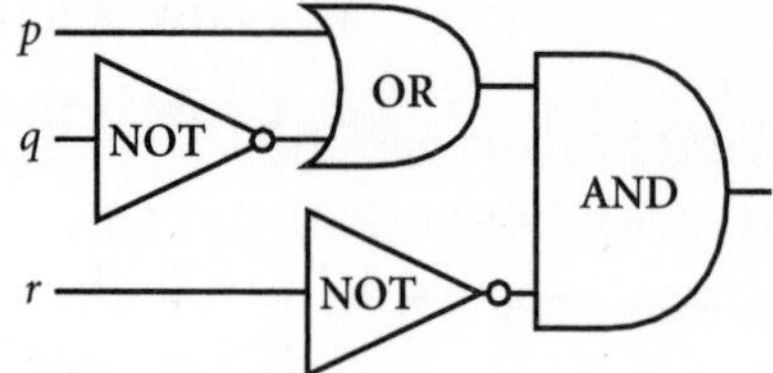

______ 13. If p is 1 and q is 0, which of the following outputs is equal to 1?

a. $\sim p$ AND q b. $\sim p$ OR q c. p AND $\sim q$ d. $\sim(p$ OR $q)$

In Exercises 14–16, choose the term from the list below that best matches each definition.

a. biconditional
b. fallacy
c. counterexample
d. indirect proof
e. binary number system

______ 14. an "if and only if" statement

______ 15. a common logical mistake of affirming the consequent or denying the antecedent

______ 16. an argument that assumes the negation of what is to be proven

NAME ______________________ CLASS __________ DATE __________

Chapter Assessment

Chapter 12, Form B, page 1

Write a valid conclusion from the given premises.

1. If Jay is in London, then Jay is in Great Britain.
   Jay is not in Great Britain.

   ____

2. If May is in Quebec, then she is in Canada.
   May is in Quebec.

   ____

Use the following conditional to write each of the related forms:

If a quadrilateral is a square, then it is a rhombus.

3. Converse ______________________
4. Inverse ______________________
5. Contrapositive ______________________

Indicate whether each statement expressed by the symbols is true or false according to the rules of logic. Use the statements given below.

p: A square is a rectangle.

q: A square is a rhombus.

r: A square does not have congruent sides.

6. $\sim p$ ________ 7. $\sim r$ ________ 8. p OR r ________

9. $p \Rightarrow r$ ________ 10. p AND q ________ 11. $q \Rightarrow \sim r$ ________

12. $\sim q$ OR r ________ 13. p AND $\sim r$ ________ 14. $\sim p \Rightarrow r$ ________

15. (p AND r) $\Rightarrow q$ ________

16. Write a logical expression for this network.

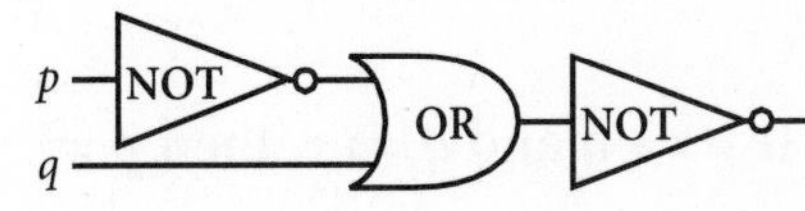

17. Construct a network diagram for [($\sim p$) OR q] AND r.

NAME ______________________ CLASS ____________ DATE ____________

Chapter Assessment

Chapter 12, Form B, page 2

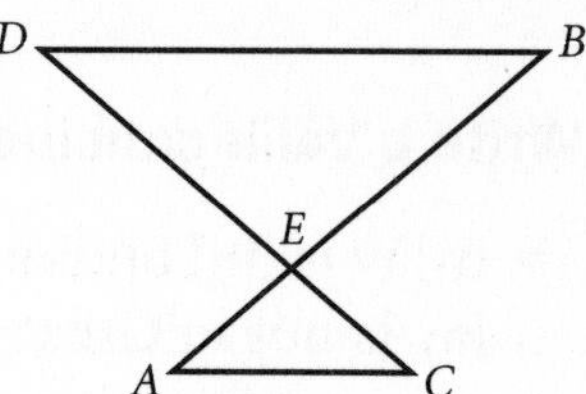

Fill in the missing statements and reasons in the following proof:

Given: $\overline{DB} \ncong \overline{AC}$

Prove: $\overline{AB}$ **and** $\overline{CD}$ **do not bisect each other.**

Statements	Reasons
18. ______________________	By assumption
$\overline{AE} \cong \overline{EB}$ and $\overline{CE} \cong \overline{ED}$	19. ______________________
20. ______________________	Vertical angles are congruent.
$\triangle AEC \cong \triangle BED$	21. ______________________
22. ______________________	CPCTC
This contradicts the fact that $\overline{DB} \ncong \overline{AC}$.	23. ______________________
24. ______________________ ______________________	The assumption is false in a proof by contradiction.

25. Complete the input-output table for this network.

p	q			
1	1	________	________	________
1	0	________	________	________
0	1	________	________	________
0	0	________	________	________

26. "If $p \Rightarrow q$ and $q \Rightarrow r$, then $p \Rightarrow r$" is an example of the If-Then ______________ Property.

27. Computers use a system of on and off "switches" represented by 1s and 0s respectively, called a ______________ number system.

Alternative Assessment

Who's the Thief? Chapter 12, Form A

TASK: Use logical reasoning to determine which suspects are thieves.

HOW YOU WILL BE SCORED: As you work through the task, your teacher will be looking for the following:

- whether or not you can determine who is telling the truth and solve the problem
- how well you support your conclusions and describe the logic you used

Three people suspected of theft were questioned by a judge. The thieves always lied, and those who were not thieves always told the truth. The judge asked each suspect the question, "Are you a thief?" The judge could not hear the first suspect's reply. The second suspect said, "He says he is not a thief. He is not a thief, and I am not a thief either." The third suspect said, "They are both lying. They are both thieves. I am not a thief."

1. Determine whether or not each person is a thief.

   ____

2. Write an argument to support your conclusions.

   ____

   ____

   ____

3. What type of argument did you use? Describe how your argument illustrates the features of this type of argument.

   ____

   ____

   ____

4. Write a problem that can be solved by using another type of argument. Provide a solution. Then exchange problems with another student.

   ____

   ____

   ____

SELF-ASSESSMENT: Make truth tables to summarize the truth values for the situation described above.

NAME ______________________ CLASS ____________ DATE ____________

Alternative Assessment

Exploring Consumer Advertising Using Truth Tables, Chapter 12, Form B

TASK: Translate advertising "promises" to and from truth tables.

HOW YOU WILL BE SCORED: As you work through the task, your teacher will be looking for the following:

- whether you can correctly create and fill in a truth table
- how well you interpret the truth table results within the context of the ad

Suppose that you saw the following ad:

LIMITED TIME ONLY! Save $\frac{1}{3}$ on ALL brand name watches purchased before December 24!

Create a truth table at right, using 1 for true and 0 for false, with at least 4 rows and any of the following columns: *p*, *q*, *r*, *s*, OR, AND, ? ⇒ *s*, *s* ⇒ ?. You may want to use those statements listed to help you.

1. Write row 1 of your truth table as a conditional statement. ______________________

____

2. Write row 2 of your truth table as a conditional statement. ______________________

____

____

3. Write row 3 of your truth table as a conditional statement. ______________________

____

____

4. Write row 4 of your truth table as a conditional statement. ______________________

____

____

SELF-ASSESSMENT: Clip a newspaper or magazine ad, and write it as a conditional. Try to prove that it is valid by using an indirect proof.

Chapter 1

Quick Warm-Up 1.1

Answers will vary.

1. compass point 2. pencil
3. corner of room
4. strips dividing ceiling tiles
5. notebook paper

Lesson Quiz 1.1

1. $\overline{TS}$ and $\overline{TU}$ 2. $\overrightarrow{TS}$ and $\overrightarrow{TU}$
3. $\angle STU$, $\angle UTS$, and $\angle T$
4. Check students' sketches.
5. line 6. plane 7. plane 8. point
9. endpoints 10. vertex 11. planes

Quick Warm-Up 1.2

Answers will vary. Sample answer: $\angle EDF$, $\overline{EH}$, $\overrightarrow{HE}$, $\overleftrightarrow{GF}$, point E.

Lesson Quiz 1.2

1. 10 2. 6 3. 8 4. 1
5. $\overline{QR}$, $\overline{RS}$, and $\overline{ST}$; $\overline{QS}$ and $\overline{RT}$
6. $\overline{AB}$, $\overline{AD}$, and $\overline{AE}$; $\overline{BC}$ and $\overline{CD}$
7. $\overline{QR} \cong \overline{RS} \cong \overline{ST}$; $QR = RS = ST$
   $\overline{QS} \cong \overline{RT}$; $QS = RT$
8. $\overline{AB} \cong \overline{AD} \cong \overline{AE}$; $AB = AD = AE$
   $\overline{BC} \cong \overline{CD}$; $BC = CD$

Quick Warm-Up 1.3

1. $\angle JKL$, $\angle EDF$, $\angle ABC$, $\angle GHI$
2. $\angle JKL$ and $\angle EDF$ (159° and 104°); $\angle ABC$ and $\angle GHI$ (62° and 13°)
3. no

Lesson Quiz 1.3

1. 35° 2. 75° 3. 140° 4. 40°
5. 105° 6. 80° 7. 45° 8. 54°
9. 64° 10. x 11. $x - 37°$

Quick Warm-Up 1.4

Answers will vary.

1. $\overline{AC}$ and $\overline{BF}$ 2. $\angle ACE$
3. $\angle ACG$ 4. $\overline{CG}$

Lesson Quiz 1.4

1. Fold the paper through points A and B.
2. Fold the paper so that line ℓ folds onto itself.
3. Lines m and p are parallel to one another.
4. An infinite number of lines can be constructed perpendicular to line ℓ.

Mid-Chapter Assessment—Chapter 1

1. b 2. c 3. a 4. d
5. False; the statement describes a ray, but a line extends forever in both directions.
6. False; the rays must have a common endpoint.
7. If the distances from any point on the line to the sides of the angle are equal, then the line is an angle bisector.
8. 25° 9. 35°

Quick Warm-Up 1.5

Check students' drawings.

Lesson Quiz 1.5

1. Construct the bisector of each angle. Using the point where the three bisectors intersect, construct a perpendicular to one side of the triangle. Use this distance as the radius of the inscribed circle.
2. Construct the perpendicular bisector of each side. Using the point where the three bisectors intersect as a center, and the distance from the center to one vertex as the radius, draw a circle through the three vertices of the triangle.

3.

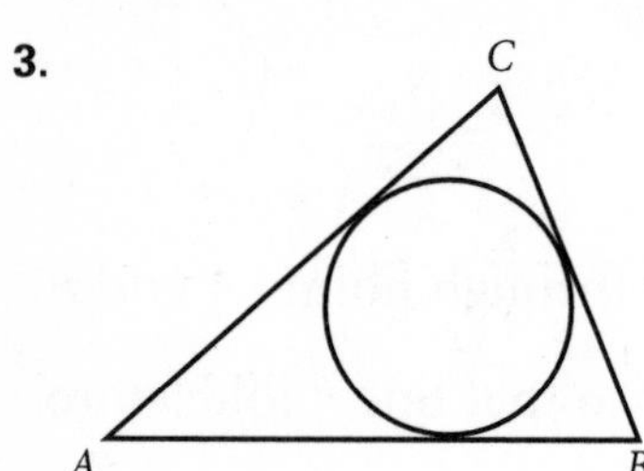

4.

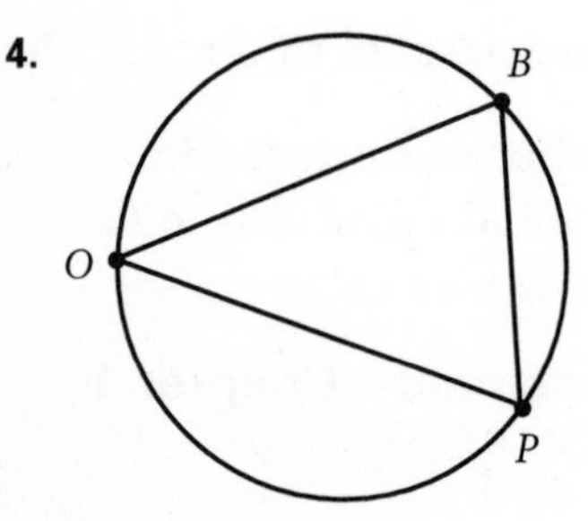

Quick Warm-Up 1.6

1–4. Check students' drawings.

Lesson Quiz 1.6

1. translation 2. reflection 3. neither

4. 5.

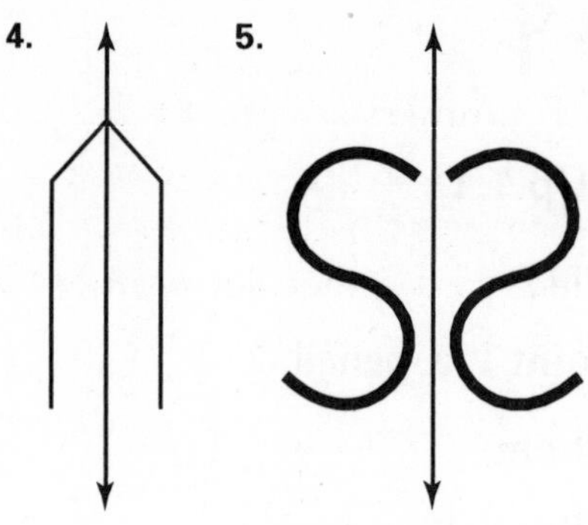

6.

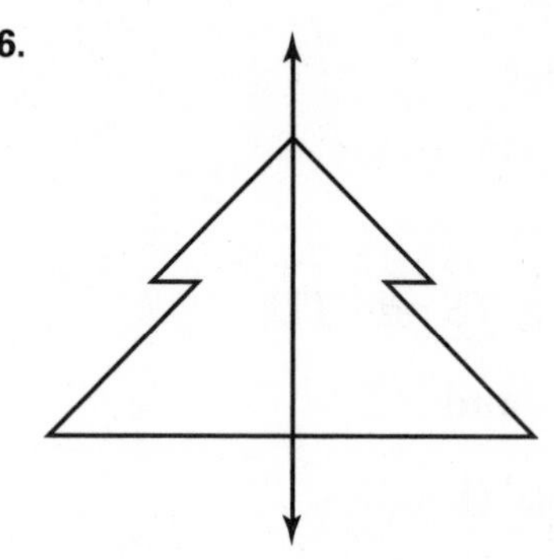

Quick Warm-Up 1.7

1. a vertical line crossing the x-axis at -3

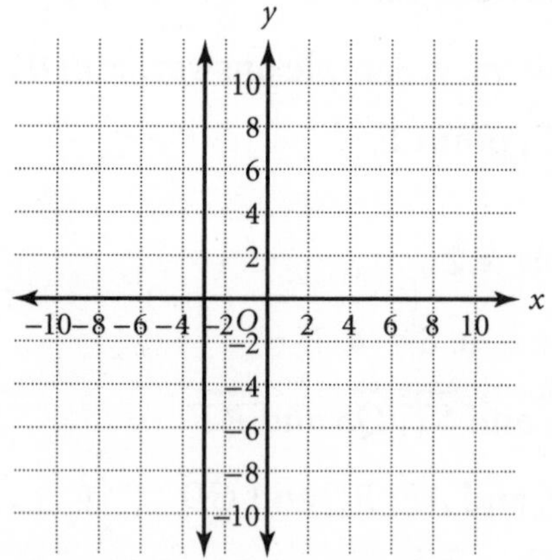

2. Check students' graphs. Two vertices have the same x-coordinate, and two have the same y-coordinate.

Answers

Lesson Quiz 1.7

1. It is a pair of numbers assigned to a point on a coordinate plane. The first number determines the distance from the *x*-axis; the second number determines the distance from the *y*-axis.

2. *x*-coordinate

3. $D(1, -1), E(3, -4), F(5, -1)$

4.

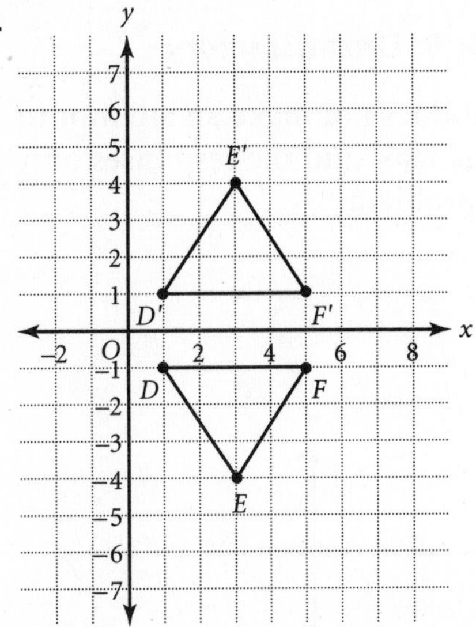

5. $D'(1, 1), E'(3, 4), F'(5, 1)$

6. $D'(4, -2), E'(6, -5), F'(8, -2)$

7. $D'(7, -3), E'(9, -6), F'(11, -3)$

5. They are perpendicular

6. 54 7. 16.1 8. 65 9. none

10. An inscribed circle touches each side of the triangle and is located inside the triangle. A circumscribed circle passes through each vertex and is located outside the circle.

11. 49° 12. 48° 13. 30°

14. Exercise 12 15. translation

16. reflection 17. rotation

18.

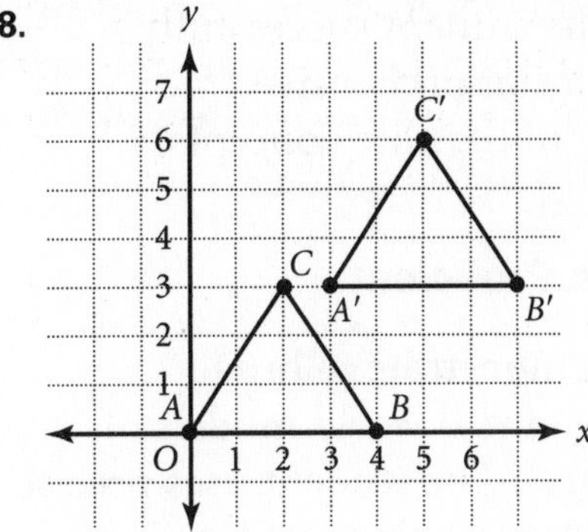

(3, 3), (5, 6), (7, 3)

19. translation

20. $R(x, y) = (x - 3, y - 3)$

21. $(-3, -3), (-1, 0), (1, -3)$

Chapter Assessment—Chapter 1

Form A

1. d 2. a 3. c 4. a 5. c 6. b 7. b

8. a 9. d 10. b 11. d 12. a 13. d

Form B

1. $\overline{GH}, \overline{HI}, \overline{IG}$ 2. $\angle G, \angle H, \angle I$

3. plane *GHI* 4. $\overrightarrow{XB}, \overrightarrow{XE}, \overrightarrow{XC}, \overrightarrow{XD}, \overrightarrow{XW}, \overrightarrow{XR}$

Alternative Assessment—Chapter 1

Form A

1–9. Answers will vary.

1. pencil point 2. pen 3. sheet of paper

4. book; point: corner of cover; line: edge of cover; plane: cover

5. two corners on same edge of cover

6. two points on cover

7. opposite edges of cover

8. intersecting edges of cover

9. shelves in a bookcase

Score Point 4: Distinguished

The student demonstrates a comprehensive understanding of real world models of geometric figures. The student uses perceptive, creative, and complex mathematical reasoning as well as precise and appropriate mathematical language throughout the task. Theoretical knowledge is apparent and is applied to concrete situations as the student successfully demonstrates a comprehensive understanding of the core concepts.

Score Point 3: Proficient

The student demonstrates a broad understanding of real world models of geometric figures. The student uses precise and appropriate mathematical language most of the time. Theoretical knowledge is apparent and is applied to concrete situations as the student attempts to draw conclusions based on the investigations.

Score Point 2: Apprentice

The student demonstrates an understanding of real world models of geometric figures. The student uses mathematical reasoning and appropriate mathematical language some of the time. The student attempts to apply theoretical knowledge to the task but may not be able to draw conclusions based on the investigations.

Score Point 1: Novice

The student demonstrates a basic understanding of real world models of geometric figures. The student uses little mathematical reasoning or appropriate mathematical language. Theoretical knowledge may appear weak, and many responses may be illogical because directions were followed incorrectly.

Score Point 0: Unsatisfactory

The student does not make an attempt to complete the task, and the responses only restate the problem.

Form B

1–2.

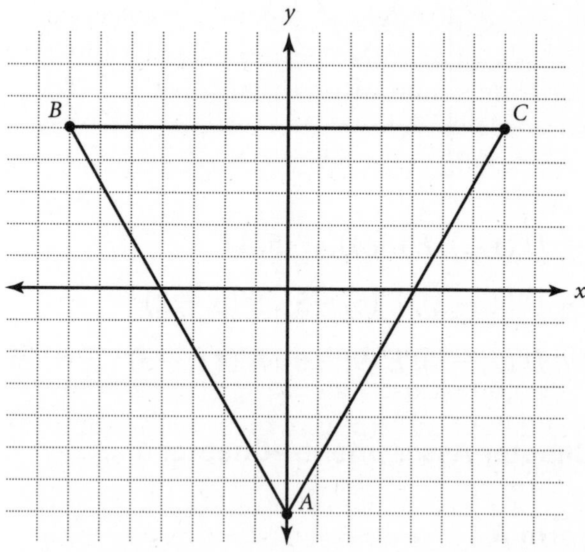

3. Find the midpoint of one side of the triangle, and connect this point with the vertex of the opposite side. Follow the same steps for the other two sides of the triangle.

4. centroid

5. Answers will vary.
6. Find the point of intersection of the three angle bisectors.
7. Answers will vary. When rolling a poster to fit in a box shaped like a triangular prism, you would need to find the incenter in order to calculate the size of the inscribed circle.
8. Yes; the angles formed by the angle bisectors are congruent to each other.

Score Point 4: Distinguished

The student demonstrates a comprehensive understanding of constructions in a triangle. The student uses perceptive, creative, and complex mathematical reasoning as well as precise and appropriate mathematical language throughout the task. Theoretical knowledge is apparent and is applied to concrete situations as the student successfully demonstrates a comprehensive understanding of the core concepts.

Score Point 3: Proficient

The student demonstrates a broad understanding of constructions in a triangle. The student uses precise and appropriate mathematical language most of the time. Theoretical knowledge is apparent and is applied to concrete situations as the student attempts to draw conclusions based on the investigations.

Score Point 2: Apprentice

The student demonstrates an understanding of constructions in a triangle. The student uses mathematical reasoning and appropriate mathematical language some of the time. The student attempts to apply theoretical knowledge to the task but may not be able to draw conclusions based on the investigations.

Score Point 1: Novice

The student demonstrates a basic understanding of constructions in a triangle. The student uses little mathematical reasoning or appropriate mathematical language. Theoretical knowledge may appear weak, and many responses may be illogical because directions were followed incorrectly.

Score Point 0: Unsatisfactory

The student does not make an attempt to complete the task, and the responses only restate the problem.

Chapter 2

Quick Warm-Up 2.1

1. 5 cm 2. 15 3. $x = 2$

Lesson Quiz 2.1

1.

n	Segment names	Number of segments, S
2	$\overline{DE}$	1
3	$\overline{DE}, \overline{DF}, \overline{EF}$	3
4	$\overline{DE}, \overline{DF}, \overline{EF}$ $\overline{DG}, \overline{EG}, \overline{FG}$	6
5	$\overline{DE}, \overline{DF}, \overline{DG}, \overline{DH}, \overline{EF},$ $\overline{EG}, \overline{EH}, \overline{FG}, \overline{FH}, \overline{GH}$	10

2. $\frac{n(n-1)}{2}$

3. Methods may vary. Make a table.

n	Odd numbers	Sum
1	1	1
2	1, 3	4
3	1, 3, 5	9
4	1, 3, 5, 7	16
5	1, 3, 5, 7, 9	25

Each number in the third column is the square of the number in the same row of the first column. Letting S represent the sum of the squares of the first n odd integers, $S = n^2$.

Quick Warm-Up 2.2

1. true 2. true 3. false

4. a diagram that is used to show logical relationships

Lesson Quiz 2.2

1. If a TV program is a comedy, then it is fictional.

2. Hypothesis: A TV program is a comedy.

   Conclusion: It is fictional.

3. 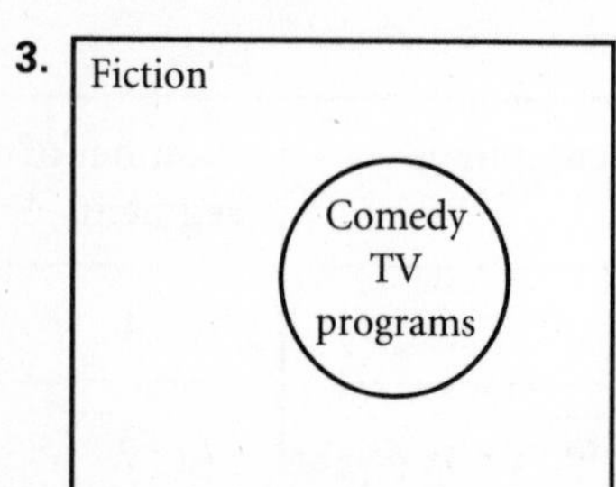

4. If a television program is fictional, then it is a comedy; False; there are many serious fictional dramas on television.

Quick Warm-Up 2.3

1. If a figure is a triangle, then it has three sides.

2. If B, then A.

3. not necessarily

Lesson Quiz 2.3

1. Conditional: If a flower is a rose, then it is red.
   Converse: If a flower is red, then it is a rose.
   Biconditional: A flower is a rose if and only if it is red.
   Conclusion: No, it is not a definition.

2. Conditional: If a city is Phoenix, then it is in Arizona.
   Converse: If a city is in Arizona, then it is Phoenix.
   Biconditional: A city is Phoenix if and only if it is in Arizona.
   Conclusion: No, it is not a definition.

3. Conditional: If a person is a high school senior, then he or she is in 12th grade.
   Converse: If a person is in 12th grade, then he or she is a high school senior.
   Biconditional: A person is a high school senior if and only if he or she is in 12th grade.
   Conclusion: Yes, it is a definition.

4. No; the converse is false.

Mid-Chapter Assessment—Chapter 2

1. c 2. b 3. false 4. true

5. If it is a rabbit, then it is white.

6. Hypothesis: It is a rabbit.
   Conclusion: It is white.

7. If it is white, then it is a rabbit. False; some white animals are not rabbits.

8. A person is a baseball player if and only that person is an athlete. False; some athletes are not baseball players.

Quick Warm-Up 2.4

1. none 2. $x = 20$

3. the segment itself 4. $AB = 5$

Lesson Quiz 2.4

1. Subtraction Property of Equality; Simplify

2. Transitive Property of Congruence

3. Symmetric Property of Congruence

4. Addition Property of Equality; Segment Addition Postulate

5. Transitive Property of Equality

6. Let A, B, and C represent three angles. If $A \cong B$ and $B \cong C$, then $A \cong C$.

Quick Warm-Up 2.5

1. true 2. false 3. false

Lesson Quiz 2.5

1. The sum of the measures of the angles of a linear pair is 180°.

2. Substitution 3. 14 4. 168° 5. 12°

6. $m\angle PMQ = m\angle SMF = 150°$
$m\angle SMP = m\angle FMQ = 30°$

Chapter Assessment—Chapter 2

Form A

1. b 2. d 3. c 4. a 5. c 6. d

7. c 8. a 9. c 10. b 11. d 12. c

Form B

1. $\frac{1}{200}, \frac{1}{150}, \frac{1}{100}, \frac{1}{50}$

2. The denominators are decreasing by 50.

3. If a person is a Knicks fan, then the person lives in New York.

4. Hypothesis: A person is a Knicks fan.
Conclusion: The person lives in New York.

5. If the sun is shining, then people go swimming.
If people go swimming, then it is hot.
If it is hot, then it is summer.
Conditional: If the sun is shining, then it is summer.

6. Conditional: If an angle is 45°, then it is an obtuse angle.
Converse: If an angle is obtuse, then it is a 45° angle.
Biconditional: An angle is obtuse if and only if it is a 45° angle.
Conclusion: The biconditional is false. In fact, a 45° angle is acute, and acute angles are never obtuse.

7. true 8. true 9. false

10. true 11. false

12. Def. of perp. rays 13. Given

14. Substitution 15. Def. of perp. rays

16. Angle Addition Postulate

17. Subtraction Property of Equality

18. Substitution

Alternative Assessment—Chapter 2

Form A

1–5. Answers will vary.

6. angles that have equal measure

7. Two angles have equal measure if and only if they are congruent.

8. and 9. Answers will vary.

Score Point 4: Distinguished

The student demonstrates a comprehensive understanding of conditional statements. The student uses perceptive, creative, and complex mathematical reasoning as well as precise and appropriate mathematical language throughout the task. Theoretical knowledge is apparent and is applied to concrete situations as the student successfully demonstrates a comprehensive understanding of the core concepts.

Score Point 3: Proficient

The student demonstrates a broad understanding of conditional statements. The student uses precise and appropriate mathematical reasoning most of the time. Theoretical knowledge is apparent and is applied to concrete situations as the student attempts to draw conclusions based on the investigations.

Score Point 2: Apprentice

The student demonstrates an understanding of conditional statements. The student uses mathematical reasoning and appropriate mathematical reasoning some of the time. The student attempts to apply theoretical knowledge to the task but may not be able to draw conclusions based on the investigations.

Score Point 1: Novice

The student demonstrates a basic understanding of conditional statements. The student uses little mathematical reasoning or appropriate mathematical language. Theoretical knowledge may appear weak, and many responses may be illogical because directions were followed incorrectly.

Score Point 0: Unsatisfactory

The student does not make an attempt to complete the task, and the responses only restate the problem.

Form B

1. Answers will vary. Check students' proofs. The statement is true.

2. Answers will vary. Check students' proofs. The statement is false.

Score Point 4: Distinguished

The student demonstrates a comprehensive understanding of writing proofs. The student uses perceptive, creative, and complex mathematical reasoning as well as precise and appropriate mathematical language throughout the task. Theoretical knowledge is apparent and is applied to concrete situations as the student successfully demonstrates a comprehensive understanding of the core concepts.

Score Point 3: Proficient

The student demonstrates a broad understanding of writing proofs. The student uses precise and appropriate mathematical reasoning most of the time. Theoretical knowledge is apparent and is applied to concrete situations as the student attempts to draw conclusions based on the investigations.

Score Point 2: Apprentice

The student demonstrates an understanding of writing proofs. The student uses mathematical reasoning and appropriate mathematical language some of the time. The student attempts to apply theoretical knowledge to the task but may not be able to draw conclusions based on the investigations.

Score Point 1: Novice

The student demonstrates a basic understanding of writing proofs. The student uses little mathematical reasoning or appropriate mathematical language. Theoretical knowledge may appear weak, and many responses may be illogical because directions were followed incorrectly.

Score Point 0: Unsatisfactory

The student does not make an attempt to complete the task, and the responses only restate the problem.

Chapter 3

Quick Warm-Up 3.1

1. Any of these: $\overrightarrow{DE}$, $\overrightarrow{DB}$, $\overrightarrow{DF}$, $\overrightarrow{BD}$

2. $\overleftrightarrow{BD}$ 3. $\overrightarrow{DF}$

Lesson Quiz 3.1

1. both 2. reflectional

3. reflectional

4.

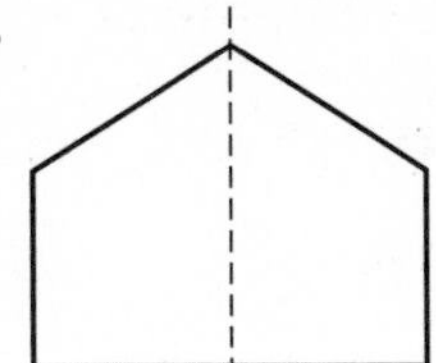

5.

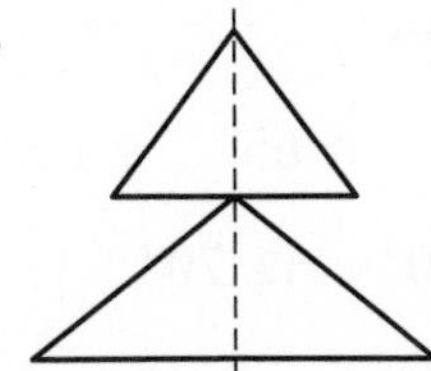

6. 90° 7. 60° 8. 36°

9. Check students' drawings.

Quick Warm-Up 3.2

1. $\angle BAD$, $\angle BCD$ 2. $\overline{AC}$

3. $\triangle ABC$, $\triangle DAB$, $\triangle BCD$, and $\triangle ADC$

Lesson Quiz 3.2

1. trapezoid 2. rhombus

3. square 4. rectangle

5. 120° 6. 8 7. 62° 8. 5 9. 10

10. 58° 11. 6 12. 15° 13. 12 14. 15°

15. 90° 16. 6 17. 9.5 18. 27° 19. 19

20. 27° 21. 63° 22. 9.5

Quick Warm-Up 3.3

1. Yes; vertical angles are congruent (Vertical Angle Theorem).

2. $\angle 8$ is congruent to $\angle 1$

3. They are a linear pair and are supplementary.

Answers

Lesson Quiz 3.3

1. alternate interior angles
2. corresponding angles
3. same-side interior angles
4. alternate exterior angles

5. 65° 6. 115° 7. 115° 8. 65° 9. 65°

10. 115° 11. 110° 12. 70° 13. 70°

14. 110° 15. 70° 16. 110°

Quick Warm-Up 3.4

1. If Rosalia can vote, then she is 18 years old.
2. 4 pairs
3. They are congruent.

Lesson Quiz 3.4

1. alternate interior angles are congruent
2. corresponding angles are congruent
3. same-side interior angles are supplementary

4. $m_1 \parallel m_2$ 5. $k_1 \parallel k_2$ 6. none

7. $m_1 \parallel m_2$ 8. $k_1 \parallel k_2$ 9. none 10. none

11. $\angle 1 \cong \angle 12$, $\angle 3 \cong \angle 10$, $\angle 5 \cong \angle 16$, $\angle 7 \cong \angle 14$

12. $\angle 1 \cong \angle 8$, $\angle 2 \cong \angle 7$, $\angle 9 \cong \angle 16$, $\angle 10 \cong \angle 15$

Mid-Chapter Assessment—Chapter 3

1. a 2. c 3. b 4. b 5. 6 6. 90°

7. 100° 8. 50° 9. 110° 10. 70°

11. 70° 12. 70°

13. Rotate n 70° clockwise or 110° counterclockwise about the point on n half the distance between k_1 and k_2.

Quick Warm-Up 3.5

1. $\ell \parallel m$ only if corresponding angles $\angle 1$ and $\angle 5$ are congruent. $m\angle 1 \neq m\angle 5$, so $\ell \nparallel m$.
2. They are parallel.

Lesson Quiz 3.5

1. They are exterior angles.

2. $\angle 1$ and $\angle 3$ 3. $\angle 1$ and $\angle 2$

4. It is equal to the sum of the remote interior angles' measures.

5. 45° 6. 60° 7. 65°

Quick Warm-Up 3.6

1. 110° 2. 60°

Lesson Quiz 3.6

1. 540° 2. 135° 3. 360° 4. 60° 5. 60°

6. 45° 7. 155° 8. 120° 9. 142°

10. 90° 11. 15

Quick Warm-Up 3.7

1. 14; 7 2. 26 3. $x = 12.5$

Lesson Quiz 3.7

1. 15 2. 50 3. 21 4. 18

5. cannot be determined

6. 7 7. 20 8. 40 9. 3 10. 5

Answers

Quick Warm-Up 3.8

1. 6; 11 **2.** 3 **3.** 1

Lesson Quiz 3.8

1. 1 **2.** -3 **3.** -2 **4.** perpendicular

5. neither **6.** parallel

7.

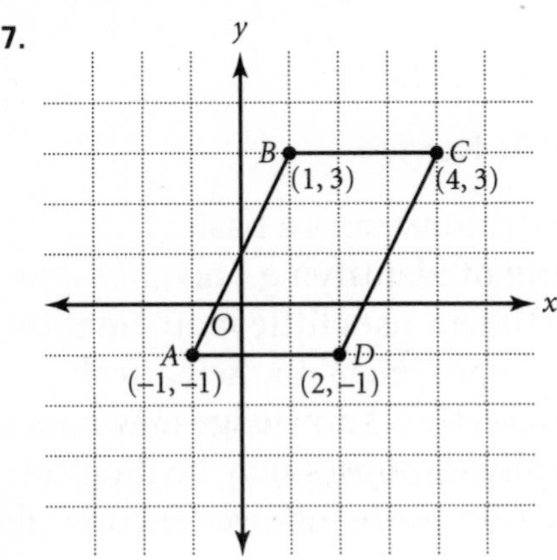

8. 2 **9.** 0 **10.** 2 **11.** 0

12. Parallelogram; the slopes of the opposite sides are equal, and the opposite sides are equal in length.

13.

Rectangle; the slopes of adjacent sides are negative reciprocals, and the opposite sides are equal in length.

Chapter Assessment—Chapter 3

Form A

1. c **2.** c **3.** b **4.** c **5.** c **6.** a

7. d **8.** d **9.** c **10.** b **11.** d **12.** a

13. b **14.** d **15.** c **16.** c **17.** b **18.** b

19. a **20.** c **21.** c **22.** d **23.** a

Form B

1. rotational **2.** both **3.** neither

4. both **5.** rotation

6. reflection across y-axis

7. rotation **8.** both **9.** rotation

10. True; a square is a rectangle whose sides are all congruent.

11. False; a rhombus does not need to have right angles.

12. 70° **13.** 10 **14.** 16 **15.** 50° **16.** 8

17. 110° **18.** 150° **19.** 30° **20.** 150°

21. 30° **22.** 30° **23.** 150°

24. the alternate interior angles are congruent

25. the same-side interior angles are supplementary

26. the corresponding angles are congruent

27. 70° **28.** 70° **29.** 2 centimeters

30. 0.75 centimeters **31.** 70° **32.** 70°

33. 2.25 centimeters **34.** 110° **35.** 110°

36. 1080° **37.** 140° **38.** 360° **39.** 36°

40. 10 **41.** 58 **42.** $-\frac{3}{2}$; (1, 1)

43. $-\frac{3}{2}$; $(-1, -1)$

Answers

Alternative Assessment—Chapter 3

Form A

1. rectangle or square
2. same-side interior angles or same-side exterior angles
3. rhombus
4. transversal
5. Answers will vary.
6. Lines *m* an *n* are parallel. If line ℓ is perpendicular to both *m* and *n*, then the corresponding angles must both be right angles and thus congruent. So the two lines must be parallel by the converse of the Corresponding Angles Postulate.

Score Point 4: Distinguished

The student demonstrates a comprehensive understanding of identifying polygons and angles. The student uses perceptive, creative, and complex mathematical reasoning as well as precise and appropriate mathematical language throughout the task. Theoretical knowledge is apparent and is applied to concrete situations as the student successfully demonstrates a comprehensive understanding of the core concepts.

Score Point 3: Proficient

The student demonstrates a broad understanding of identifying polygons and angles. The student uses precise and appropriate mathematical language most of the time. Theoretical knowledge is apparent and is applied to concrete situations as the student attempts to draw conclusions based on the investigations.

Score Point 2: Apprentice

The student demonstrates an understanding of identifying polygons and angles. The student uses mathematical reasoning and appropriate mathematical language some of the time. The student attempts to apply theoretical knowledge to the task but may not be able to draw conclusions based on the investigations.

Score Point 1: Novice

The student demonstrates a basic understanding of identifying polygons and angles. The student uses little mathematical reasoning or appropriate mathematical language. Theoretical knowledge may appear weak, and many responses may be illogical because directions were followed incorrectly.

Score Point 0: Unsatisfactory

The student does not make an attempt to complete the task, and the responses only restate the problem.

Form B

1. Answers will vary.

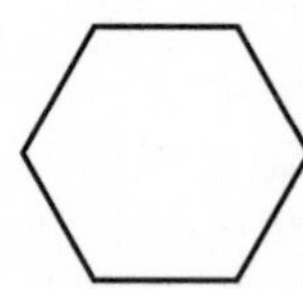

regular hexagon with 120° interior angles and 6 congruent sides and angles

2. Answers will vary.

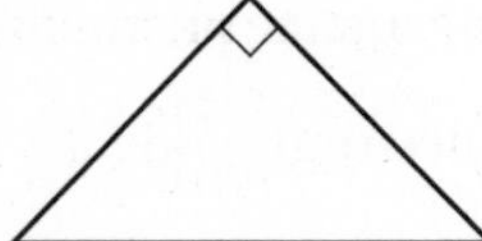

isosceles right triangle with 45°, 45°, and 90° angles

3. Answers will vary.

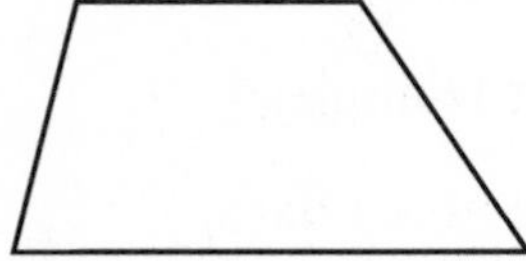

trapezoid with angles adding up to 360°

4. Answers will vary.

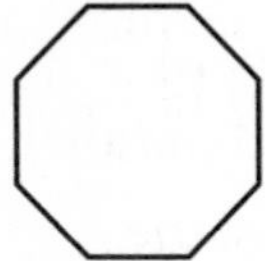

regular octagon with 135° interior angles and 8 congruent sides and angles

Score Point 4: Distinguished

The student demonstrates a comprehensive understanding of identifying and describing polygons. The student uses perceptive, creative, and complex mathematical reasoning as well as precise and appropriate mathematical language throughout the task. Theoretical knowledge is apparent and is applied to concrete situations as the student successfully demonstrates a comprehensive understanding of the core concepts.

Score Point 3: Proficient

The student demonstrates a broad understanding of identifying and describing polygons. The student uses precise and appropriate mathematical language most of the time. Theoretical knowledge is apparent and is applied to concrete situations as the student attempts to draw conclusions based on the investigations.

Score Point 2: Apprentice

The student demonstrates an understanding of identifying and describing polygons. The student uses mathematical reasoning and appropriate mathematical language some of the time. The student attempts to apply theoretical knowledge to the task but may not be able to draw conclusions based on the investigations.

Score Point 1: Novice

The student demonstrates a basic understanding of identifying and describing polygons. The student uses little mathematical reasoning or appropriate mathematical language. Theoretical knowledge may appear weak, and many responses may be illogical because directions were followed incorrectly.

Score Point 0: Unsatisfactory

The student does not make an attempt to complete the task, and the responses only restate the problem.

Chapter 4

Quick Warm-Up 4.1

1. $\angle B$ 2. $\overline{YW}$ 3. 60 degrees

Answers

Lesson Quiz 4.1

1. Yes; the angles have equal measures, so they are congruent.
2. No; the lengths of the corresponding sides are not equal, so the quadrilaterals are not congruent.
3. Yes; the segments are of equal length, so they are congruent.

4. $\overline{PQ}$ 5. $\angle P$ 6. $\overline{BC}$ 7. $\angle B$ 8. $\overline{SR}$

9. $\angle R$ 10. $\overline{AD}$ 11. $\angle D$

Quick Warm-Up 4.2

$\overline{DF}$ and $\overline{RQ}$, $\overline{RS}$ and $\overline{DE}$, $\overline{EF}$ and $\overline{SQ}$, $\angle D$ and $\angle R$, $\angle Q$ and $\angle F$, $\angle S$ and $\angle E$.

Lesson Quiz 4.2

1. $\triangle ABC \cong \triangle \text{FDE}$; Angle-Side-Angle
2. No; the three angles are congruent, but none of the sides are congruent.
3. $\triangle LNO \cong \triangle NLM$; Side-Angle-Side. ($\overline{LM} \parallel \overline{ON}$, so $\angle MLN \cong \angle ONL$; and by the Reflexive Property $\overline{LN} \cong \overline{LN}$.)
4. $\overline{BC} \cong \overline{CE}$
5. $\angle ACB \cong \angle DCE$ because vertical angles are congruent.
6. Answers will vary. $\angle B \cong \angle E$ because they are both right angles.

Quick Warm-Up 4.3

1. yes; any rhombus without right angles
2. no
3. yes; any 45-45-90 triangle
4. no

Lesson Quiz 4.3

1. No; *SSA* is not a congruence postulate.
2. $\triangle HJK \cong \triangle KLH$; Side-Angle-Side
3. $\triangle MPQ \cong \triangle ONQ$; Hypotenuse-Leg
4. Given
5. Given
6. Definition of perpendicular lines
7. Definition of right triangles
8. Reflexive Property
9. Hypotenuse-Leg

Quick Warm-Up 4.4

Answers will vary. Opposite sides of a parallelogram are congruent and parallel, so $\overline{AB} \cong \overline{DC}$. $\angle BAC \cong \angle DCA$ by the Alternate Interior Angles Theorem. Since $m\angle B = m\angle D = 90°$, $\angle B$ and $\angle D$ are congruent. Therefore the triangles are congruent by ASA.

Lesson Quiz 4.4

1. Two sides are congruent, and the angles opposite those sides are congruent.
2. Given
3. Reflexive Property
4. SAS
5. CPCTC

6. Sample Proof:

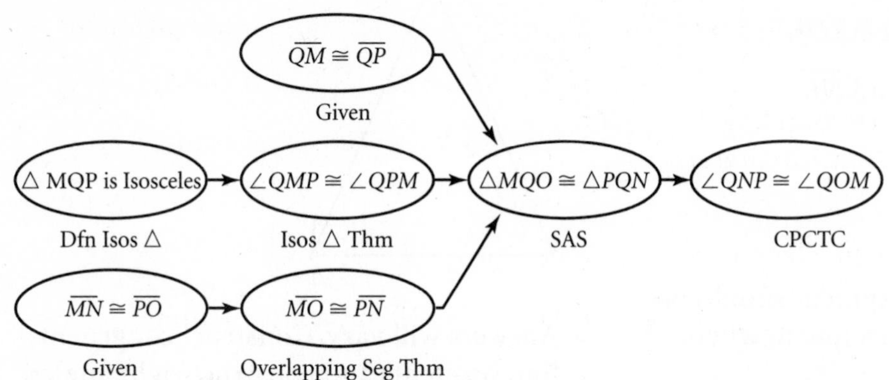

Mid-Chapter Assessment—Chapter 4

1. d 2. b 3. c 4. d 5. no

6. $\Delta FGH \cong \Delta KJI$; Angle-Side-Angle

7. $\Delta PQS \cong \Delta RSQ$; Hypotenuse-Leg

8. Sample Proof:

Statements	Reasons
1. $\overline{AD} \cong \overline{EC}$; $\overline{AB} \cong \overline{BC}$	Given
2. ΔABC is isosceles	Definition of isosceles triangle
3. $\angle BAC \cong \angle BCA$	Isosceles Triangle Theorem
4. $\overline{AC} \cong \overline{AC}$	Reflexive Property
5. $\Delta DAC \cong \Delta ECA$	SAS
6. $\overline{AE} \cong \overline{DC}$	CPCTC

Quick Warm-Up 4.5

1. Alike: opposite sides are parallel and congruent. Different: a square has four right angles and four congruent sides.

2. Alike: four congruent sides and opposite angles are congruent; diagonals are perpendicular. Different: a square has four right angles.

Lesson Quiz 4.5

1. They are alternate interior angles of $\overline{PS}$ and $\overline{QR}$ which are parallel.

2. four 3. eight 4. two 5. given

6. The sides of a square are congruent.

7. The diagonals of a parallelogram bisect each other.

8. Reflexive property 9. SSS 10. CPCTC

11. Congruent linear pairs are 90° angles.

Quick Warm-Up 4.6

1. $\overline{WX}$ and $\overline{ZY}$, $\overline{WZ}$ and $\overline{XY}$

2. $\angle X$ and $\angle Z$, $\angle W$ and $\angle Y$

3. $\overline{WX}$ and $\overline{XY}$, $\overline{YZ}$ and $\overline{XY}$, $\overline{YZ}$ and $\overline{ZW}$, $\overline{ZW}$ and $\overline{WX}$

4. $\angle W$ and $\angle X$, $\angle X$ and $\angle Y$, $\angle Y$ and $\angle Z$, $\angle Z$ and $\angle W$

Lesson Quiz 4.6

1. rectangle 2. rhombus 3. square

4. true 5. true 6. true 7. $\overline{KM}$ and $\overline{NL}$

8. $\overline{OL}$ 9. $\overline{OM}$ 10.–11. $\angle NOK$ and $\angle KOL$

12. right 13. 90°

14–16. $\triangle MON$, $\triangle NOK$, and $\triangle KOL$

17. SAS 18–20. $\overline{LM}$, $\overline{MN}$, and $\overline{NK}$

21. CPCTC 22. rhombus

Quick Warm-Up 4.7

1. Check students' work: segment should be from a starting point on a line to where the arc crosses the line.
2. Check students' work for same arc length as in Ex. 1, dragged around the starting point 360 degrees.
3. Draw a line segment. From one endpoint draw an arc with length a little more than half the length of the segment. Using the same compass setting, draw another arc from the other endpoint. The arcs should intersect at two points, and the line connecting the two intersection points is the perpendicular bisector.

Lesson Quiz 4.7

1.

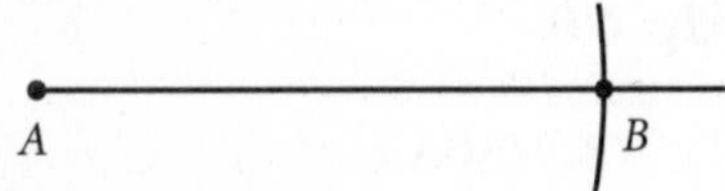

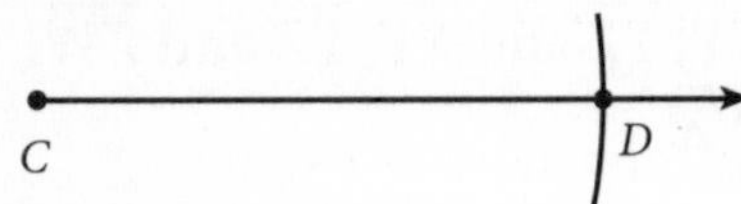

2.

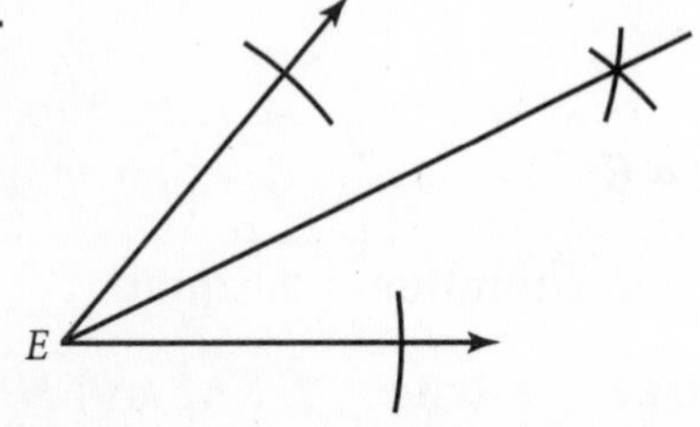

3.

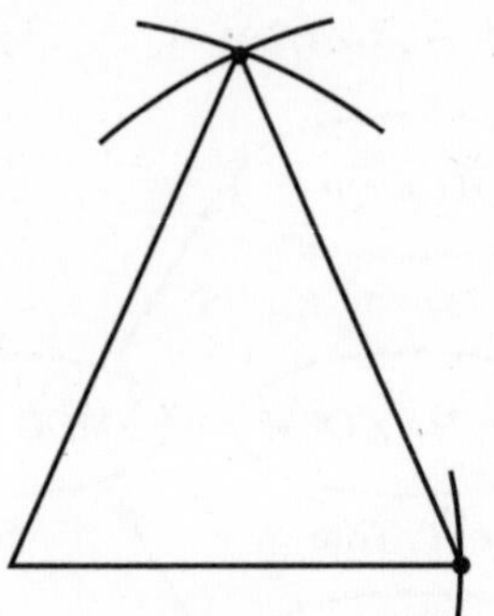

4. Answers will vary. Construct congruent line segments and construct right angles.
5. A single compass setting gives segments that are equal in length. Such line segments are congruent.

Quick Warm-Up 4.8

1. b 2. d 3. a 4. c

Lesson Quiz 4.8

1. The points are collinear since $7.2 + 8.08 = 15.28$, so $PR + RQ = PQ$ with R between P and Q.
2. yes; $3 + 4 > 5$, $4 + 5 > 3$, and $3 + 5 > 4$
3. no; $2 + 5 < 9$ 4. no; $4 + 4 = 8$
5. yes; $3 + 3 > 5$, $3 + 5 > 3$ 6. $\angle E$
7. $\overline{FG}$ 8. $\angle G$ 9. $\overline{HE}$ 10. $\angle F$ 11. $\overline{HG}$
12. Construct perpendiculars from A and B through line m. Name the points of intersection C and D. Construct $\overline{CX} \cong \overline{AC}$ and $\overline{DY} \cong \overline{BM}$. Draw $\overline{XY}$ which is congruent to $\overline{AB}$.

Chapter Assessment—Chapter 4

Form A

1. b 2. d 3. c 4. b 5. a 6. b

7. d 8. a 9. c 10. c 11. d 12. a

13. b 14. d 15. a 16. a 17. c

Form B

1. No; only corresponding angles are congruent.
2. Yes; when a figure is rotated, it has the same size and shape.
3. Given
4. Definition of isosceles triangle
5. Reflexive Property
6. Definition of perpendicular lines
7. Hypotenuse-Leg 8. CPCTC
9. $\angle K \cong \angle P, \angle L \cong \angle Q, \angle M \cong \angle R, \angle N \cong \angle S, \angle O \cong \angle T$
10. $\overline{KL} \cong \overline{PQ}, \overline{LM} \cong \overline{QR}, \overline{MN} \cong \overline{RS}, \overline{NO} \cong \overline{ST}, \overline{OK} \cong \overline{TP}$
11. 6
12. yes; by the Overlapping Segments Theorem
13. SAS.
14. Sample proof: By definition of a rectangle, $\overline{KJ} \cong \overline{LM}$ and, $\angle KJM$ and $\angle LMJ$ are both right angles. By the Reflexive Property, $\overline{JM} \cong \overline{JM}$. Therefore, $\Delta KJM \cong \Delta LMJ$ by SAS. The diagonals $\overline{KM}$ and $\overline{JL}$ are congruent by CPCTC.
15.

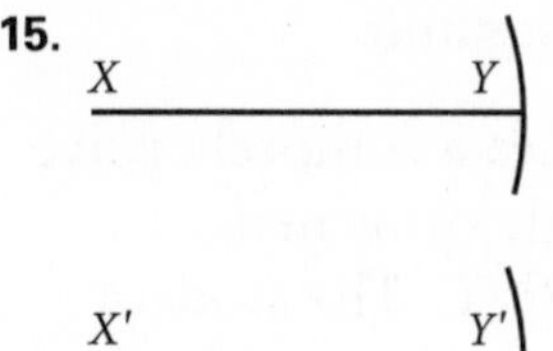

16.

E

17. yes 18. no 19. yes

Alternative Assessment—Chapter 4

Form A

1. There are six different pairs of triangles that can be proved congruent. See answers to Exercise 2.
2. $\Delta LPO \cong \Delta NPM$; SSS, SAS, ASA, AAS
   $\Delta LPM \cong \Delta NPO$; SSS, SAS, ASA, AAS;
   $\Delta LNM \cong \Delta OMN$; SSS, SAS
   $\Delta MLO \cong \Delta NOL$; SSS, SAS
   $\Delta LOM \cong \Delta NMO$; SSS, SAS, ASA, AAS
   $\Delta LON \cong \Delta MNO$: SSS, SAS, ASA, AAS
3. Yes; the information provided by the definition of a rectangle is enough to prove all of the triangles listed.
4. Yes. The information provided by P being the midpoint of $\overline{OM}$ and $\overline{LN}$ is enough to prove two pairs of triangles congruent. CPCTC can be used to prove the remaining pairs of triangles congruent.
5. Answers will vary. The proof should be one of the listed possibilities from Exercise 2.

Score Point 4: Distinguished

The student demonstrates a comprehensive understanding of triangle congruence theorems and using CPCTC. The student uses perceptive, creative, and complex mathematical reasoning as well as precise and appropriate mathematical language throughout the task. Theoretical knowledge is apparent and is applied to concrete situations as the student successfully demonstrates a comprehensive understanding of the core concepts.

Score Point 3: Proficient

The student demonstrates a broad understanding of triangle congruence theorems using CPCTC. The student uses precise and appropriate mathematical language most of the time. Theoretical knowledge is apparent and is applied to concrete situations as the student attempts to draw conclusions from the investigations.

Score Point 2: Apprentice

The student demonstrates an understanding of triangle congruence theorems and using CPCTC. The student uses mathematical reasoning and appropriate mathematical language some of the time. The student attempts to apply theoretical knowledge to the task but may not be able to draw conclusions from the investigations.

Score Point 1: Novice

The student demonstrates a basic understanding of triangle congruence theorems and using CPCTC. The student uses little mathematical reasoning or appropriate mathematical language. Theoretical knowledge may appear weak, and many responses may be illogical because directions were followed incorrectly.

Score Point 0: Unsatisfactory

The student does not make an attempt to complete the task, and the responses only restate the problem.

Form B

1. $\triangle WXZ \cong \triangle WYZ$, SSS, SAS; $\triangle XZV \cong \triangle YZV$, ASA; $\triangle WXV \cong \triangle WYV$, ASA

Statements	Reasons
1. $\overline{XW} \cong \overline{YW}$	Definition of equilateral triangle
2. $\overline{XZ} \cong \overline{YZ}$	Definition of isosceles triangle
3. $\overline{WZ} \cong \overline{WZ}$	Reflexive Property
4. $\triangle ZXW \cong \triangle ZYW$	SSS
5. $\angle XZV \cong \angle YZV$	CPCTC
6. $\angle VXZ \cong \angle VYZ$	Base angles of isosceles triangle are congruent.
7. $\triangle VXZ \cong \triangle VYZ$	ASA
8. $\angle XVZ \cong \angle YVZ$	CPCTC

Score Point 4: Distinguished

The student demonstrates a comprehensive understanding of triangle congruence theorems and construction steps. The student uses perceptive, creative, and complex mathematical reasoning as well as precise and appropriate mathematical language throughout the task. Theoretical knowledge is apparent and is applied to concrete situations as the student successfully demonstrates a comprehensive understanding of the core concepts.

Score Point 3: Proficient

The student demonstrates a broad understanding of triangle congruence theorems and construction steps. The student uses precise and appropriate mathematical language most of the time. Theoretical knowledge is apparent and is applied to concrete situations as the student attempts to construct the diagram and prove figures congruent.

Score Point 2: Apprentice

The student demonstrates an understanding of triangle congruence theorems and construction steps. The student uses mathematical reasoning and appropriate mathematical language some of the time. The student attempts to apply theoretical knowledge to the task but may not be able to draw conclusions based on the investigations.

Score Point 1: Novice

The student demonstrates a basic understanding of polygons. The student uses little mathematical reasoning or appropriate mathematical language. Theoretical knowledge may appear weak, and many responses may be illogical because directions were followed incorrectly.

Score Point 0: Unsatisfactory

The student does not make an attempt to complete the task, and the responses only restate the problem.

Chapter 5

Quick Warm-Up 5.1

1. 28 units **2.** 48 square units

3. Answers will vary. a 10×4 rectangle and a 13×1 rectangle

4. Answers will vary. a 16×3 rectangle and a 12×4 rectangle

Lesson Quiz 5.1

1. 50 units **2.** 35 units **3.** 16 units

4. 55 units **5.** 50 units **6.** 55 units

7. 32 square units **8.** 4 square units

9. 2 square units **10.** 28 square units

11. $P = 14$ units; $A = 12$ square units

12. $P = 30$ units; $A = 56$ square units

Quick Warm-Up 5.2

1. 3 sides, 1 right angle
2. 3 sides, all angles less than 90°
3. 4 sides, opposides sides parallel and congruent

Lesson Quiz 5.2

1. 32 square centimeters
2. 42 square meters 3. 96 square inches
4. Yes; two rectangles could both have an area of 24 square units with one having dimensions of 3 units by 8 units and the other 4 units by 6 units.
5. 180 square centimeters
6. 60 square centimeters
7. 240 square centimeters
8. 16 centimeters

Quick Warm-Up 5.3

1. 25 2. 44.89 3. 6 4. 6.93 5. 3.14

6. 6.28 7. 10.72 8. 28.27 9. 88.83

10. Check student drawings for labels: capital letter for the center, *r* for radius, and *d* for diameter.

Lesson Quiz 5.3

1. 75.4 2. 15.7 3. 201.1 4. 63.6

5. 373.85 square centimeters
6. 13.73 square inches
7. about 28 feet

Quick Warm-Up 5.4

1. 8.49 2. 3.16 3. 10.95 4. 7.21

5. 13.08

Lesson Quiz 5.4

1. 10 2. 17 3. 11.2 4. 16.6

5. 12 centimeters 6. 7 in. 7. $\sqrt{95}$ m

8. $8\sqrt{2} \approx 11.3$ centimeters
9. 50 square centimeters
10. $4\sqrt{2} \approx 5.7$ centimeters
11. obtuse 12. not possible 13. right

14. obtuse 15. not possible 16. acute

Mid-Chapter Assessment—Chapter 5

1. c 2. b 3. b 4. a

5. 12 centimeters 6. 26 inches

7. approximately 12.57 meters
8. 450 square centimeters
9. 120 square inches
10. 162 square centimeters
11. $924.48

Quick Warm-Up 5.5

1. $a = 16, b = 4$ 2. $a = 9, b = 3$

1. $a = 9, b = 3$

Lesson Quiz 5.5

1. $x = 8\sqrt{3}; y = 16$ 2. $x = 5; y = 10$

3. $x = 4; y = 4\sqrt{2}$ 4. $x = 7; y = 7$

5. $5\sqrt{2}$ centimeters 6. 6 centimeters

7. 72 square centimeters

8. $8\sqrt{3} \approx 13.9$ square inches

9. $54\sqrt{3} \approx 93.5$ square centimeters

10. $36\sqrt{3} \approx 62.4$ square centimeters

Quick Warm-Up 5.6

1. 4 2. 25 3. 5

Lesson Quiz 5.6

1. 5 2. 13 3. $\sqrt{10} \approx 3.16$

4. $4\sqrt{2} \approx 5.66$ 5. 5 6. 4

7. Yes; $(AC)^2 + (BC)^2 = 3^2 + 4^2 = 5^2 = (AB)^2$

8. Yes; $AD = DE = EF = FA = \sqrt{18}$, so all sides are congruent.

9. (2.5, 1) 10. 6 square units

11. 3 square units

Quick Warm-Up 5.7

1. -1 2. 10 3. $(4, -2)$

Lesson Quiz 5.7

1. $C(4, 4)$; $D(4, 0)$ 2. $C(6, 0)$ 3. 5

4. $(a + m, b + n)$ 5. $-\frac{t}{s}$

6. The coordinates of the fourth vertex C are (9, 6). By the distance formula, $AC = \sqrt{89}$ and $BD = \sqrt{89}$.

Quick Warm-Up 5.8

Answers will vary. A circle of radius 3 inside a 10 × 7 rectangle has an area ratio of 9π to 70, or about 2 to 5.

Lesson Quiz 5.8

1. 20% 2. 60% 3. 62.5% 4. $\frac{3}{8}$ 5. $\frac{3}{4}$

6. $\frac{1}{5}$ 7. $\frac{6}{11}$ 8. $\frac{9}{11}$ 9. $\frac{2}{11}$ 10. $\frac{4}{25}$ 11. $\frac{3}{4}$

12. $\frac{2}{9}$

Chapter Assessment—Chapter 5

Form A

1. d 2. c 3. a 4. c 5. a 6. b 7. b

8. d 9. b 10. b 11. c 12. b 13. a

14. b 15. b 16. c 17. b 18. c 19. c

Form B

1. $P = 24$ centimeters; $A = 28$ square centimeters

2. $P = 38$ inches; $A = 42$ square inches

3. $P = 24$ centimeters; $A = 24\sqrt{3} \approx 41.6$ centimeters

4. 120 square centimeters

5. base = 18 centimeters; height = 6 centimeters

6. $x = 4.2$ 7. $x = 8.9$ inches

8. $p = 3.77$; $a = 1.13$

9. $p = 47.12$ inches; 176.71 square inches

10. right 11. obtuse 12. not possible

13. acute 14. 10.27 15. $b = 12$

16. $s = 5$ 17. $x = 3\sqrt{3}$; $h = 9$

18. about 4 revolutions 19. $75\sqrt{2}$

20. $AB = 3\sqrt{10}$; $BC = 2\sqrt{10}$; $AC = \sqrt{130}$

21. $(AB)^2 + (BC)^2 = 90 + 40 = 130 = (AC)^2$

22. 30 square units

Answers

23. P = 42 feet; A ≈ 127.31 square feet

24. $P = 28\sqrt{3} \approx 48.50$; $A = \frac{196\sqrt{3}}{3} \approx 113.16$

25. $\frac{18}{60} = 30\%$.

Alternative Assessment—Chapter 5

Form A

1. The rectangular space in the fair location.
2. The rectangular space in the fair location.
3. Any space except the circular space because it is slightly over $48,000.
4. Answers will vary. Answers should be supported by data.
5. Answers will vary depending on the student's response. The perimeters are as follows: circle = 100.53 feet, rectangle = 132 feet, isosceles trapezoid = 126.6 feet, and trapezoid = 122.5 feet.

Score Point 4: Distinguished

The student demonstrates a comprehensive understanding of perimeter and area. The student uses perceptive, creative, and complex mathematical reasoning as well as precise and appropriate mathematical language throughout the task. Theoretical knowledge is apparent and is applied to concrete situations as the student successfully demonstrates a comprehensive understanding of the core concepts.

Score Point 3: Proficient

The student demonstrates a broad understanding of perimeter and area. The student uses precise and appropriate mathematical language most of the time. Theoretical knowledge is apparent and is applied to concrete situations as the student attempts to calculate the cost of each store space.

Score Point 2: Apprentice

The student demonstrates an understanding of perimeter and area. The student uses mathematical reasoning and appropriate mathematical language some of the time. The student attempts to apply theoretical knowledge to the task but may not be able to calculate the cost of each store space.

Score Point 1: Novice

The student demonstrates a basic undertanding of perimeter and area. The student uses little mathematical reasoning or appropriate mathematical language. Theoretical knowledge may appear weak, and many responses may be illogical because directions were followed incorrectly.

Score Point 0: Unsatisfactory

The student does not make an attempt to complete the task, and the responses only restate the problem.

Answers

Form B

The figure is a trapezoid with a height of 7 units and an area of 87.5 square units. The figure consists of two right triangles and a rectangle if vertical lines are drawn at the endpoints of the top base. The area of the triangle on the left is 14 square units. By the Pythagorean Theorem, the sides of this triangle are 4 units, 7 units, and about 8.06 units. Using the 45°-45°-90° pattern, the isosceles right triangle with sides of 7 units and hypotenuse of $7\sqrt{2}$ units has an area of 24.5 units. The perimeter of the entire figure is about 40.96 units.

The area can also be determined by the distance formula. The area of the rectangle is 42 square units and has a perimeter of 26 units. The midpoint of the bottom base of the entire trapezoid is (9.5, 1). The midpoint of the right side is (14.5, 4.5). The midpoint of the top base is (8, 8). The midpoint of the left side is (3, 4.5)

Score Point 4: Distinguished

The student demonstrates a comprehensive understanding of triangle concepts–perimeter, area, the Pythagorean Theorem, special triangle patterns, and the distance formula. The student uses perceptive, creative, and complex mathematical reasoning as well as precise and appropriate mathematical language throughout the task. Theoretical knowledge is apparent and is applied to concrete situations as the student successfully demonstrates a comprehensive understanding of the core concepts.

Score Point 3: Proficient

The student demonstrates a broad understanding of perimeter, area, the Pythagorean Theorem, special triangle patterns, and the distance formula. The student uses precise and appropriate mathematical language most of the time. Theoretical knowledge is apparent and is applied to concrete situations as the student attempts to describe the figure.

Score Point 2: Apprentice

The student demonstrates an understanding of perimeter, area, the Pythagorean Theorem, special triangle patterns, and the distance formula. The student uses mathematical reasoning and appropriate mathematical language some of the time. The student attempts to apply theoretical knowledge to the task but may not be able to draw conclusions about the figure.

Score Point 1: Novice

The student demonstrates a basic understanding of perimeter, area, the Pythagorean Theorem, special triangle patterns, and the distance formula. The student uses little mathematical reasoning or appropriate mathematical language. Theoretical knowledge may appear weak, and many responses may be illogical because directions were followed incorrectly.

Score Point 0: Unsatisfactory

The student does not make an attempt to complete the task, and the responses only restate the problem.

Chapter 6

Quick Warm-Up 6.1

1. 6 2. 3 3. 1 cm^2

Lesson Quiz 6.1

1. 13 cubic units 2. 54 square units

3–4.

front and back

5–6.

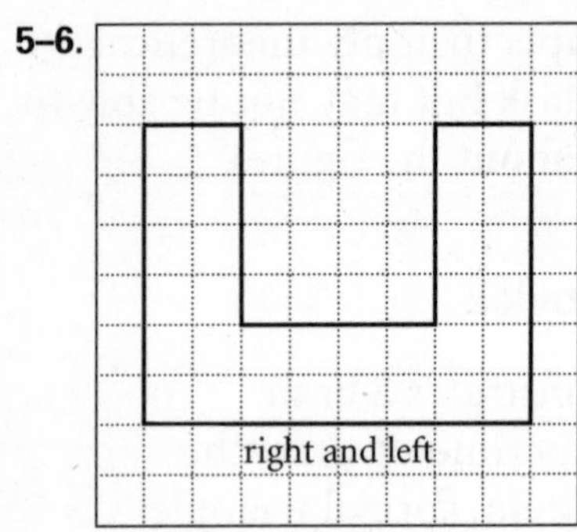

7–8.

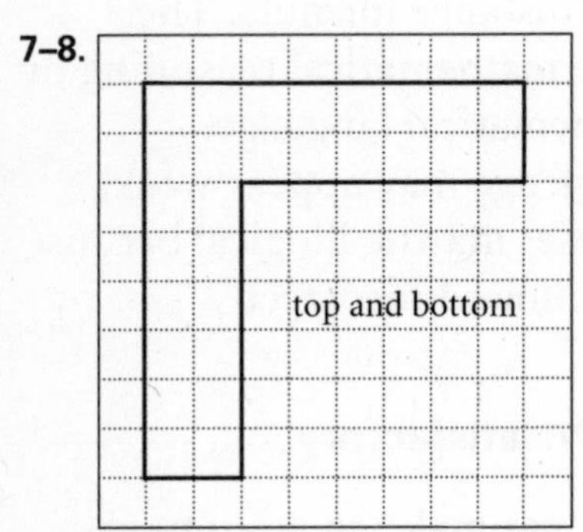

Quick Warm-Up 6.2

1. false 2. true 3. true 4. true

Lesson Quiz 6.2

1. *EFGH* 2. *ABEH, BCFE, CDGF, ADGH*

3. $\overline{AD}$, $\overline{EF}$, $\overline{HG}$, 4. *CDFG, ABEH*

5. $\overline{AH}$, $\overline{BE}$, $\overline{CF}$, $\overline{DG}$ 6. yes; plane *IPKN*

7. no 8. sample answer: $\overline{JK}$, $\overline{MN}$, $\overline{IJ}$, $\overline{MP}$

9. *JKMN* and *MNOP* 10. $\overline{JK}$

11. *MNOP* and *ILOP*

Quick Warm-Up 6.3

1. *ABC, EFD, BCDF, ACDE,* and *ABFE*

2. *ABC* and *EFD*

Lesson Quiz 6.3

1. right triangular prism 2. cube

3. oblique rectangular prism

4. $\overline{VB}$, $\overline{WC}$, $\overline{XD}$, $\overline{YE}$, $\overline{ZF}$ 5. *UVWXYZ*

6. *AUVB, BVWC, CWXD, DXYE, EYZF, FZUA*

7. regular hexagon 8. 7 9. 11

10. 12 faces; 20 vertices; 30 edges

Mid-Chapter Assessment—Chapter 6

1. c 2. a 3. b 4. b 5. c 6. d 7. 11

8. 13 9. 17 10. $6\sqrt{3}$

Quick Warm-Up 6.4

1. (5.5, −3.5) 2. *x*-axis 3. (5, 7)

Answers

Lesson Quiz 6.4

1–3.

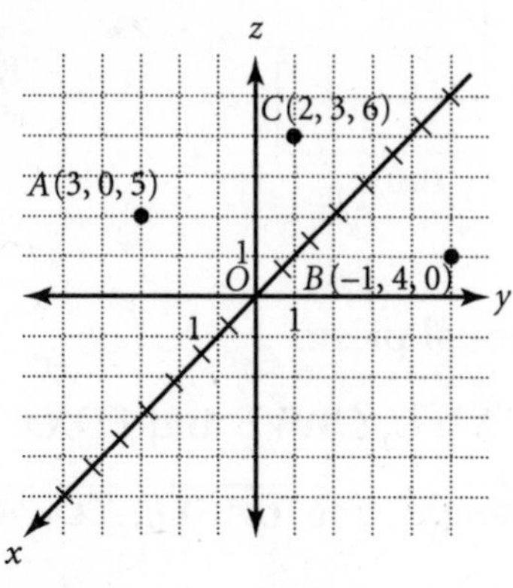

4. (4, 0, 0) 5. (4, 0, 3) 6. (4, 6, 3)
7. (0, 6, 3) 8. 3; (6, 2.5, 1) 9. 7; (4, 3, 1.5)
10. 7; (0, 3.5, 1) 11. $2\sqrt{14}$; (2, −2, 0)
12. *xy* plane; first octant
13. *yz* plane; bottom-front-left octant

Quick Warm-Up 6.5

1. *x*-intercept (3, 0); *y*-intercept (0, 2)
2. no 3. $y = -3$

Lesson Quiz 6.5

1.

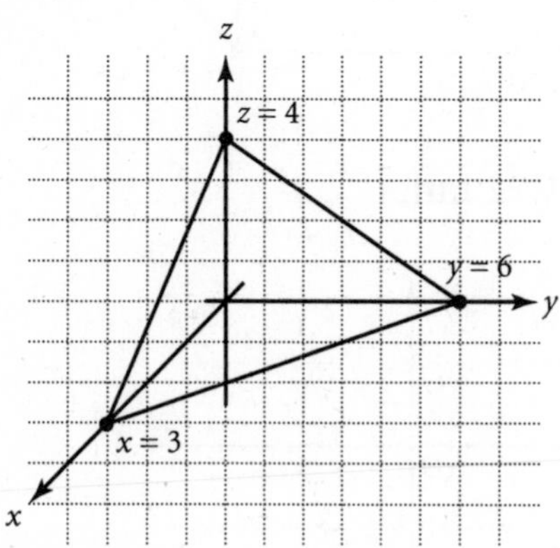

2.

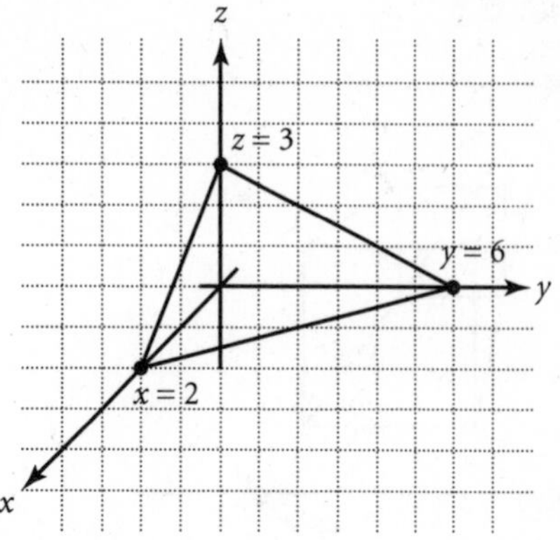

3.

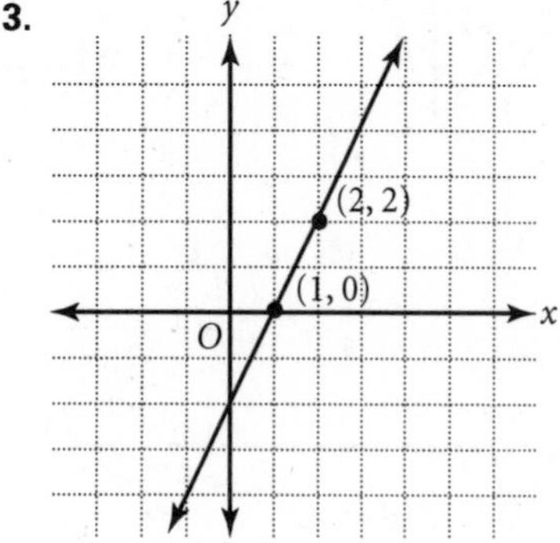

4.

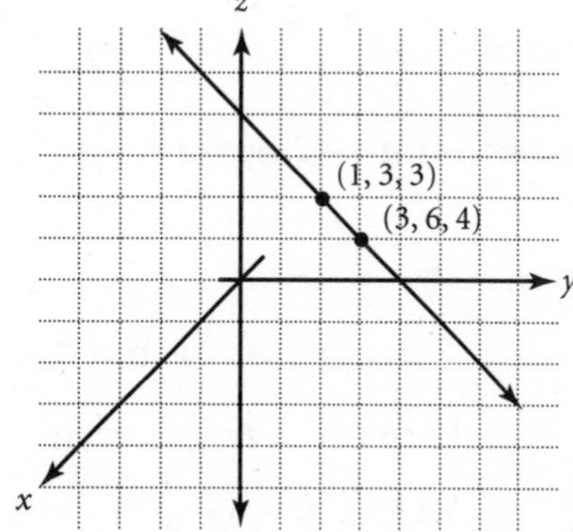

Quick Warm-Up 6.6

1. Answers will vary.
2. Answers will vary. Sample answer: Draw parallelograms for sides, like the ones shown, instead of rectangles.

Lesson Quiz 6.6

1\.

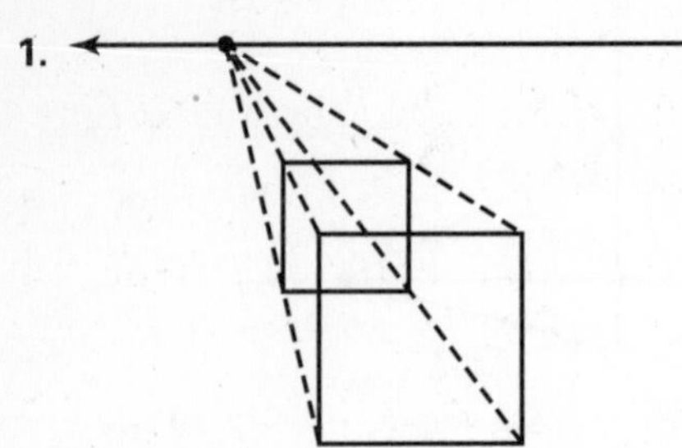

2\.

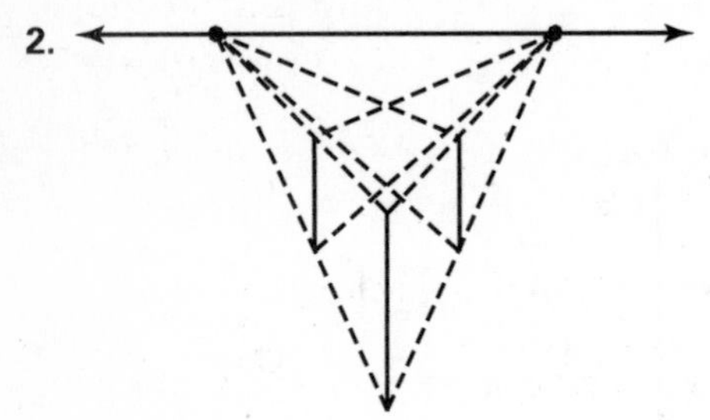

3\.

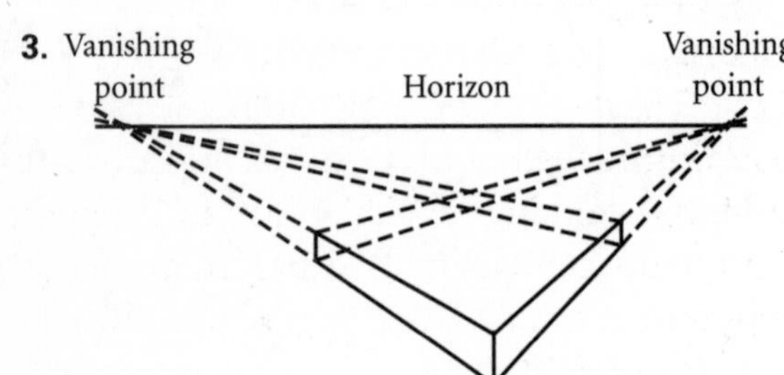

4–5.

front left

6\. right pentagonal prism

7\. *AUYE* and *DXWC*, *CWVB* and *EDXY*

8\. 90°   9. rectangle   10. $\overline{UA}$, $\overline{YE}$, $\overline{XD}$, $\overline{WC}$

11\. 11 units   12. 17 units

13\. about 9.54 units; (1.5, 4.5, −5.5)

14\. about 11.58 units; (5, 1.5, 1.5)

15\. (6, 0, 5), (0, 0, 5), (0, 9, 5) and (0, 9, 0)

16\.

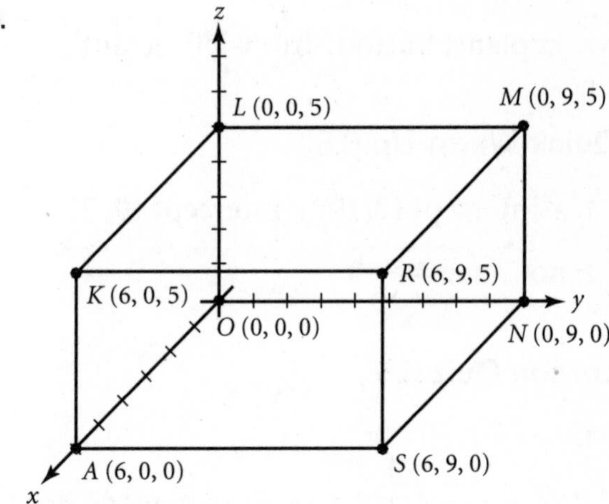

17\. about 11.92 units

18\.

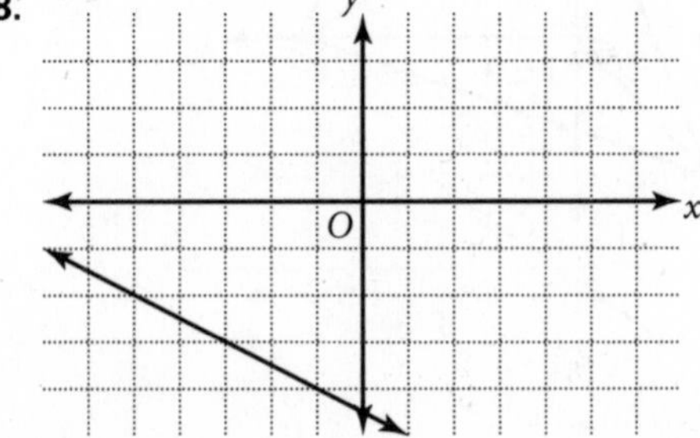

Chapter Assessment—Chapter 6

Form A

1\. b   2. c   3. a   4. c   5. a   6. b   7. d

8\. c   9. b   10. d   11. a   12. c   13. a

14\. d   15. a   16. b   17. b   18. a

Form B

1\. 10; 10 cubic units   2. 12

3\. 40 square units

19.
20.
21.
22.

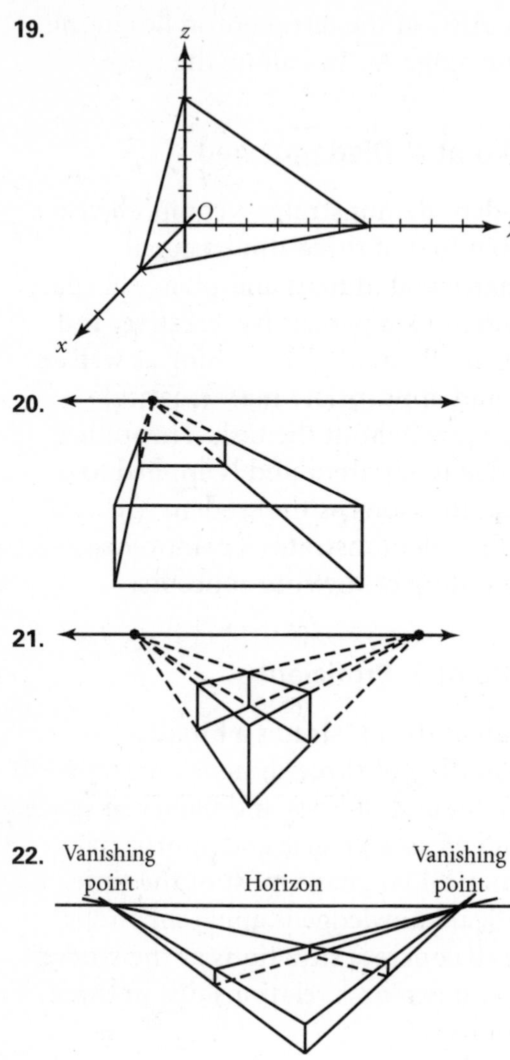

Alternative Assessment—Chapter 6

Form A

1. Sample answer:

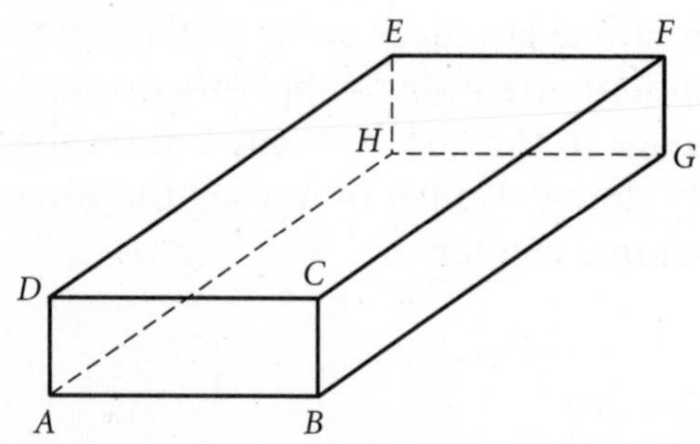

2. *ABGH* and *DCFE*; rectangles
3. *ABCD* and *HGFE*, *BCFG* and *ADEH*; rectangles
4. *ABGH* and *DCFE*, *ABCD* and *HGFE*, *BCFG* and *ADEH*
5. *ABCD* and *BCFG*, *ABGH* and *HGFE*, *DCFE* and *AHED*
6. All interior angles measure 90°. The prism is right and rectangular, so the bases and faces meet at right angles.
7. Use the formula $d = \sqrt{\ell^2 + w^2 + h^2}$.

Score Point 4: Distinguished

The student demonstrates a comprehensive understanding of prisms and their properties. The student uses perceptive, creative, and complex mathematical reasoning as well as precise and appropriate mathematical language throughout the task. Theoretical knowledge is apparent and is applied to concrete situations as the student successfully demonstrates a comprehensive understanding of the core concepts.

Score Point 3: Proficient

The student demonstrates a broad understanding of prisms and their properties. The student uses precise and appropriate mathematical language most of the time. Theoretical knowledge is apparent and is applied to concrete situations as the student attempts to draw a prism and identify its features and properties.

Score Point 2: Apprentice

The student demonstrates an understanding of prisms and their properties. The student uses mathematical reasoning and appropriate mathematical language some of the time. The student attempts to apply theoretical knowledge to the task but may not be able to draw conclusions from the investigation.

Score Point 1: Novice

The student demonstrates a basic understanding of prisms and their properties. The student uses little mathematical reasoning or appropriate mathematical language. Theoretical knowledge may appear weak, and many responses may be illogical because directions were followed incorrectly.

Score Point 0: Unsatisfactory

The student does not make an attempt to complete the task, and the responses only restate the problem.

Form B

1., 2., and 5.

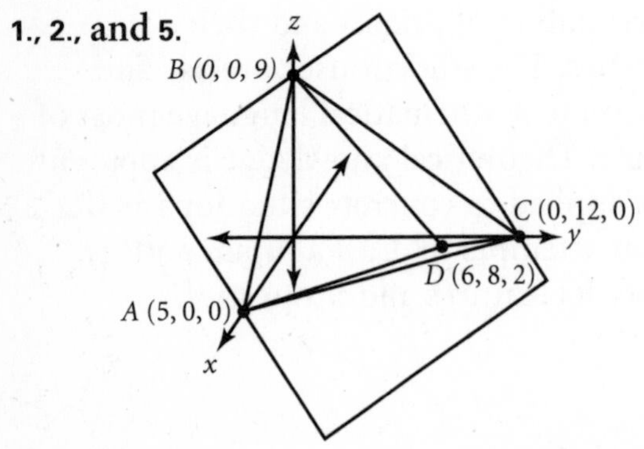

2. a tetrahedron 3. 13 4. about 12.8

5. about 8.3 6. See figure.

7. $12x + 5y = 60$

8. Side ABC of the tetrahedron lies in the plane. Edge $\overline{AC}$ lies along the trace.

Score Point 4: Distinguished

The student demonstrates a comprehensive understanding of three-dimensional coordinates and of lines and planes in space. The student uses perceptive, creative, and complex mathematical reasoning as well as precise and appropriate mathematical language throughout the task. Theoretical knowledge is apparent and is applied to concrete situations as the student successfully demonstrates a comprehensive understanding of the core concepts.

Score Point 3: Proficient

The student demonstrates a broad understanding of three-dimensional coordinates and of lines and planes in space. The student uses precise and appropriate mathematical language most of the time. Theoretical knowledge is apparent and is applied to concrete situations as the student locates and explores relationships in three-dimensions.

Score Point 2: Apprentice

The student demonstrates an understanding of three-dimensional coordinates and of lines and planes in space. The student uses mathematical reasoning and appropriate mathematical language some of the time. The student attempts to apply theoretical knowledge to the task but may not be able to identify the relationships among the three-dimensional elements.

Score Point 1: Novice

The student demonstrates a basic understanding of three-dimensional coordinates and of lines and planes in space. The student uses little mathematical reasoning or appropriate mathematical language. Theoretical knowledge may appear weak, and many responses may be illogical because directions were followed incorrectly.

Score Point 0: Unsatisfactory

The student does not make an attempt to complete the task, and the responses only restate the problem.

Chapter 7

Quick Warm-Up 7.1

1. surface area: 96 in.^{2}; volume: 64 in.^{3}
2. surface area: 94 square inches; volume: 60 cubic inches
3. 8:12 or 2:3

Lesson 7.1

1. 2:1 2. 8:3 3. 10:3 4. 6:5 5. $\frac{6}{x}$

6.

Side (in.)	Length (in.)	Width (in.)	Height (in.)	Volume (in.^{3})
1	8	6	1	48
1.5	7	5	1.5	52.5
2	6	4	2	48
2.5	5	3	2.5	37.5
3	4	2	3	24

7. the 1.5-inch square

Quick Warm-Up 7.2

1. 12 square cm
2. $\sqrt{57}$
3. surface area: 384 square inches; volume: 512 cubic inches

Lesson 7.2

1.

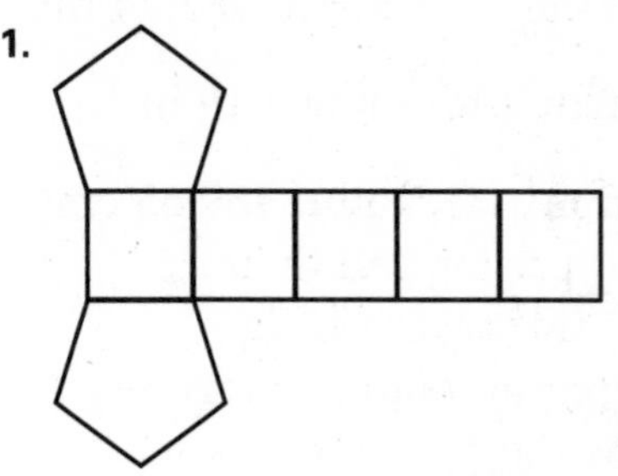

2. 112 in.^{2}; 64 in.^{3} 3. 85.2 m^2; 36 m^3

4. 12 ft 5. 360 cm^2

6. $432\sqrt{3} \approx 748.25$ ft^3

7. $108\sqrt{3} \approx 187.06$ ft^3

Quick Warm-Up 7.3

1. 384 square units 2. 7.42 or $\sqrt{55}$

3. 150 square cm

Lesson 7.3

1.

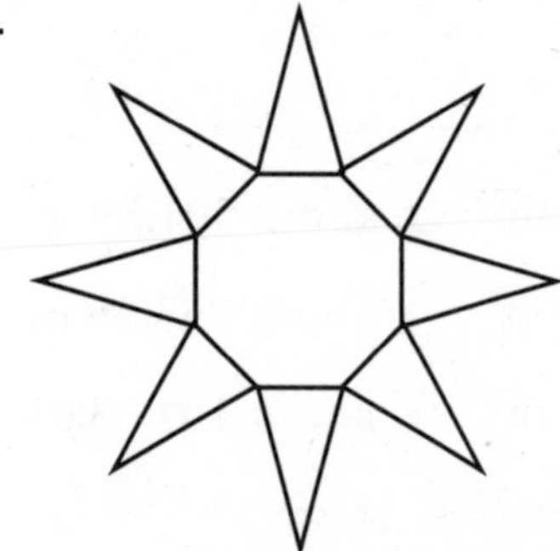

2. 360 in.^{2} 3. $64\sqrt{3} = 110.85$ cm^2

4. 4 ft 5. 96 m^3 6. 72 cm^3

Answers

Quick Warm-Up 7.4

1. circumference: 37.7 units; area: 113.1 square units
2. surface are: 312 square units; volume: 360 cubic units

Lesson 7.4

1. about 414.69 cm^2 2. about 282.74 m^2
3. about 8.25 ft^2 4. about 603.19 $in.^3$
5. about 502.65 m^3 6. about 169.65 cm^3
7. 4 cm 8. 3 cm
9. about 1,430,000 gallons
10. 4π; 4 times as much
11. about 15,039 gallons

Mid-Chapter Assessment—Chapter 7

1. d 2. d 3. c 4. b 5. b 6. 3:2
7. 156 m^2 8. 48 ft^3 9. 9 cm
10. about 251.3 cm^2 11. about 1526.8 $in.^3$
12. 18 labels

Quick Warm-Up 7.5

1. true 2. true 3. true 4. false

Lesson 7.5

1. about 282.74 cm^2 2. about 301.59 $in.^2$
3. about 1507.96 m^2 4. about 753.98 m^3
5. about 1005.31 m^3 6. about 1017.88 $in.^3$
7. 2.5 cm 8. 6 cm 9. 24 cm

Quick Warm-Up 7.6

1. 64π, or 201.1 square cm
2. 288π, or 904.8 cubic inches
3. 32π, or 100.5 cubic cm

Lesson 7.6

1. The surface area is divided by 4.
2. Student proofs should include the facts that $LA = 2\pi rh$ and $h = 2r$. Thus, $LA = 2\pi r(2r) = 4\pi r^2$.
3. 201.06 $in.^2$; 268.08 $in.^3$
4. 5026.55 cm^2; 33,510.32 cm^3
5. 314.16 m^2; 523.60 m^3
6. 3.14 cm^2; 0.52 cm^3
7. 452.39 $in.^2$; 904.78 $in.^3$
8. 113.10 ft^2; 113.10 ft^3
9. about 706.86 $in.^2$ 10. about 34 $in.^3$

Quick Warm-Up 7.7

1. (2, 4) 2. (3, 4) 3. yz-plane

Lesson 7.7

1.

2. (3, 5, −6) 3. (3, −5, 6) 4. (−3, 5, 6)

5a. cylinder with a radius of 5 and height of 10

5b. circle with a radius of 10 and its interior

6. about 471.24 square units; about 314.16 square units

Chapter Assessment—Chapter 7

Form A

1. c **2.** d **3.** c **4.** b **5.** a **6.** b **7.** b

8. a **9.** a **10.** d **11.** c **12.** c **13.** c

14. b **15.** d **16.** c **17.** d **18.** a **19.** b

Form B

1. 13:6 **2.** 600 cm^2 **3.** about 113.10 $in.^2$

4. 184 m^2 **5.** about 408.41 ft^2 **6.** 144 cm^2

7. about 301.59 cm^2 **8.** 6 yd **9.** 8 in.

10. 5 m **11.** about 6.2 $in.^2$ **12.** sphere

13. 768 cm^2 **14.** 235,000,000 square miles

15. 90 m^3 **16.** about 603.19 ft^3

17. 400 cm^3 **18.** about 904.78 ft^3

19. 1005.31 cm^3 **20.** 48 cm^3 **21.** 64 ft^3

22. 4.0 mi^3 **23.** 800 $in.^3$ **24.** (2, 3, 1)

25. $(-2, 3, -1)$ **26.** $(2, -3, -1)$

27. cylinder with a radius of 3 and height of 5; about 141.37 cubic units

28. 35 is the area of the trapezoid

Alternative Assessment—Chapter 7

Form A

1. Sample answer: to determine package costs, size of shipping containers, and transportation costs

2. right cylinder

Number of balls	Diam. (in.)	Height (in.)	SA ($in.^2$)	Volume ($in.^3$)
3	1.5	4.5	24.74	7.95
4	1.5	6	31.81	10.6
5	1.5	7.5	38.88	13.25
6	1.5	9	45.95	15.9

3. right square prism

Number of balls	Diam. (in.)	Height (in.)	SA ($in.^2$)	Volume ($in.^3$)
3	1.5	4.5	31.5	10.13
4	1.5	6	40.5	13.5
5	1.5	7.5	49.5	16.88
6	1.5	9	58.5	20.25

Score Point 4: Distinguished

The student demonstrates a comprehensive understanding of the properties of prisms and cylinders. The student uses perceptive, creative, and complex mathematical reasoning as well as precise and appropriate mathematical language throughout the task. Theoretical knowledge is apparent and is applied to concrete situations as the student successfully demonstrates a comprehensive understanding of the core concepts.

Score Point 3: Proficient

The student demonstrates a broad understanding of the properties of prisms and cylinders. The student uses precise and appropriate mathematical language most of the time. Theoretical knowledge is apparent and is applied to concrete situations as the student attempts to calculate the size of each package.

Score Point 2: Apprentice

The student demonstrates an understanding of the properties of prisms and cylinders. The student uses mathematical reasoning and appropriate mathematical language some of the time. The student attempts to apply theoretical knowledge to the task but may not be able to calculate the size of each package.

Score Point 1: Novice

The student demonstrates a basic understanding of the properties of prisms and cylinders but is unable to complete the task. The student uses little mathematical reasoning or appropriate mathematical language. Theoretical knowledge may appear weak, and many responses may be illogical because directions were followed incorrectly.

Score Point 0: Unsatisfactory

The student does not make an attempt to complete the task, and the responses only restate the problem.

Form B

1. $5.99 \approx 6$ m 2. $23.87 \approx 24$ m

3. $10.61 \approx 11$ m

4. sample answer: $h = 9$ m, $\ell = 8.82$ m

5. sample answer: $h = 9$ m, $r = 9.77$ m

6. There is no other sphere with the same volume.

Score Point 4: Distinguished

The student demonstrates a comprehensive understanding of spheres, cones, and pyramids and their properties. The student uses perceptive, creative, and complex mathematical reasoning as well as precise and appropriate mathematical language throughout the task. Theoretical knowledge is apparent and is applied to concrete situations as the student successfully demonstrates a comprehensive understanding of the core concepts.

Score Point 3: Proficient

The student demonstrates a broad understanding of spheres, cones, and pyramids and their properties. The student uses precise and appropriate mathematical language most of the time. Theoretical knowledge is apparent and is applied to concrete situations as the student attempts to describe the figure.

Score Point 2: Apprentice

The student demonstrates an understanding of spheres, cones, and pyramids and their properties. The student uses mathematical reasoning and appropriate mathematical language some of the time. The student attempts to apply theoretical knowledge to the task but may not be able to make mathematical determinations about the figure.

Score Point 1: Novice

The student demonstrates a basic understanding of spheres, cones, and pyramids and their properties but is unable to complete the task. The student uses little mathematical reasoning or appropriate mathematical language. Theoretical knowledge may appear weak, and many responses may be illogical because directions were followed incorrectly.

Score Point 0: Unsatisfactory

The student does not make an attempt to complete the task, and the responses only restate the problem.

Chapter 8

Quick Warm-Up 8.1

1. $\sqrt{40}$, or $2\sqrt{10}$ 2. -3 3. $(-2, -1)$

4. $y = -x + 4$

Lesson Quiz 8.1

1. $(-6, 4)$ 2. $(-9, 6)$ 3. $\frac{1}{2}$

4.

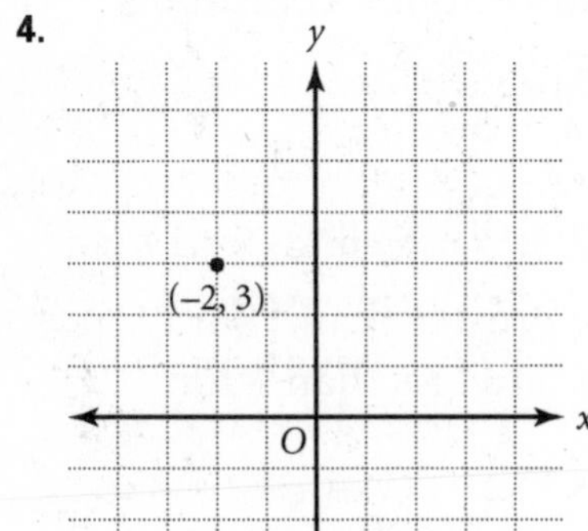

5. $y = 0$ 6. $x = -2y$

Quick Warm-Up 8.2

1. $\angle ABC \cong \angle DEF$; $\angle ACB \cong \angle DFE$; $\angle BAC \cong \angle EDF$; $\overline{AB} \cong \overline{DE}$; $\overline{AC} \cong \overline{DF}$; $\overline{BC} \cong \overline{EF}$

2. 1.5 3. 7.5

Lesson Quiz 8.2

1. yes 2. no 3. 6 4. 10 5. -5

6. 10 7. 7.5 8. true

9. false; $\frac{1}{2} = \frac{2}{4}$ but $\frac{1+4}{2} \neq \frac{2+2}{4}$

Quick Warm-Up 8.3

1. SSS, SAS, ASA 2. 1 to 1 3. 1 to 1

Lesson Quiz 8.3

1. If two angles of one triangle are congruent to two angles of another triangle, then the triangles are similar.

2. If two sides of one triangle are proportional to two sides of another triangle and their included angles are congruent, then the triangles are similar.

3. yes; AA Similarity Postulate

4. yes; SSS Similarity Theorem

5. 6 6. $\frac{3}{4}$

Mid-Chapter Assessment—Chapter 8

1. d 2. a 3. b 4. a

5. 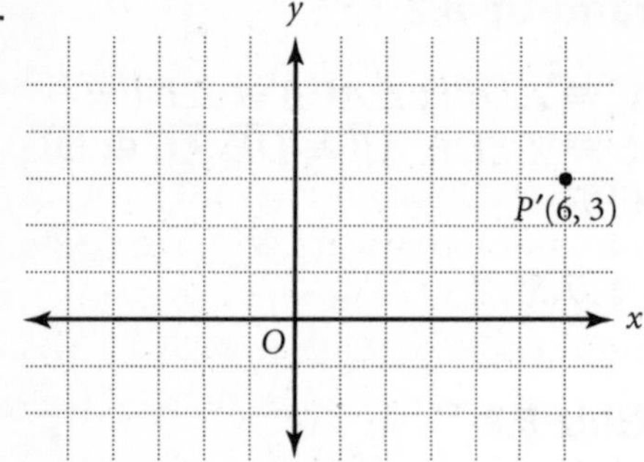

6. 20 7. 10 8. no

9. yes; SAS Similarity Theorem

Quick Warm-Up 8.4

AA Triangle Similarity Postulate, SSS and SAS Triangle Similarity Theorems

Lesson 8.4

1. parallel; proportionally 2. parallel lines
3. If a segment divides two sides of a triangle proportionally, then the segment is parallel to the third side.

4. 4 5. 1.5 6. 18 7. 25 8. 3 9. 6

Quick Warm-Up 8.5

1. $x = 4$ 2. $\frac{AB}{DE} = \frac{AC}{DF} = \frac{BC}{EF}$ 3. 4

Lesson Quiz 8.5

1. the same ratio as the corresponding sides
2. angle bisector 3. $\frac{8}{3}$, or $2\frac{2}{3}$ 4. 10 feet
5. 9 6. 8 7. about 10.20
8. $\frac{2}{3}$ 9. 4.5 cm

Quick Warm-Up 8.6

1. 16π sq cm 2. 168 cubic cm
3. 150π cubic cm

Lesson Quiz 8.6

1. $\frac{1}{2}$ or $\frac{2}{1}$ 2. $\frac{1}{4}$ or $\frac{4}{1}$ 3. $\frac{1}{8}$ or $\frac{8}{1}$ 4. $\frac{1}{3}$ 5. $\frac{1}{9}$
6. 3 : 4 7. 3 : 4 8. about 171.9 in.^{2}
9. 24 in.^{2}

Chapter Assessment—Chapter 8

Form A

1. d 2. b 3. c 4. c 5. c 6. d 7. c
8. b 9. b 10. a 11. a 12. b 13. b
14. c 15. c 16. d

Form B

1. $\frac{1}{3}$ 2. (0, 1) 3. (3, 0) 4. (−6, 0)
5. 5 6. 31.25 7. yes; $\frac{3}{9} = \frac{5}{15} = \frac{4}{12}$
8. 15 9. 1.5 10. 27 11. $x = 9$; $y = 10.8$
12. $x = 10.5$; $y = 15$ 13. $x = 12$; $y = 8$
14. 20 15. 6.5 16. 48 17. 27 cm
18. 9 feet 19. 195 cm^2 20. 9 gallons

Alternative Assessment—Chapter 8

Form A

1. Check students' drawings.
2. Ratios will be greater than 1 for enlargements and less than 1 for reductions.

3. Students many have measured the drawing, selected a ratio, and used the ratio to find lengths for their drawings. They may also have drawn a section of the aircraft, the size they wanted, measured it and the drawing to determine the ratio, and then used the ratio to find the remaining lengths.
4. approximately 70 to 1
5. Answers will vary. 6. about 22 feet
7. Answers will depend on the ratio used.

Score Point 4: Distinguished

The student demonstrates a comprehensive understanding of ratios and scale drawings. The student uses perceptive, creative, and complex mathematical reasoning as well as precise and appropriate mathematical language throughout the task. Theoretical knowledge is apparent and is applied to concrete situations as the student successfully demonstrates a comprehensive understanding of the core concepts.

Score Point 3: Proficient

The student demonstrates a broad understanding of ratios and scale drawings. The student uses precise and appropriate mathematical language most of the time. Theoretical knowledge is apparent and is applied to concrete situations as the student attempts to create a scale drawing and use ratios to solve problems.

Score Point 2: Apprentice

The student demonstrates an understanding of ratios and scale drawings. The student uses mathematical reasoning and appropriate mathematical language some of the time. The student attempts to apply theoretical knowledge to the task but may not be able to use ratios to solve problems.

Score Point 1: Novice

The student demonstrates a basic understanding of ratios and scale drawings but is unable to complete the task. The student uses little mathematical reasoning or appropriate mathematical language. Theoretical knowledge may appear weak, and many responses may be illogical because directions were followed incorrectly.

Score Point 0: Unsatisfactory

The student does not make an attempt to complete the task, and the responses only restate the problem.

Form B

1. Sample answer: Techniques might employ a mirror or shadows. Compare the shadow or mirror image of a known object to that of the tree.
   Properties and theorems used: triangle similarity postulates and properties of proportions
2. Sample answer: Find one-half of the length of a frame pole. Use the proportion $\frac{\text{base pole}}{\text{frame pole}} = \frac{\text{brace}}{0.5 \text{ frame pole}}$.
   Properties and theorems used: triangle similarity postulates, properties of proportions, and Proportional Medians Theorem

Score Point 4: Distinguished

The student demonstrates a comprehensive understanding of indirect measurement using ratios. The student uses perceptive, creative, and complex mathematical reasoning as well as precise and appropriate mathematical language throughout the task. Theoretical knowledge is apparent and is applied to concrete situations as the student successfully demonstrates a comprehensive understanding of the core concepts.

Score Point 3: Proficient

The student demonstrates a broad understanding of indirect measurement using ratios. The student uses precise and appropriate mathematical language most of the time. Theoretical knowledge is apparent and is applied to concrete situations as the student attempts to explain indirect measuring techniques and write appropriate ratios.

Score Point 2: Apprentice

The student demonstrates an understanding of indirect measurement using ratios. The student uses appropriate mathematical language some of the time. The student attempts to apply theoretical knowledge to the task but may not be able to explain techniques or find the appropriate ratios.

Score Point 1: Novice

The student demonstrates a basic understanding of indirect measurement using ratios but is unable to complete the task. The student uses little mathematical reasoning or appropriate mathematical language. Theoretical knowledge may appear weak, and many responses may be illogical because directions were followed incorrectly.

Score Point 0: Unsatisfactory

The student does not make an attempt to complete the task, and the responses only restate the problem.

Chapter 9

Quick Warm-Up 9.1

1. a set of points in a plane equidistant from a given point in that plane
2. a line segment connecting the center of a circle with any point on the circle
3. a line segment that connects two points on a circle and passes through the center

Lesson Quiz 9.1

1. congruent 2. $\frac{A°}{360°} \times 2\pi r$ 3. 42°

4. 75° 5. 150° 6. 117° 7. 180° 8. 243°

9. 18.85 m 10. 60°

11. $3 \times 94.24 \approx 283$ feet 12. 300°

Quick Warm-Up 9.2

1. 2 2. 1 3. 2, if $\theta \neq 360°n$

Lesson Quiz 9.2

1. tangent 2. bisects 3. $\overline{TR}$

4. $\overline{QR}$ or $\overline{QS}$ 5. 12 6. 15 7. 5 8. 2

9. 8 10. 9 11. 58° 12. 13 inches

13. 2.12 ft

Answers

Quick Warm-Up 9.3

1. Possible arrangements: Neither, both, or one of the rays of the angle intersects the circle.
2. neither: 1; both: 3; one: 2

Lesson Quiz 9.3

1. one-half 2. same measure

3. 90° 4. 30° 5. 60° 6. 240°

7. $m\angle C = 36°$; $m\angle D = 36°$

8. $m\widehat{AB} = 84°$; $m\angle C = 42°$

9. $m\angle A = 28°$; $m\angle B = 28°$

10. $m\widehat{DC} = 110°$; $m\angle A = 55°$

11. They intercept major and minor arcs that equal the measure of the entire circle, or 360°. Thus, the sum of the two angles is one-half 360°, or 180°.

Mid-Chapter Assessment—Chapter 9

1. e 2. d 3. a 4. b 5. c 6. b 7. a

8. c 9. a 10. c 11. 12 12. 16 13. 10.5

Quick Warm-Up 9.4

Possibilities are as follows: The circle intersects the angle at a vertex and on one ray; the circle is inside the angle with intersection points at tangents to the circle; one ray intersects the circle at two points other than the vertex; the vertex is in the interior of the circle and the circle intersects both rays at one point each.

Lesson Quiz 9.4

1. Find half the measure of the intercepted arc.
2. Find half the sum of the measures of the intercepted arcs.

3. 55° 4. 75° 5. 30° 6. 170° 7. 80°

8. 70° 9. 90° 10. 50° 11. 25° 12. 160°

Quick Warm-Up 9.5

1. Check students' constructions.
2. Check students' constructions.

Lesson Quiz 9.5

1. *CB* 2. *GC* 3. *BH*; *CB* 4. 3 5. 5

6. 8 7. 9 8. 6 9. 9

Quick Warm-Up 9.6

Solve for x.

1. $x = \pm 3$ 2. $x = \pm 5$

3. $x = \pm 3\sqrt{3}$ or $x = \pm 5.2$

4. $x = \pm\sqrt{16 - y^2}$, where $16 - y^2 \geq 0$

Lesson Quiz 9.6

1. $(7, 0), (-7, 0); (0, -7), (0, 7)$

2. $(-9, 0), (9, 0); (0, -9), (0, 9)$

3. $(-12, 0), (12, 0); (0, -12), (0, 12)$

4. $x^2 + y^2 = 9$ 5. $(x - 4)^2 + y^2 = 64$

6. $(x + 3)^2 + (y - 2)^2 = 225$

7. $(x - 3)^2 + (y - 5)^2 = 36$

8. $(x - 2)^2 + (y + 4)^2 = 4$

9. $(x + 5)^2 + (y + 1)^2 = 1$

10. $(0, 0); 6$ 11. $(0, 3); 5$ 12. $(2, -1); 10$

13. $x^2 + y^2 = 4$ 14. $(x - 2)^2 + y^2 = 4$

15. $(x - 3)^2 + (y - 1)^2 = 4$

16. $(x + 2)^2 + (y - 1)^2 = 9$

Chapter Assessment—Chapter 9

Form A

1. c 2. a 3. b 4. b 5. c 6. d 7. d

8. a 9. c 10. d 11. a 12. b 13. d

14. b 15. c 16. a 17. b 18. d 19. d

Form B

1. 96° 2. 42° 3. 9 4. 138° 5. 69°

6. 69° 7. 7π 8. 30π 9. 12 10. 17

11. 6 12. 10.6 13. 6.1 14. 70°

15. 55° 16. 110° 17. 95° 18. 50°

19. 56° 20. 30° 21. 135° 22. 9 23. 15

24. 20π, or 62.8 cm 25. 160°

26. 180°; the measure of an angle formed by two chords inside a circle is half the sum of the measures of the intercepted arcs.

27. $(-4, 0), (4, 0); (0, -4), (0, 4)$

28. $(-\sqrt{20}, 0), (\sqrt{20}, 0); (0, -\sqrt{20}), (0, \sqrt{20})$

29. $(0, 0); 6$ 30. $(0, 2); 7$

31. $(1, -3); \sqrt{10}$ 32. $(x - 4)^2 + y^2 = 25$

33. $(x + 1)^2 + (y - 5)^2 = 4$

34. $(x + 1)^2 + (y + 2)^2 = 4$

35. $(x - 3)^2 + (y + 2)^2 = 9$

Alternative Assessment—Chapter 9

Form A

1. Minor arcs: (1) an arc between two vertices of the triangle (outside the smaller triangle); (2) an arc between two vertices of the smaller triangle (inside the triangle); (3) an arc forming one side of a "petal" of the "flower" (one-half of (2)); major arc: an arc enclosing two sides of the smaller triangle; the arc measures are 60°, 120°, and 240°; the length measures are $\frac{1}{6}$, $\frac{1}{3}$, and $\frac{2}{3}$ times the circumference of the circle.

2. The segments inside the circle are chords. None of the lines are secants. The segments outside the circle are tangents.

3. Use a compass. Draw a circle. Place the point of the compass on the circle. Without changing the compass setting, draw a minor arc inside the circle. Move the compass so the point is on the intersection of this arc and the circle, and draw another minor arc. Repeat this for a third minor arc. Draw chords to connect the points on the circumference of the circle. Draw tangents to the circle that pass through the three points. Extend the lines to form an equilateral triangle.

Score Point 4: Distinguished

The student demonstrates a comprehensive understanding of circles, chords, arcs, and tangents. The student uses perceptive, creative, and complex mathematical reasoning as well as precise and appropriate mathematical language throughout the task. Theoretical knowledge is apparent and is applied to concrete situations as the student successfully demonstrates a comprehensive understanding of the core concepts.

Score Point 3: Proficient

The student demonstrates a broad understanding of circles, chords, arcs, and tangents. The student uses precise and appropriate mathematical language most of the time. Theoretical knowledge is apparent and is applied to concrete situations as the student attempts to analyze the logo.

Score Point 2: Apprentice

The student demonstrates an understanding of circles, chords, arcs, and tangents. The student uses mathematical reasoning and appropriate mathematical language some of the time. The student attempts to apply theoretical knowledge to the task but may not be able to identify the parts of the logo.

Score Point 1: Novice

The student demonstrates a basic understanding of circles, chords, arcs, and tangents but is unable to complete the task. The student uses little mathematical reasoning or appropriate mathematical language. Theoretical knowledge may appear weak, and many responses may be illogical because directions were followed incorrectly.

Score Point 0: Unsatisfactory

The student does not make an attempt to complete the task, and the responses only restate the problem.

Form B

Sample answer: The measure of $\widehat{EF}$ is 60°, twice the measure of $\angle EDF$. Since $\overline{EB}$ is a diameter, the measure of $\widehat{EFB}$ and $\widehat{EDB}$ are each 180°. $\widehat{BF}$ is 120°, found by subtracting $m\widehat{EFB} - m\widehat{EF}$.

$$m\angle C = \frac{120° - 80°}{2} = 20°$$

$$m\angle A = \frac{180° - 120°}{2} = 30°$$

$$m\angle DEO = \frac{80°}{2} = 40°$$

Thus, $m\angle EGD = 110°$. Since $\overline{OB} \perp \overline{AC}$, $m\angle ABO = m\angle OBC = 90°$.

$$m\angle EFD = \frac{100°}{2} = 50°$$

$$m\angle EGF = 180° - 110° = 70°$$

$$m\angle FEG = 180° - (50° + 70°) = 60°$$

$$m\angle AED = \frac{200°}{2} = 100°$$

The remaining angles can be found by using the Vertical Angles Theorem and Linear Pair Property.

Score Point 4: Distinguished

The student demonstrates a comprehensive understanding of arcs and angles and their properties. The student uses perceptive, creative, and complex mathematical reasoning as well as precise and appropriate mathematical language throughout the task. Theoretical knowledge is apparent and is applied to concrete situations as the student successfully demonstrates a comprehensive understanding of the core concepts.

Score Point 3: Proficient

The student demonstrates a broad understanding of arcs and angles and their properties. The student uses precise and appropriate mathematical language most of the time. Theoretical knowledge is apparent and is applied to concrete situations as the student attempts to find the measures of the arcs and angles in the figure.

Score Point 2: Apprentice

The student demonstrates an understanding of arcs and angles and their properties. The student uses appropriate mathematical language some of the time. The student attempts to apply theoretical knowledge to the task but may not be able to find the measure of an arc or angle from the known information.

Score Point 1: Novice

The student demonstrates a basic understanding of arcs and angles and their properties but is unable to complete the task. The student uses little mathematical reasoning or appropriate mathematical language. Theoretical knowledge may appear weak, and many responses may be illogical because directions were followed incorrectly.

Score Point 0: Unsatisfactory

The student does not make an attempt to complete the task, and the responses only restate the problem.

Chapter 10

Quick Warm-Up 10.1

1. $\overline{AB}$ 2. $\overline{BC}$ 3. $\overline{AC}$ 4. $\overline{AC}$ 5. $\overline{BC}$

Lesson Quiz 10.1

1. the opposite side to the adjacent side, or $\frac{BC}{AC}$
2. increases
3. 1.00 4. 0.58 5. 1.73 6. 42° 7. 54°
8. 85° 9. 42.89 10. 22.57 11. 56.31°
12. about 14.3 feet

Quick Warm-Up 10.2

The other acute angle decreases in 5° increments.

Lesson Quiz 10.2

1. $\frac{9}{15}$, or $\frac{3}{5}$ 2. $\frac{9}{15}$, or $\frac{3}{5}$ 3. $\frac{12}{15}$, or $\frac{4}{5}$
4. 53° 5. 0.97 6. 0.98 7. 0.05
8. 26° 9. 82° 10. 58° 11. 14.56
12. 31.06 13. 38.25 14. 30.16
15. no less than ≈ 2.3 feet

Quick Warm-Up 10.3

Check students' drawings.

Lesson Quiz 10.3

1. 0.500 2. −0.866 3. −0.574
4. −0.819 5. −0.819 6. −0.574
7. −0.707 8. 0.707 9. −0.500
10. (0.707, 0.707) 11. (−.707, 0.707)
12. (−0.500, −0.866) 13. (0, −1)
14. (0.707, −0.707) 15. (0.866, −0.500)
16. 15°, 165° 17. 25°, 155° 18. 30°, 150°
19. 60°, 120° 20. 75°, 105° 21. 80°, 100°
22. 55°, 125° 23. 50°, 130° 24. 45°, 135°
25. y

Answers

Mid-Chapter Assessment—Chapter 10

1. a **2.** a **3.** c **4.** c **5.** 23.8 **6.** 28.4
7. 28.1° **8.** about 2.1 meters **9.** 50°
10. −0.866 **11.** 0.5 **12.** 23°, 157°

Quick Warm-Up 10.4

1. Check students' drawings.

2. *A, B, a; A, B, b; A, B, c; A, C, a; A, C, b; A, C, c; C, B, a; C, B, b; C, B, c*

Lesson Quiz 10.4

1. two angles and a side of the triangle

2. two sides and an angle that is opposite one of the sides

3. 4.7 **4.** 25.4° **5.** 12.4 **6.** 9.7; 3.7

7. the station forming the 100° angle; 60 kilometers

Quick Warm-Up 10.5

1. Check students' drawings.

2. *a, b, C; a, c, B; b, c, A*

Lesson Quiz 10.5

1. two sides and the included angle

2. all three sides

3. 84.3° **4.** 12.4 **5.** 14.8
6. 36.3° **7.** 436 feet

Quick Warm-Up 10.6

1. Check students' drawings.

2. $a\sqrt{2}$

Lesson Quiz 10.6

1. vector sum **2.** diagonal
3. parallel **4.** perpendicular

5.

$\vec{m}+\vec{n}$

$\vec{m}$

$\vec{n}$

6.

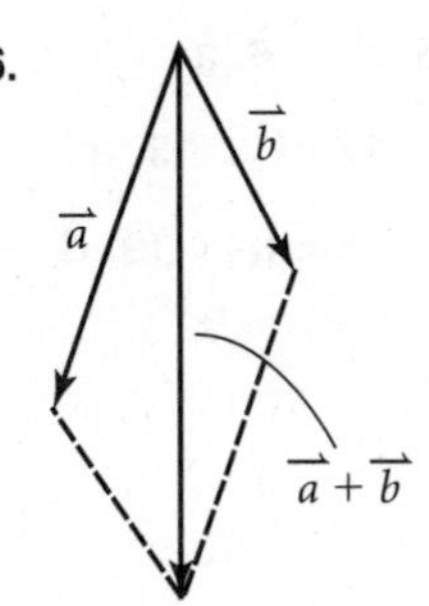

7.

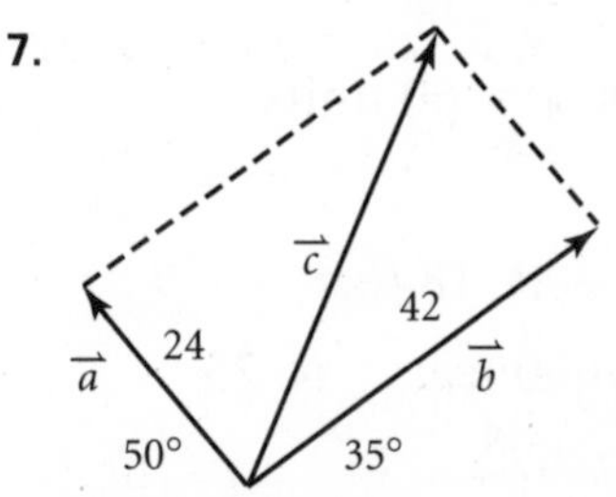

8. 95°; 85° **9.** 47 **10.** 31°

Quick Warm-Up 10.7

1. Check students' drawings.

2. Check students' drawings.

Lesson Quiz 10.7

1. (1, 2) **2.** (2, −3) **3.** (0, 3)
4. (−1, −1) **5.** (0, 2) **6.** (3, 2)
7. (6.4, 7.8) **8.** (−4.1, −9.2)
9. (−13.7, −3.7) **10.** (6.1, −3.5) **11.** 90°
12. 270° **13.** 180° **14.** 360° **15.** 90°

16. $\begin{bmatrix} \cos 45° & -\sin 45° \\ \sin 45° & \cos 45° \end{bmatrix} \times \begin{bmatrix} 2 \\ -1 \end{bmatrix} =$

$\begin{bmatrix} 2\cos 45° & \sin 45° \\ 2\sin 45° & -\cos 45° \end{bmatrix}$

Chapter Assessment—Chapter 10

Form A

1. a 2. c 3. b 4. c 5. d 6. a 7. c
8. c 9. b 10. c 11. a 12. d 13. a
14. c 15. b 16. c 17. b 18. a 19. d

Form B

1. $\frac{8}{17}$ 2. $\frac{8}{17}$ 3. $\frac{15}{8}$
4. 28° 5. 65° 6. −0.174
7. $x = 16$ units 8. $x = 60$ units
9. $x = 62°$
10. 45 square units 11. 18 feet
12. 1860 feet 13. 5600 feet 14. 25
15. 40° 16. 28 17. 67°
18. 93 feet 19. 19 inches
20.

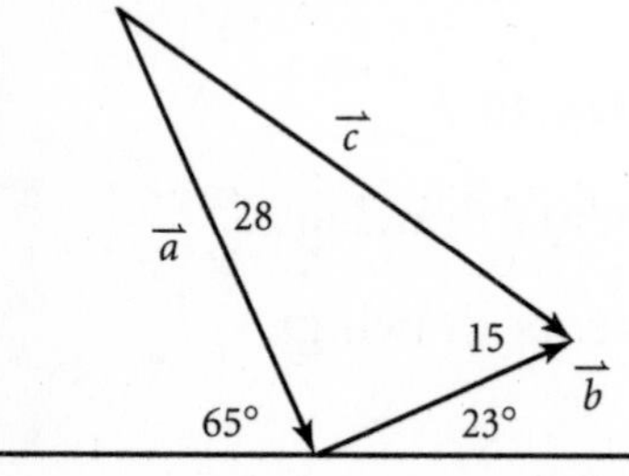

21. 92° 22. 32 23. 28° 24. (1, −5)
25. (−2, −4) 26. (−3, 2) 27. 270° 28. 90°
29. 180°

Alternative Assessment—Chapter 10

Form A

1. Sample answer: Find the height from sea level to the top of the lighthouse. Estimate the angle between the boat and the top of the lighthouse, and find the tangent of that angle. Divide the height by the tangent to find the distance from the boat to the shore.

2. Use an instrument to measure the angle or estimate the angle.

3. about 260 feet

4. about 10.6°

Score Point 4: Distinguished

The student demonstrates a comprehensive understanding of the use of trigonometry to find unknown distances. The student uses perceptive, creative, and complex mathematical reasoning as well as precise and appropriate mathematical language throughout the task. Theoretical knowledge is apparent and is applied to concrete situations as the student successfully demonstrates a comprehensive understanding of the core concepts.

Score Point 3: Proficient

The student demonstrates a broad understanding the use of trigonometry to find unknown distances. The student uses precise and appropriate mathematical language most of the time. Theoretical knowledge is apparent and is applied to concrete situations as the student attempts to use trigonometry to find the unknown distances.

Answers

Score Point 2: Apprentice

The student demonstrates an understanding of the use of trigonometry to find unknown distances. The student uses mathematical reasoning and appropriate mathematical language some of the time. The student attempts to apply theoretical knowledge to the task but may not be able to use trigonometry to find the unknown distance.

Score Point 1: Novice

The student demonstrates a basic understanding of the use of trigonometry to find unknown distances but is unable to complete the task. The student uses little mathematical reasoning or appropriate mathematical language. Theoretical knowledge may appear weak, and many responses may be illogical because directions were followed incorrectly.

Score Point 0: Unsatisfactory

The student does not make an attempt to complete the task, and the responses only restate the problem.

Form B

1. Sample answer:

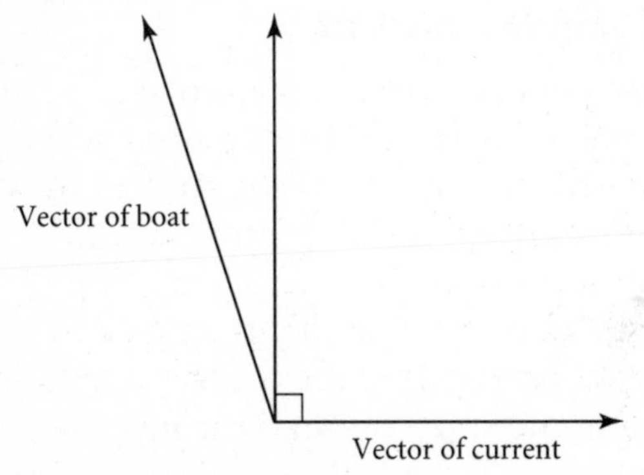

2. river speed, rowing speed

3. Sample answer: From experience the rowing speed in still water is about 5 mph. Estimate the water current flows at approximately 3 mph. Use the parallelogram method of vector addition. Since $5^2 = 3^2 + x^2, x = 4$ mph. The resulting vector at 90° across the river is 4 mph.
   $\theta = \tan^{-1}\left(\frac{4}{3}\right) \approx 53.1°$

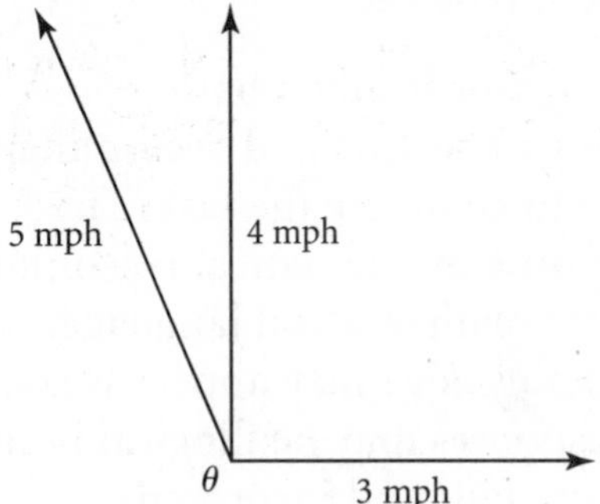

Score Point 4: Distinguished

The student demonstrates a comprehensive understanding of vectors and vector analysis. The student uses perceptive, creative, and complex mathematical reasoning as well as precise and appropriate mathematical language throughout the task. Theoretical knowledge is apparent and is applied to concrete situations as the student successfully demonstrates a comprehensive understanding of the core concepts.

Score Point 3: Proficient

The student demonstrates a broad understanding of vectors and vector analysis. The student uses precise and appropriate mathematical language most of the time. Theoretical knowledge is apparent and is applied to concrete situations as the student attempts to analyze the problem and assign appropriate values.

Score Point 2: Apprentice

The student demonstrates an understanding of vectors and vector analysis. The student uses appropriate mathematical language some of the time. The student attempts to apply theoretical knowledge to the task but may not be able to analyze the problem or assign appropriate values.

Score Point 1: Novice

The student demonstrates a basic understanding of vectors and vector analysis but is unable to complete the task. The student uses little mathematical reasoning or appropriate mathematical language. Theoretical knowledge may appear weak, and many responses may be illogical because directions were followed incorrectly.

Score Point 0: Unsatisfactory

The student does not make an attempt to complete the task, and the responses only restate the problem.

Chapter 11

Quick Warm-Up 11.1

1. $x = 4$ 2. $x = \pm\sqrt{5} + 1$ 3. $x \approx 1.618$

Lesson Quiz 11.1

1. A rectangle is a golden rectangle with the property that if a square is cut off one end of the rectangle, the remaining piece is similar to the original rectangle.

2. $\frac{a}{b} = \frac{b}{a-b}$ 3. $\frac{1+\sqrt{5}}{2} \approx 1.618$

4. 25.9 5. 11.1 6. 5.6

Quick Warm-Up 11.2

1. square 2. $2\sqrt{13}$, or 7.21 3. 10 units

Lesson Quiz 11.2

1. the least number of grid units that a taxi must travel to move from one point to another
2. the set of all points *r* units from a given point

3. 16 4. 16 5. 4 units

6. 8 units 7. 5 units

Quick Warm-Up 11.3

1. Answers will vary. One common use of the term *network* refers to a television network.
2. They are equal.
3. Answers will vary. Traverse means "to pass through."

Lesson Quiz 11.3

1. A graph is composed of points, called vertices, and segments or curves linking the vertices, called edges.
2. An Euler path travels along each edge of a graph exactly once.
3. 2

4.

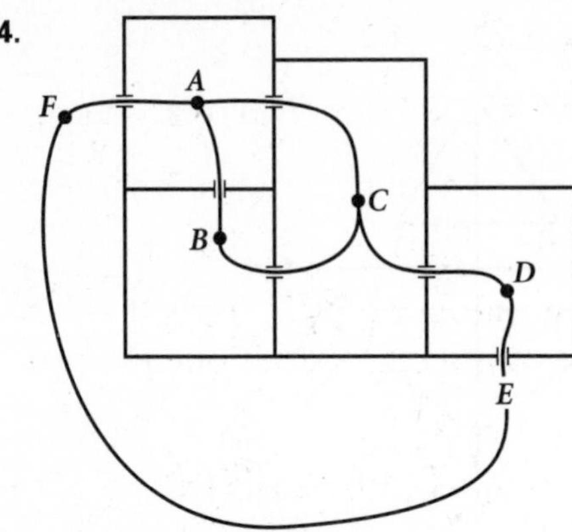

5. No; not all doorways are even vertices.

6. yes 7. yes 8. no

Quick Warm-Up 11.4

1. The line is "straight" because it is the shortest distance.

2. Answers will vary. The message will be smaller and not as "stretched out."

Lesson Quiz 11.4

1. when one of the shapes can be stretched, shrunk, or otherwise distorted into the other without cutting, tearing, or intersecting itself

2. Every line that connects a point on the inside to a point on the outside must intersect the curve.

3. $V - E + F = 2$

4. Yes; the triangle can be made into a circle without cutting or tearing it.

5. No; the circle has no intersections, but the figure eight has one intersection.

6. No; Jordan's Theorem is not true for a torus but is true for a sphere.

7. Yes; Euler's formula is true for both the pyramid and the sphere.

Mid-Chapter Assessment—Chapter 11

1. c 2. d 3. d 4. c 5. d 6. 5

7. Yes; there are exactly two odd vertices and the other vertices are even.

8. No; the prism satisfies Jordan's Theorem but the torus does not.

Quick Warm-Up 11.5

1. greater than

2. Yes; on a saddle, the sides of a triangle are concave and the sum of its angles is less than 180°.

Lesson Quiz 11.5

1. Two statements are logically equivalent if each can be derived from the other.

2. A line is a great circle of the sphere.

3. infinitely many

4. 180°

5. always greater than 180°

6. always less than 180°

7. 300°

8. $\overline{AB}$ and $\overline{AC}$ or $\overline{AB}$ and $\overline{BC}$

9. It becomes smaller.

Quick Warm-Up 11.6

Answers will vary. Students with previous experience with fractals will be likely to mention fractal qualities.

Lesson Quiz 11.6

1. a geometric object that is self-similar and can be created by doing a procedure over and over

2. creating a mathematical object by repetitive application of the same rule

3.

Original line

First iteration

Second iteration

Third iteration

4.

1
1 2 1
1 3 3 1
1 4 6 4 1
1 5 10 10 5 1

5.

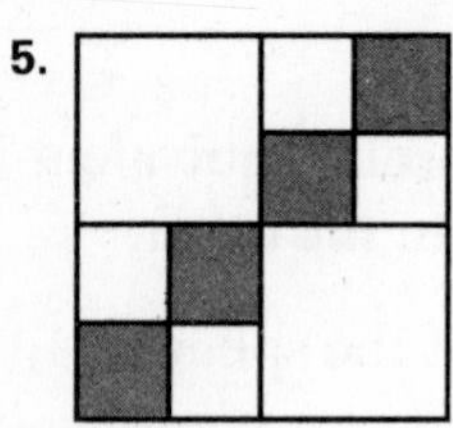

6. 12

Quick Warm-Up 11.7

1. reflections, rotations, and translations
2. shape
3. translations, dilations, rotations, and reflections

Lesson Quiz 11.7

1. true 2. false 3. true 4. true

5.

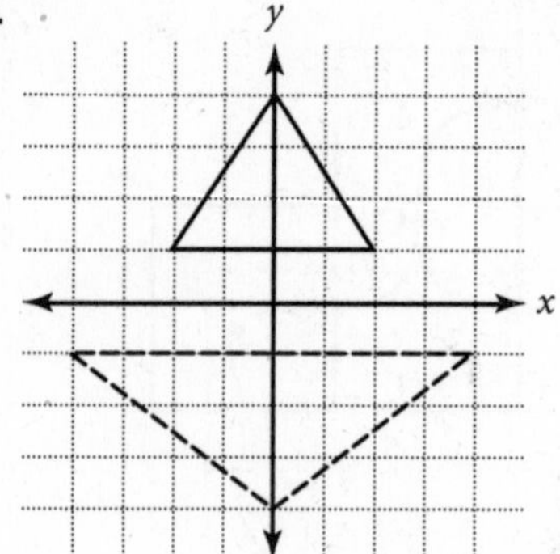

6.

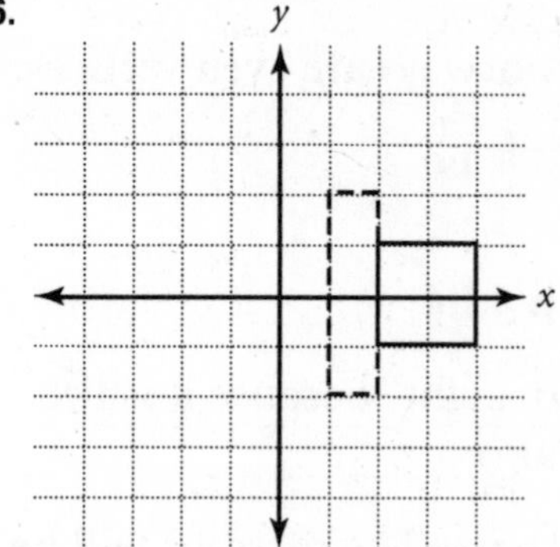

7.

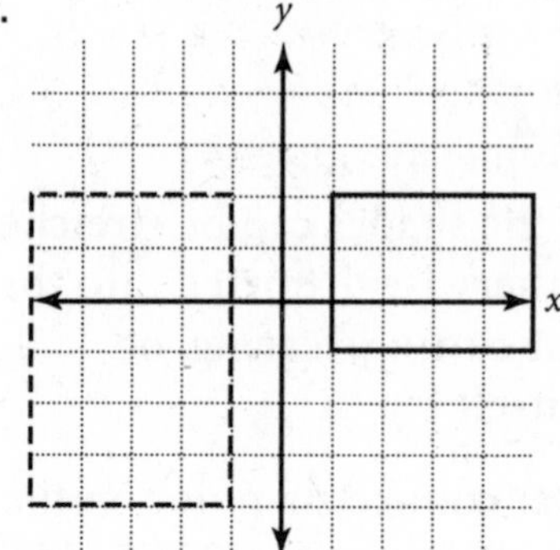

8. *D*

Chapter Assessment—Chapter 11

Form A

1. a 2. b 3. d 4. b 5. d
6. c 7. b 8. b 9. c 10. c
11. d 12. c 13. d 14. b 15. d

Form B

1. 16 2. 9 3. 15 4. 6

5. 5 6. 11 7. 11 8. 40

9. 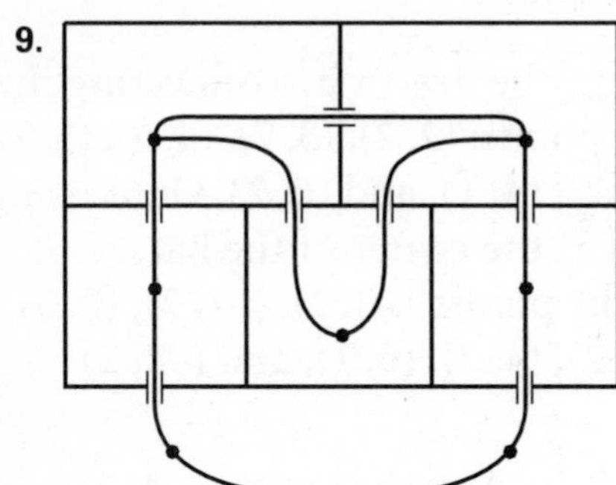

10. 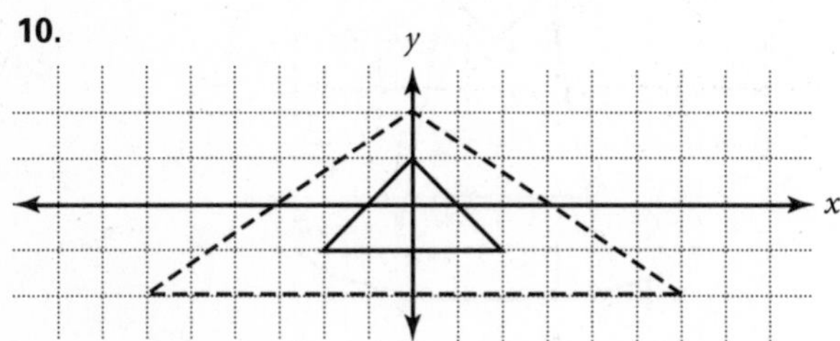

11. Yes; both are simple closed curves that can be transformed into each other without cutting or tearing.
12. No; a torus can not be changed into a sphere.
13. hyperbolic geometry
14. spherical geometry
15. Euclidean geometry
16. No; there are more than 2 odd vertices.
17. Yes; all vertices are even.
18. Yes; there are exactly 2 odd vertices.

19. *C* 20. *E* 21. *D* 22. *F*

Alternative Assessment—Chapter 11

Form A

1. Sample answer.

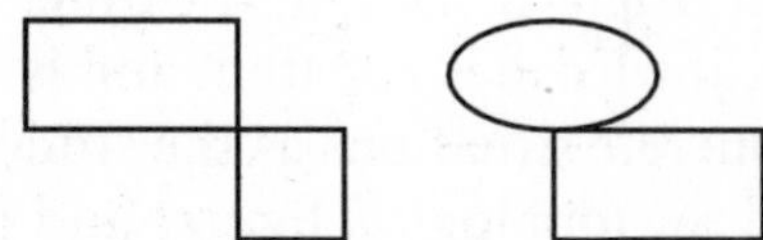

2. One shape can be stretched, shrunk, or otherwise distorted to form the other.
3. Sample answer:

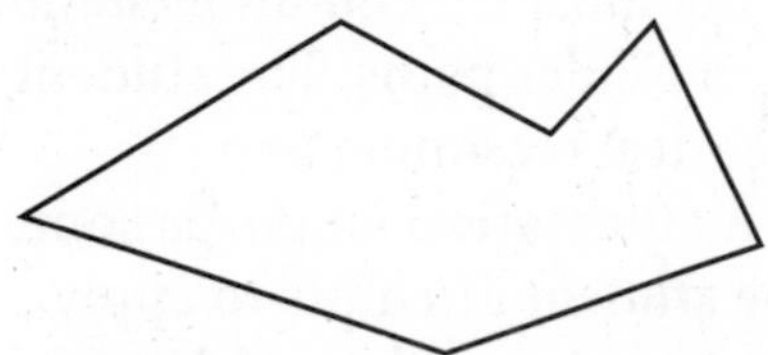

4. The neighborhood can be patrolled by walking each street only once if the beginning and ending points are the cul-de-sac and the first street that intersects the cul-de-sac. This is the odd vertex. All other vertices have an even number of paths from them.

Score Point 4: Distinguished

The student demonstrates a comprehensive understanding of topology and Euler paths. The student uses perceptive, creative, and complex mathematical reasoning as well as precise and appropriate mathematical language throughout the task. Theoretical knowledge is apparent and is applied to concrete situations as the student successfully demonstrates a comprehensive understanding of the core concepts.

Score Point 3: Proficient

The student demonstrates a broad understanding of topology and Euler paths. The student uses precise and appropriate mathematical language most of the time. Theoretical knowledge is apparent and is applied to concrete situations as the student attempts to draw topological figures and find an Euler path through a graph.

Score Point 2: Apprentice

The student demonstrates an understanding of topology and Euler paths. The student uses mathematical reasoning and appropriate mathematical language some of the time. The student attempts to apply theoretical knowledge to the task but may not be able to draw topological figures and find an Euler path through a graph.

Score Point 1: Novice

The student demonstrates a basic understanding topology and Euler paths but is unable to complete the task. The student uses little mathematical reasoning or appropriate mathematical language. Theoretical knowledge may appear weak, and many responses may be illogical because directions were followed incorrectly.

Score Point 0: Unsatisfactory

The student does not make an attempt to complete the task, and the responses only restate the problem.

Form B

1. Sample answer: in or under the hat
2. Redraw the figure in the coordinate plane.
3. He transformed the figure onto a polar plane.
4. The hat may be drawn by connecting the following points: (2, 2), (3, 7), (4, 6), (5, 7), (6, 2), (8, 2), (4, 1), and (0, 2). Or, if using the y-axis as the center of the hat, connect the points (–2, 2), (–1, 7), (0, 6), (1, 7), (2, 2), (4, 2), (0, 1), and (–4, 2).

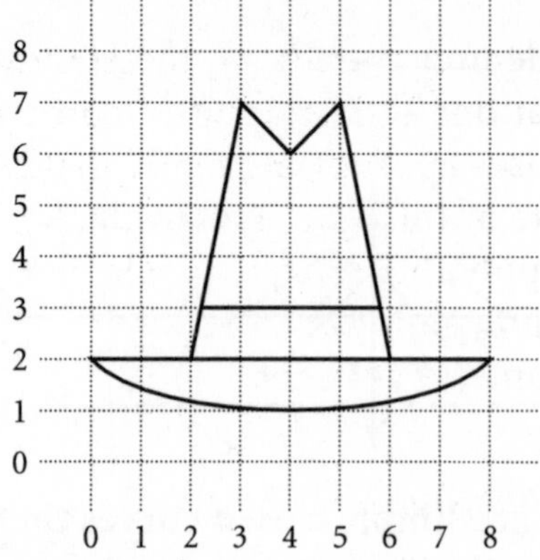

5. Sample answer: Multiply the x-coordinates by 2 and the y-coordinates by 3.

Score Point 4: Distinguished

The student demonstrates a comprehensive understanding of transformations and projective geometry. The student uses perceptive, creative, and complex mathematical reasoning as well as precise and appropriate mathematical language throughout the task. Theoretical knowledge is apparent and is applied to concrete situations as the student successfully demonstrates a comprehensive understanding of the core concepts.

Score Point 3: Proficient

The student demonstrates a broad understanding of transformations and projective geometry. The student uses precise and appropriate mathematical language most of the time. Theoretical knowledge is apparent and is applied to concrete situations as the student attempts to apply projective geometry concepts.

Score Point 2: Apprentice

The student demonstrates an understanding of transformations and projective geometry. The student uses appropriate mathematical language some of the time. The student attempts to apply theoretical knowledge to the task, but may not always be able to apply projective geometry concepts.

Score Point 1: Novice

The student demonstrates a basic understanding of transformations and projective geometry but is unable to complete the task. The student uses little mathematical reasoning or appropriate mathematical language. Theoretical knowledge may appear weak, and many responses may be illogical because directions were followed incorrectly.

Score Point 0: Unsatisfactory

The student does not make an attempt to complete the task, and the responses only restate the problem.

Chapter 12

Quick Warm-Up 12.1

1. If an animal is a frog, then it is an amphibian.
2. If Jamie gets an A in the course, then she scored 95% on the test.

Lesson Quiz 12.1

1. It is cold. Therefore, Juanita will wear a coat; valid
2. Juanita will not wear a coat. Therefore, it is not cold; valid
3. Juanita will wear a coat. Therefore, it is cold; invalid
4. It is not cold. Therefore, Juanita will not wear a coat; invalid

5. valid 6. not valid

7. valid 8. valid

Quick Warm-Up 12.2

1. It is raining today *and* school is in session.
2. It is raining today *or* school is in session.
3. Albert did *not* get permission to see a movie.

Lesson Quiz 12.2

1. when both of its parts are true
2. when both of its parts are false
3. Omar did not fail math. (Or Omar passed math.)
4. The light is not on. (Or The light is off.)

5. true 6. true 7. false 8. true

9. false 10. false 11. true 12. true

Quick Warm-Up 12.3

1. If an animal is warm-blooded, then it is a mammal.

2. If an animal is a mammal, then it is warm-blooded.

Lesson Quiz 12.3

1. If pq is an even number, then p and q are even numbers.

2. If p and q are not even numbers, then pq is not an even number.

3. If pq is not an even number, then p and q are not even numbers.

4. false 5. false 6. true

7. If you practice the piano everyday, then you will improve your skill.

8. If the temperature is 30°F, then the lake freezes.

9. If the student is Kate, then she is a member of the student organization.

Mid-Chapter Assessment—Chapter 12

1. b 2. c 3. a

4.

p	q	p OR q
T	T	T
T	F	T
F	T	T
F	F	F

5. If the sun is shining, then Joe goes swimming.

6. The sun is shining and Joe goes swimming.

7. The sun is shining or Joe is not wet.

8. If the sun is not shining, then Joe is wet.

9. If Joe does not go swimming, then the sun is not shining.

Quick Warm-Up 12.4

1. If the defendant was not in an automobile accident, then he was not in his car at 10:00 P.M.

2. No; it is not possible because if the sum of the angles is 180°, then it cannot contain both an angle of 90° and an angle greater than 90°.

Lesson Quiz 12.4

1. $\overline{CD} \perp \overline{AB}$

2. Definition of perpendicular lines

3. $\angle CDA \cong \angle CDB$

4. Given

5. Definition of a median

6. Reflexive Property

7. SAS 8. CPCTC

9. In a scalene triangle, no two sides are congruent.

10. $\overline{CD}$ is not perpendicular to $\overline{AB}$.

Quick Warm-Up 12.5

1. sample: VCR, CD-player

2.

p	q	p AND q
T	T	T
T	F	F
F	T	F
F	F	F

3.

p	$\sim p$
T	F
F	T

Lesson Quiz 12.5

1. ~(p AND q) 2. ~p OR q

3. 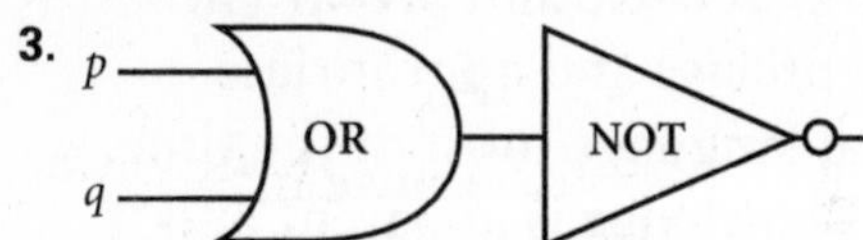

4. 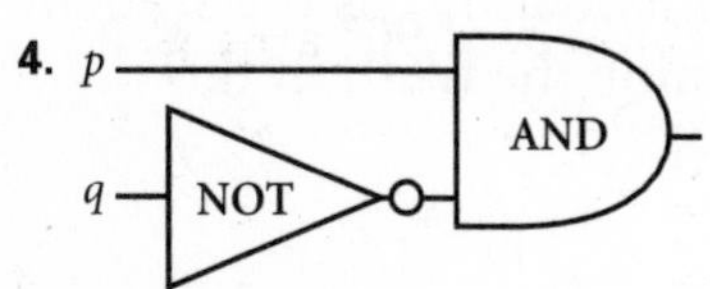

5. 0; 1 6. 1; 1 7. 0; 0 8. 1; 1

9.

p	q	(~p)	~p AND q
1	1	0	0
1	0	0	0
0	1	1	1
0	0	1	0

p	q	~(~p) AND q	
1	1	0	0
1	0	0	0
0	1	1	1
0	0	1	0

Chapter Assessment—Chapter 12

Form A

1. a 2. b 3. b 4. b 5. c 6. b 7. d

8. b 9. c 10. d 11. c 12. b 13. c

14. a 15. b 16. d

Form B

1. Jay is not in London.

2. May is in Canada.

3. If a quadrilateral is a rhombus, then it is a square.

4. If a quadrilateral is not a square, then it is not a rhombus.

5. If a quadrilateral is not a rhombus, then it is not a square.

6. false 7. true 8. true 9. false

10. true 11. true 12. false 13. true

14. true 15. true

16. ~(~p OR q)

17. 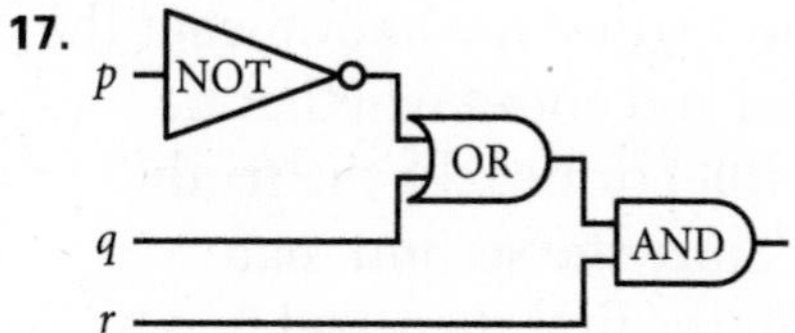

18. $\overline{AB}$ and $\overline{CD}$ bisect each other.

19. Definition of a bisector

20. $\angle AEC \cong \angle BED$

21. SAS

22. $\overline{DB} \cong \overline{AC}$

23. Given

24. $\overline{AB}$ and $\overline{CD}$ do not bisect each other.

25.

p	q	(~q)	p AND (~q)
1	1	0	0
1	0	1	1
0	1	0	0
0	0	1	0

p	q	~(p AND ~q)
1	1	1
1	0	0
0	1	1
0	0	1

26. Transitive 27. binary

Alternative Assessment—Chapter 12

Form A

1. The first and second suspects are not thieves. The third suspect is a thief.
2. Assume that the first suspect will say he is not a thief whether he is or not. The second suspect either lies about what the first man said and about whether he himself is a thief or he tells the truth about both. Since the second man confirms that the first man said he was not a thief, he is telling the truth; so the first and second suspects are both not thieves. When the third man says they are both thieves, he is not telling the truth. Therefore, the first two are not thieves, and the third is the thief.
3. Indirect argument. It was assumed the first suspect said he was not a thief. Reasoning proceeded until a conclusion was drawn.
4. Check students' work.

Score Point 4: Distinguished

The student demonstrates a comprehensive understanding of logic and proof. The student uses perceptive, creative, and complex mathematical reasoning as well as precise and appropriate mathematical language throughout the task. Theoretical knowledge is apparent and is applied to concrete situations as the student successfully demonstrates a comprehensive understanding of the core concepts.

Score Point 3: Proficient

The student demonstrates a broad understanding of logic and proof. The student uses precise and appropriate mathematical language most of the time. Theoretical knowledge is apparent and is applied to concrete situations as the student attempts to determine the truth in the situation presented.

Score Point 2: Apprentice

The student demonstrates an understanding of logic and proof. The student uses mathematical reasoning and appropriate mathematical language some of the time. The student attempts to apply theoretical knowledge to the task, but may not be able to proceed step-by-step through the proof.

Score Point 1: Novice

The student demonstrates a basic understanding logic and proof but is unable to complete the task. The student uses little mathematical reasoning or appropriate mathematical language. Theoretical knowledge may appear weak, and many responses may be illogical because directions were followed incorrectly.

Score Point 0: Unsatisfactory

The student does not make an attempt to complete the task, and the responses only restate the problem.

Form B

Check students' truth tables.

1. Answers may vary. If I buy a brand name watch, then I do not necessarily save $\frac{1}{3}$.

2. Answers may vary. If I buy a watch before December 24, then I do not necessarily save $\frac{1}{3}$.
3. Answers may vary. If I buy a brand name watch and I buy it before December 24, then I save $\frac{1}{3}$.
4. Answers may vary. If I do not buy a brand name watch and do not buy it before December 24, then I do not save $\frac{1}{3}$.

Score Point 4: Distinguished

The student demonstrates a comprehensive understanding of truth tables. The student uses perceptive, creative, and complex mathematical reasoning as well as precise and appropriate mathematical language throughout the task. Theoretical knowledge is apparent and is applied to concrete situations as the student successfully demonstrates a comprehensive understanding of the core concepts.

Score Point 3: Proficient

The student demonstrates a broad understanding of truth tables. The student uses precise and appropriate mathematical language most of the time. Theoretical knowledge is apparent and is applied to concrete situations as the student attempts to create, complete, and translate a truth table's symbolic logic.

Score Point 2: Apprentice

The student demonstrates an understanding of truth tables. The student uses appropriate mathematical language some of the time. The student attempts to apply theoretical knowledge to the task but may not be able to complete and translate a truth table's symbolic logic.

Score Point 1: Novice

The student demonstrates a basic understanding of truth tables but is unable to complete the task. The student uses little mathematical reasoning or appropriate mathematical language. Theoretical knowledge may appear weak, and many responses may be illogical because directions were followed incorrectly.

Score Point 0: Unsatisfactory

The student does not make an attempt to complete the task, and the responses only restate the problem.